# Cloud-Based Cyber-Physical Systems in Manufacturing

Lihui Wang · Xi Vincent Wang

# Cloud-Based Cyber-Physical Systems in Manufacturing

Springer

Lihui Wang
Department of Production Engineering
KTH Royal Institute of Technology
Stockholm
Sweden

Xi Vincent Wang
Department of Production Engineering
KTH Royal Institute of Technology
Stockholm
Sweden

ISBN 978-3-319-67692-0 ISBN 978-3-319-67693-7 (eBook)
https://doi.org/10.1007/978-3-319-67693-7

Library of Congress Control Number: 2017957672



Printed on acid-free paper

This Springer imprint is published by Springer Nature
The registered company is Springer International Publishing AG
The registered company address is: Gewerbestrasse 11, 6330 Cham, Switzerland

# Preface

New technologies in manufacturing are tightly connected to innovation. They have thus been the key factors that support and influence a nation's economy since the eighteenth century, from stream engines to Industry 4.0. As the primary driving force behind economic growth and sustainable development, manufacturing serves as the foundation of and contribute to other industries, with products ranging from heavy-duty machinery to hi-tech home electronics. In the past centuries, they have contributed significantly to modern civilisation and created the momentum that still drives today's economy and society. Despite many achievements, we are still facing challenges due to growing complexity and uncertainty in manufacturing, such as adaptability to uncertainty, resource and energy conservation, ageing workforce, and secure information sharing. Researchers and engineers across organisations often find themselves in situations that demand advanced new technologies when dealing with new challenges in daily activities related to manufacturing, which cannot be addressed by existing approaches.

Targeting the challenges in solving daily problems, over the past a few years, researchers have focused their efforts on innovative approaches to improving the adaptability to complex situations on shop floors and energy efficiency along the life cycle of products. New technologies and innovations include cyber-physical system (CPS), cloud manufacturing (CM), Internet of Things (IoT), big data analytics, which are related to embedded systems and system of systems. These new technologies are now driving industry towards yet another revolution and are referred to the German initiative Industry 4.0. While these efforts have resulted in a large volume of publications recently and impacted both present and future practices in factories and beyond, there still exists a gap in the literature for a focused collection of knowledge dedicated to cloud-based CPS in manufacturing. To bridge this gap and present the state of the art to a much broader readership, from academic researchers to practicing engineers, is the primary motivation behind this book.

The first three chapters form Part 1 of this book on the literature surveys and trends. Chapter 1 begins with a clear definition of cloud computing (CC) versus cloud manufacturing (CM). CC and CM represent the latest advancement and applications of the cloud technologies in computing and manufacturing, respectively. The aim of Chap. 1 is to provide a comprehensive introduction to both CC and CM and to present their status and advancement. The discussions on CC and CM are extended in Chap. 2 to cover the latest advancement of CPS and IoT, especially in manufacturing systems. To comprehensively understand CPS and IoT, a brief introduction to both of them is given, and the key enabling technologies related to CPS and IoT are outlined. Key features, characteristics, and advancements are explained, and a few applications are reported to highlight the latest advancement in CPS and IoT. Chapter 3 then provides an overview of cybersecurity measures being considered to ensure the protection of data being sent to physical machines in a cybernetic system. While common to other cybernetic systems, security issues in CM are focused in this chapter for brevity.

Part 2 of this book focuses on cloud-based monitoring, planning, and control in CPS and is constituted from four chapters. Targeting distributed manufacturing, the scope of Chap. 4 is to present an Internet- and web-based service-oriented system for machine availability monitoring and process planning. Particularly, this chapter introduces a tiered system architecture and introduces IEC 61499 function blocks for prototype implementation. It enables real-time monitoring of machine availability and execution status during metal-cutting operations, both locally or remotely. The closed-loop information flow makes process planning and monitoring two feasible services for the distributed manufacturing. Based on the machine availability and the execution status, Chap. 5 introduces Cloud-DPP for collaborative and adaptive process planning in cloud environment. Cloud-DPP supports parts machining with a combination of milling and turning features and offers process planning services for multi-tasking machining centres with special functionalities to minimise the total number of set ups. In Chap. 6, the Cloud-DPP is linked to physical machines by function blocks to form a CPS. Within the CPS, function blocks run at control level with embedded machining information such as machining sequence and machining parameters to facilitate adaptive machining. To utilise the machines properly, right maintenance strategies are required. Chapter 7 reviews the historical development of prognosis theories and techniques and projects their future growth in maintenance enabled by the cloud infrastructure. Techniques for cloud computing are highlighted, as well as their influence on cloud-enabled prognosis for manufacturing.

Sustainable robotic assembly in CPS settings is covered in Chaps. 8 through 11 and organised into Part 3 of this book. Chapter 8 explains how to minimise a robot's energy consumption during assembly. Given a trajectory and based on the inverse kinematics and dynamics of the robot, a set of attainable configurations for the robot can be determined, perused by calculating the suitable forces and torques on the joints and links of the robot. The energy consumption is then calculated for each configuration and based on the assigned trajectory. The ones with the lowest energy consumption are chosen for robot motion control. This approach becomes instrumental and can be wrapped as a cloud service for energy-efficient robotic assembly. Another robotic application is for human–robot collaborative assembly. Chapter 9 addresses safety issues in human–robot collaboration. This chapter first reviews the traditional safety systems and then presents the latest accomplishments in active collision avoidance through immersive human–robot collaboration by using two depth cameras installed carefully in a robotic assembly cell. A remote robotic assembly system is then introduced in the second half of the chapter as one cloud robotic application. In Chap. 10, recent cloud robotics approaches are reviewed. Function block-based integration mechanisms are introduced to integrate various types of manufacturing facilities including robots. By combining cloud with robots in form of cloud robotics, it contributes to a ubiquitous and integrated cloud-based CPS system in robotic assembly. Chapter 11 further explores the potential of establishing context awareness between a human worker and an industrial robot for physical human–robot collaborative assembly. The context awareness between the human worker and the industrial robot is established by applying gesture recognition, human motion recognition, and augmented reality (AR)-based worker instruction technologies. Such a system works in a cyber-physical environment, and its results are demonstrated through case studies.

In Part 4 of this book, the aspect of CPS systems design and lifecycle analysis is shared by Chaps. 12–15. Chapter 12 presents the architecture design of cloud CPS in manufacturing. Manufacturing resources and capabilities are discussed in terms of cloud services. A service-oriented, interoperable CM system is introduced. Service methodologies are developed to support two types of cloud users, customer user versus enterprise user, along with standardised data models describing cloud service and relevant features. Two case studies are revealed to evaluate the system. System design is extended in Chap. 13 to cover lifecycle analysis and management of products. In this chapter, CM is extended to the recovery and recycling of waste electrical and electronic equipment (WEEE). Cloud services are used in the recovery and recycling processes for WEEE tracking and management. These services include all the stakeholders from the beginning to the end of life of the electrical and electronic equipment. A product tracking mechanism is also introduced with the help of the quick response (QR) code method. Chapter 14 focuses on big data analytics. In order to minimise machining errors in advance, a big data analytics-based fault prediction approach is presented for shop-floor job scheduling, where machining jobs, machining resources, and machining processes are represented by data attributes. Based on the available data on the shop floor, the potential fault/error patterns, referring to machining errors, machine faults, maintenance states, etc., are mined to discover unsuitable scheduling arrangements before machining as well as the prediction of upcoming errors during machining. Chapter 15 presents a summary of the current status and the latest advancement of CM, CPS, IoT, and big data in manufacturing. Cloud-based CPS shows great promise in factories of the future in the areas of future trends as identified at the end of this chapter. It also offers an outlook of research challenges and directions in the subject areas.

All together, the fifteen chapters provide an overview of some recent R&D achievements of cloud-based CPS applied to manufacturing, especially machining and assembly. We believe that this research field will continue to be active for years to come.

Finally, the authors would like to express their appreciations to Springer for supporting this book project and would especially like to thank Anthony Doyle, Senior Editor for Engineering, and Amudha Vijayarangan, Project Coordinator, for their patience, constructive assistance, and earnest cooperation, both with the publishing venture in general and with the editorial details. We hope that readers find this book informative and useful.

Stockholm, Sweden
September 2017

Lihui Wang
Xi Vincent Wang

# Acknowledgements

It took a long time for the authors to prepare the book materials. Several people contributed to this book directly or indirectly during the process. The authors would like to take this opportunity to express their sincere appreciations to Dr. Abdullah Alhusin Alkhdur, Dr. Wei Ji, Dr. Yongkui Liu, Mr. Mohammad Givehchi, Mr. Hongyi Liu, and Mr. Sichao Liu for their fine contributions to this book. Their commitment, enthusiasm, and assistance are what made this book possible.

The authors are also grateful to Profs. Robert Gao, Mauro Onori, Ihab Ragai, and Martin Törngren for their cooperative work and technical expertise, the results of which added a great value to this book to complete the coverage of new knowledge.

Finally, the authors are deeply thankful to their families with ultimate respect and gratitude for their continuous support during the process of this book preparation in evenings and weekends. Without their understanding and encouragement, this book would never be materialised.

# Contents

# Part I
# Literature Survey and Trends

# Chapter 1
# Latest Advancement in Cloud Technologies

## 1.1 Introduction to Cloud Computing

During the past decade, a new computing paradigm—cloud computing has emerged as a result of the availability of high-performance networks, low-cost computers and storage devices as well as the widespread adoption of hardware virtualisation, Service-Oriented Architecture (SOA), and autonomic and utility computing. Cloud computing is a model of service delivery and access where dynamically scalable and virtualised resources are provided as a service with high reliability, scalability and availability over the Internet. Cloud computing introduces a new operating and business model that allows customers to pay only for resources they actually use instead of making heavy upfront investments. It creates a brand new opportunity for enterprises with the advantages of higher performance, lower cost, high scalability, availability and accessibility, and reduced business risks and maintenance expenses. Cloud computing relies on sharing of resources to achieve coherence and economy of scale.

The objective of this section is to give a brief but comprehensive introduction to cloud computing. Specifically, Sect. 1.1.1 presents the historical evolution and background of cloud computing. Section 1.1.2 gives a comprehensive introduction to the concept of cloud computing, including its definition, operation pattern, architecture, service delivery models, deployment models, characteristics, and architectural requirements with respect to cloud providers, enterprises that use the cloud as a platform, and end users. Section 1.1.3 devotes to the core and related technologies of cloud computing. Section 1.1.4 introduces a number of typical cloud computing infrastructure and platforms. Some tools for implementing cloud computing are presented in Sect. 1.1.5. Finally, in Sect. 1.1.6, challenges of cloud computing to be addressed in the future are discussed.



L. Wang and X. V. Wang, *Cloud-Based Cyber-Physical Systems in Manufacturing*,
https://doi.org/10.1007/978-3-319-67693-7_1

### 1.1.1 Historical Evolution and Background

The idea of cloud computing is not completely new. In fact, as early as in the 1960s, John McCarthy already envisioned that computing facilities could be provided to the general public as a utility. In 1969, Leonard Kleinrock [1], one of the chief scientists of the original Advanced Research Projects Agency Network (ARPANET) project, also anticipated that computing services could be obtained on demand as conveniently as obtaining other utility services such as water, electricity, gas, and telephony available in today's society in the 21st century [2].

The advent of the Internet provides an important basis for achieving that vision. Over the past decades, with the emergence of the Internet, many new computing paradigms such as grid computing, peer-to-peer (P2P) computing, service computing, market-oriented computing, and utility computing have been proposed and adopted to edge closer towards achieving the vision of cloud computing. Grid computing [3, 4] made it possible to share, select and aggregate a wide variety of geographically distributed resources such as supercomputers, storage systems, data sources and dedicated devices from different organisations for solving large-scale problems in science, engineering and commerce. The idea of P2P computing [5] is to allows peer nodes (i.e. computers) to share content directly with each other in a decentralised environment. Services computing [6] establishes a linkage between business processes and Information Technology (IT) services to enable seamless automation of business processes by making use of IT services such as SOA and Web Services. Market-oriented computing [7] views computing resources in economic terms such that users can utilise computing resources needed by paying resource providers.

The latest paradigm is cloud computing, in which computing resources are transformed into services that are commoditised and delivered in a similar manner that traditional utilities such as water, electricity, gas and telephony are delivered. In such a model, users can access services based on their requirements without needing to know where the services are hosted or how they are delivered. In fact, cloud computing emerges as a result of the evolution and convergence of several computing trends such as Internet delivery, "pay-as-you-go/use" utility computing, elasticity, virtualisation, distributed computing, storage, content outsourcing, Web 2.0 and grid computing. Cloud computing possesses several salient features that differentiate it from traditional service computing, including multi-tenancy, shared resource pooling, geo-distribution and ubiquitous network access, service oriented, dynamic resource provisioning, self-organising, and utility-based pricing. Table 1.1 describes how cloud entered into the market [8].

### 1.1.2 Concept

Many definitions of cloud computing have been reported [9]. In particular, the National Institute of Standards and Technology (NIST) [10] defined cloud

**Table 1.1** Cloud retrospective, adopted from [8]

| Year | Description |
|---|---|
| 2000–2005 | Dot.com bubble burst led to introduction of cloud |
| 2006 | Amazon entered the cloud market |
| 2007–2008 | The market disagreed on the understanding of cloud |
| 2008 | Cloud market expanded as more vendors joined |
| 2008–2009 | IT attention shifted to emerging private cloud |
| 2009–2010 | The open source cloud movement took hold (e.g. Openstack) |
| 2009–2011, 2012 | Cloud computing found its way, became popular, and every organisation started implementing cloud platform. In 2011, a new deployment model called hybrid cloud was born |
| 2012–2013, 2014 | The Australian Bureau of Statistics (ABS) 2013–14 Business Characteristics Survey (BCS) showed that one in five businesses had been using some form of paid cloud computing service. The overall results showed that between 2012–13 and 2013–14, businesses using information technology increased. When examining the areas where businesses used IT to a high extent, 60% used it for accounting, and 55% used it for invoicing business processes (http://www.zdnet, ABS article, online, 24 September 2015) |
| 2014–2015 | Many IT companies moved towards adopting cloud technology because of its effectiveness and fast growth |

manufacturing as "a *model for enabling ubiquitous, on-demand access to a shared pool of configurable computing resources (e.g. computer networks, servers, storage, applications and services), which can be rapidly provisioned and released with minimal management effort or service provider interaction*".

The typical operation model of cloud computing is as follows. Large companies such as Google, Amazon and Microsoft build and manage their cloud infrastructure and platforms and lease resources to enterprises using a usage-based pricing model. In the ecosystem of cloud computing, there may also be service providers who provide services to end users by renting resources from cloud infrastructure providers.

Five essential characteristics of cloud computing have been identified by the NIST [10]:

- **On-demand self-service**. A consumer can unilaterally provision computing capabilities, such as server time and network storage, as needed automatically without requiring human interaction with each service provider.
- **Broad network access**. Capabilities are available over the network and accessed through standard mechanisms that promote use by heterogeneous thin or thick client platforms (e.g. mobile phones, tablets, laptops, and workstations).
- **Resource pooling**. The provider's computing resources are pooled to serve multiple consumers using a multi-tenant model, with different physical and virtual resources dynamically assigned and reassigned according to consumer demands. There is a sense of location independence in that the customer generally has no control or knowledge over the exact location of the provided

resources but may be able to specify location at a higher level of abstraction (e.g. country, state, or data center). Examples of resources include storage, processing, memory and network bandwidth.

- **Rapid elasticity**. Capabilities can be elastically provisioned and released, in some cases automatically, to scale rapidly outward and inward commensurate with demand. To the consumer, the capabilities available for provisioning often appear to be unlimited and can be appropriated in any quantity at any time.
- **Measured service**. Cloud systems automatically control and optimise resource use by leveraging a metering capability at some level of abstraction appropriate to the type of service (e.g. storage, processing, bandwidth, and active user accounts). Resource usage can be monitored, controlled, and reported, providing transparency for both providers and consumers of the service utilised.

Cloud computing requires an architecture as a guidance for its implementation. Generally speaking, the architecture of a cloud computing system can be divided into four layers: hardware/datacentre layer, infrastructure layer, platform layer, and application layer, as shown in Fig. 1.1. Each layer is loosely coupled with the adjacent layers. This loose coupling between different layers allows each layer to evolve separately. This layered and modularised architecture makes cloud computing able to support a wide range of application requirements while reducing management and maintenance overhead [11].

- **Hardware layer**. This layer is responsible for managing the physical resources of the cloud, including physical servers, routers, switches, power and cooling systems. In practice, the hardware layer is typically implemented in data centres. A data centre usually contains thousands of servers that are organised in racks and interconnected through switches, routers or other fabrics. Typical issues at

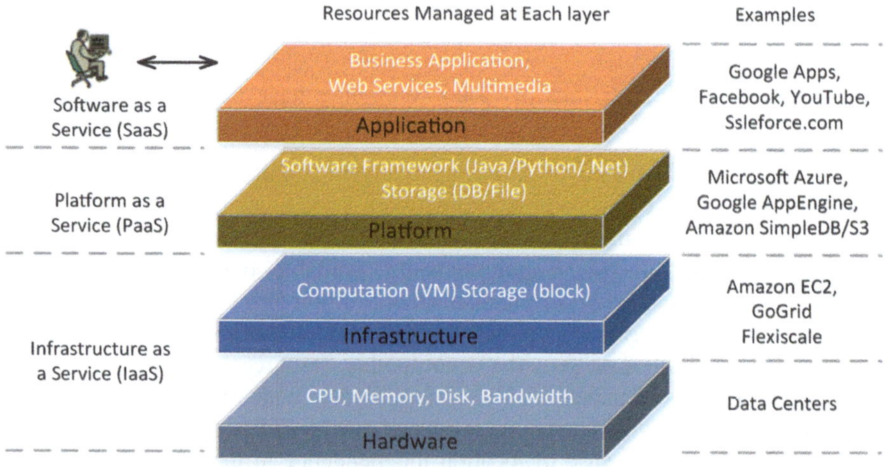


**Fig. 1.1** Cloud computing architecture, adopted from [11]

hardware layer include hardware configuration, fault tolerance, traffic management, power and cooling resource management.

- **Infrastructure layer**. This layer, also known as the virtualisation layer, creates a pool of storage and computing resources by partitioning the physical resources using the virtualisation technology. The infrastructure layer is an essential component of cloud computing, since many key features, such as dynamic resource assignment, are only made possible through virtualisation.
- **Platform layer**. Built on top of the infrastructure layer, the platform layer consists of operating systems and application frameworks. The purpose of the platform layer is to minimise the burden of deploying applications directly into virtual machine containers. For example, Google App Engine operates at the platform layer to provide Application Programming Interface (API) support for implementing storage, database and business logic of typical web applications.
- **Application layer**. The application layer consists of actual cloud applications. Different from traditional applications, cloud applications can leverage the automatic-scaling feature to achieve better performance, availability and lower operating costs.

In cloud computing, everything is treated as a service (i.e. XaaS), e.g. SaaS, PaaS, and IaaS. These services are usually delivered through industry standard interfaces such as Web services, SOA or REpresentational State Transfer (REST) services.

- **SaaS**. In this service model, software applications that run on a cloud infrastructure are delivered to consumers over the Internet. As a result, this model is sometimes referred to as Application as a Service (AaaS). Users can access SaaS applications and services from any location using various client devices through either a thin client interface, such as a web browser (e.g. web-based email), or a program interface based on subscription whenever there is an Internet access. For SaaS, consumers do not manage or control the underlying cloud infrastructure, including network, storage, servers, operating systems, or even individual application capabilities, with the possible exception of limited user-specific application configuration settings. Typical examples of SaaS are the Salesforce Customer Relationships Management system, NetSuite, Google Office Productivity application, Microsoft Office 365, Facebook, YouTube, and Twitter.
- **PaaS**. PaaS is a collection of runtime environments such as software and development tools hosted on the provider's infrastructures. It acts as the background that provides runtime environment, software deployment framework and component to facilitate the direct deployment of application level assets or web applications. Users access these tools over the Internet by means of APIs, Web portals or gateway software. Application developers, implementers, testers, and administrators can go for developing, testing and deploying their software in this platform. Users does not manage or control the underlying cloud infrastructure, including network, storage, servers, or operating systems,

but has control over the deployed applications and possibly configuration settings for the application-hosting environment. Commonly found PaaS includes Facebook F8, Salesforge App Exchange, Google App Engine, Amazon EC2, and Microsoft Azure.

- **IaaS**. IaaS provides consumers with processing, storage, networks, and other fundamental computing resources where consumers are able to deploy and run arbitrary software such as operating systems and applications. Hence, IaaS is sometimes called Hardware as a Service (HaaS). Virtualisation is the backbone behind this model where resources such as network, storage, virtualised servers, routers and so are consumed by users through virtual desktop, provided by cloud service providers (CSPs). Users are charged based on usage of CPU, storage space, value added services (e.g. monitoring, auto-scaling etc.), network bandwidth, and network infrastructure. The consumer does not manage or control the underlying cloud infrastructure but has control over operating systems, storage, and deployed application, and possibly limited control of selected networking components (e.g. host firewalls). On-demand, self-sustaining or self-healing, multi-tenant, customer segregation are the key requirements of IaaS. Examples of IaaS include Mosso/Rackspace, VMWare, and storage services provided by Amazon S3, Amazon EC2, and GoGrid.

Figure 1.2 illustrates the different service levels of cloud services in cloud computing for different service models [12, 13]. It should be noted that IaaS, PaaS, and SaaS are usually suitable for IT professionals, developers, and business end users, respectively.

- **Hybrid cloud**. The cloud infrastructure is a composition of two or more distinct cloud infrastructures (private and public) that remain unique entities, but are bound together by standard or proprietary technology that enables data and application portability (e.g. cloud bursting for load balancing between clouds). A hybrid cloud allows one to extend either the capacity or the capability of a

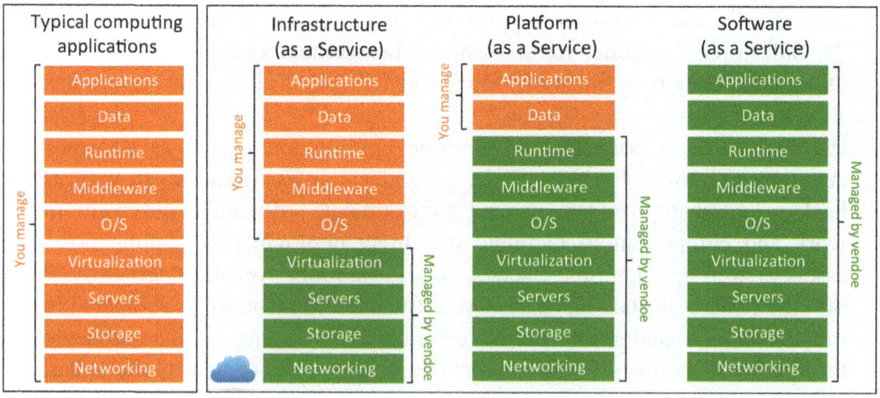


**Fig. 1.2** Different service levels of cloud services in cloud computing

cloud service, by aggregation, integration or customisation with another cloud services. Examples are Cybercon.com (US Microsoft Hybrid Cloud), Bluemix.net (IBM Cloud App Development), etc.

Overall, there are several stakeholders involved in a cloud computing system, including cloud providers, enterprises that use the cloud as a platform, and end-users. There are specific architectural requirements with respect to the participants mentioned above [14, 15]. From the service provider's perspective, highly efficient service architecture is needed to provide virtualised and dynamic services. SaaS, PaaS, and IaaS are three effective service delivery models that can satisfy the architectural requirement of cloud computing. There are also some other essential requirements, including:

- **Service-centric issues**. To fulfil the requirements of enterprise's IT management, cloud architecture needs to take a unified service-centric approach. This approach requires services to be autonomic, self-describing, etc. Autonomic means that cloud systems/applications should be able to adapt dynamically to changes with less human assistance, and self-describing means that clients will be notified exactly about how services should be called and what type of data services will return.
- **QoS**. QoS provides the guarantee of performance and availability as well as other aspects of service quality such as security, reliability and dependability etc. QoS requirements are associated with service providers and end-users. SLAs are an effective means for assuring QoS between service providers and end-users. QoS may entail systematic monitoring of resources, storage, network, virtual machine, service migration and fault tolerance. From the perspective of a cloud service provider, QoS should emphasise performance of virtualisation and monitoring tools.
- **Interoperability**. Interoperability is an essential requirement for both service providers and enterprises. It refers to the fact that applications should be able to be ported between clouds or use multiple cloud infrastructures before business applications are delivered from the cloud. In order to achieve interoperability, an agreed-upon framework/ontology, open data format or open protocols/APIs that enable easy migration and integration of applications and data between different cloud service providers as well as facilities for the secure information exchange across platforms should be created.
- **Fault-tolerance**. Fault-tolerance refers to the ability of a system to continue to operate in the event of the failure of some of its components. Fault-tolerance requires the falling components to be isolated, and the availability of reversion mode, etc. Application-specific, self-healing, and self-diagnosis mechanisms are, for example, enabling tools for cloud providers to detect failure. Cloud providers need proper tools and mechanism such as application-specific self-healing and self-diagnosis mechanism to detect failure in cloud systems/applications. Inductive systems such as classification or clustering can also be helpful for detection of failure and identification of possible causes.

- **Load balancing**. A load balancer is a key requirement for cloud computing in order to build dynamic and stable cloud architecture. Load balancing, which represents the mechanism of self-regulating workloads properly within the cloud's entities (one or more servers, hard drives, network, and IT resources), can be provided by software or hardware. Load balancing is often used to implement failover in that the service components are monitored continually and when one becomes non-responsive, the load balancer stops sending traffic, de-provisions it and provisions a new service component.
- **Virtualisation management**. Virtualisation refers to abstraction of logical resources from their underlying physical characteristics in order to improve agility, enhance flexibility and reduce cost. There are many different types of virtualisation in cloud computing, including server virtualisation, client/desktop/application virtualisation, storage virtualisation, network virtualisation, and service/application infrastructure virtualisation, etc. Handling a number of virtualisation machines on the top of operating systems and evaluating, testing servers and deployment to the targets are some of the important concerns of virtualisation. Virtualisation is indispensable for a dynamic cloud infrastructure as it brings important advantages in sharing of cloud facilities, management of complex systems as well as isolation of data/application. Quality of virtualisation determines the robustness of a cloud infrastructure.

For enterprises that use cloud computing, the critical requirements are that they should always know what services they are paying for, as well as be clear about issues like service levels, privacy matters, compliances, data ownership, and data mobility. This section describes some of the cloud deployment requirements for enterprises.

- **Cloud deployment for enterprises**. As far as enterprise users are concerned, an important requirement is how cloud is deployed because this can impact the way they access services. As mentioned above, there are four types of cloud deployment models, public, private, community and hybrid. Different types of deployment models suit different situations. Public cloud enables sharing the services and infrastructure provided by an offsite, third-party service provider in a multi-tenant environment; private cloud aims to achieve sharing services and infrastructure, which are provided either by an organisation or a specific service provider in a single-tenant environment. Community cloud provides a means for sharing resources among several organisations that have shared interests and concerns. Hybrid cloud consists of multiple internal (private) or external (public) clouds. Enterprises need to have a strategy that leverages all four options mentioned above.
- **Security**. Security is the focal concern of enterprises. When corporate information, including data of customers, consumers and employees, business know-how and intellectual properties is stored and managed by external entities on remote servers in the cloud, how to safeguard them in the shared environment will become a primary issue. Different service models provide different

security levels in the cloud environment: IaaS is the foundation of all cloud services, with PaaS built upon it and SaaS in turn built upon PaaS. Just as capabilities are inherited, so are the information security issues and risks.

- **Business Process Management (BPM)**. Typically, a business process management system concerns providing a business structure, security and consistent rules across business processes, users, organisations and territories. Cloud-based BPM (e.g. combining SaaS with a BPM application) enhances flexibility, deployability and affordability for complex enterprise applications. With cloud-based solution, the classical concept of BPM is enhanced as cloud delivers a business operating platform for enterprises such as combining SaaS and BPM applications (e.g. customer relationship management, workforce performance management, enterprise resource planning, e-commerce portals etc.) which helps achieve flexibility, deployability and affordability of complex enterprise applications. When an enterprise adopts cloud-based services or business processes, the return of investment of overall business measurement is important.

Users' requirements are the third key factor that need to be addressed for the adoption of any cloud system. For users, the trust issues are a major concern for the adoption of cloud services. In order to win users' trust, cloud should be trustworthy, stable, and secure. Stability and security can play a vital role to increase the trust between user and service providers. Furthermore, cloud-based applications should also be able to support personalisation, localisation and internationalisation to create a user-friendly environment. This section describes users' requirements from the perspectives of consumption-based billing and metering, user centric privacy, service level agreements and user experience.

- **User consumption-based billing and metering**. Users' billing and metering with respect to consumption of cloud services in a cloud system should be similar to the consumption measurement and allocation of water, gas or electricity on a consumption unit basis as users have a strong need for transparency of consumption and billings. Cost management is important for making planning and controlling decisions. Cost breakdown analysis, tracing the utilised activity, adaptive cost management, transparency of consumption and billings are also important considerations.
- **User-centric privacy**. User-centric privacy mainly concerns the storage of users' personal/enterprise sensitive data such as intellectual property at mega-data centres located around the world. There is strong resistance and reluctance for an enterprise to store any sensitive data in the cloud. Cloud providers need to make every effort to win the trust of their users. Currently, there are various technologies that can enhance data integrity, confidentiality, and security in the clouds, e.g. data compressing and encrypting at the storage level, virtual LANs and network middle-boxes (e.g. firewalls and packet filters).
- **SLAs**. SLAs are mutual contract between providers and users, representing the ability to deliver services in line with predefined agreements. Currently, many

cloud providers offer SLAs but these SLAs are rather weak on user compensations on outages. There are some important architectural issues concerning SLAs to be addressed including measurement of service delivery, method of monitoring performance, and amendment of SLA over time.

- **User Experience (UX)**. UX represents the overall feeling of users in using cloud-based application/systems. The notion of UX can provide important insights into the needs and behaviours of end-users so as to maximise the usability, desirability and productivity of application/systems. Cloud-based application/systems should be easy to use, capable of providing faster and reliable services, easily scalable, and customisable to meet the goal of localisation and standardisation. Human-Computer Interaction, ergonomics and usability engineering are some of the key technologies that can be used for designing UX-based cloud applications.

### *1.1.3 Technologies*

Cloud computing evolves from the evolution and adoption of existing technologies and paradigms. One of the most important technologies for cloud computing is virtualisation. Other technologies are concerned with the architecture of data centres, distributed file system, as well as distributed application framework. Cloud computing is also often compared to other technologies such as grid computing, utility computing, and autonomous computing, each of which shares something in common with cloud computing [11].

- **Virtualisation**. Virtualisation is a technology that abstracts details of physical hardware and provides virtualised resources for high-level applications. Virtualisation is able to separate a physical computing device into one or more virtual devices, each of which can be used and managed to perform computing tasks independently. With operating system-level virtualisation essentially creating a scalable system of multiple independent computing devices, idle computing resources can be allocated and used more efficiently. Virtualisation constitutes the foundation of cloud computing, as it provides the ability to pool computing resources from clusters of servers and dynamically assigning or reassigning virtual resources to applications on demand. Virtualisation provides the agility needed to speed up IT business operations and reducing costs by increasing infrastructure utilisation.
- **Architectural design of data centres**. A data centre is home to thousands of devices like servers, switches and routers. Proper planning of this network architecture is critical as it has a major impact on application performance and throughput in such a distributed environment. Scalability and resiliency are features that also need to be considered carefully. A common practice is adopting the layered approach for the design of network architecture of a data centre. Basically, the design of network architecture should be able to meet

objectives such as high capacity, free VM migration, resiliency, scalability, and backward compatibility.

- **Distributed file system over clouds**. There are two major file systems for cloud computing: Google File System (GFS) and Hadoop Distributed File System (HDFS). The former is a proprietary distributed file system designed especially for providing efficient, reliable access to data using large clusters of commodity servers by Google. Compared with traditional file systems, GFS is designed and optimised to provide extremely high data throughput, low latency and survive individual server failures. Inspired by GFS, the open source HDFS stores large files across multiple machines, achieving high reliability by replicating the data across multiple servers. Similar to GFS, data is stored on multiple geographically distributed nodes. The file system is built from a cluster of data nodes, each of which serves blocks of data over the network using a block protocol specific to HDFS. Data is also provided over HTTP, allowing access to all content from a web browser or other types of clients. Data nodes can talk to each other to rebalance data distribution, to move copies, and to keep the replication of data high.
- **Distributed application framework over clouds**. MapReduce is a software framework introduced by Google to support distributed computing on large datasets on clusters of computers. MapReduce consists of one Master, to which client applications submit MapReduce jobs. The Master pushes work to available task nodes in the data centre, striving to keep the tasks as close to the data as possible. The open source Hadoop MapReduce project is inspired by Google's work. Today, many organisations are using Hadoop MapReduce to run large data-intensive computations.
- **Grid computing**. Grid computing is a distributed and parallel computing model. A grid computing system is a cluster of networked, loosely coupled computers acting together to perform large tasks. The development of grid computing was originally driven by computation-intensive scientific applications. Cloud computing is similar to grid computing in that it also employs distributed resources. However, cloud computing leverages virtualisation technologies at multiple levels (hardware and application platform) which enable it to achieve resource sharing and dynamic resource provisioning. Cloud computing can be considered the business-oriented evolution of grid computing.
- **Utility computing**. Utility computing represents a model of providing resources on demand and charging customers by a pay-per-use model. Cloud computing can be perceived as a realisation of utility computing. With on-demand resource provisioning and utility-based pricing, cloud computing service providers can maximise resource utilisation and minimise their operating costs.
- **Autonomic computing**. Originally coined by IBM in 2001, autonomic computing aims at building computing systems capable of self-management, i.e. reacting to internal and external events without human intervention. The goal of autonomic computing is to overcome the management complexity of today's computer systems. Cloud computing exhibits some autonomic features such as

automatic resource provisioning. However, its objective is to lower resource costs instead of reducing system complexity.

### *1.1.4 Cloud Platforms*

A number of industrial organisations have developed their cloud computing infrastructure, among which the dominant ones include Amazon Elastic Compute Cloud (Amazon EC2), Google App Engine, and Microsoft Azure [2]. Table 1.2 describes the features of these cloud platforms from the perspectives of SLA, reliability, auto-scaling, virtualisation, privacy, storage, and security.

- **Amazon EC2**. Amazon EC2 is a web service that provides secure, resizable compute capacity in the cloud. It creates a virtual computing environment for users to launch and manage server instances in data centres using APIs or available tools and utilities. Users can either create a new Amazon Machine Image (AMI) containing the applications, libraries, data and associated configuration settings, or select from a library of globally available AMIs. Users then need to upload the created or selected AMIs to Amazon Simple Storage Service (S3) before they can perform some activities such as starting, stopping,

**Table 1.2** Comparison of the representative cloud platforms, adapted from [2]

| Property | System | | |
|---|---|---|---|
| | Amazon EC2 | Google App Engine | Microsoft Azure |
| Focus | Infrastructure | Platform | Platform |
| Service type | Compute, storage (Amazon S3) | Web application | Web and non-web |
| Virtualisation | OS level running on a Xen hypervisor | Application container | OS level through fabric controller |
| Dynamic negotiation of QoS parameters | None | None | None |
| User access interface | Amazon EC2 command-line tools | Web-based administration console | Microsoft Windows Azure portal |
| Web APIs | Yes | Yes | Yes |
| Value-added service providers | Yes | No | Yes |
| Programming framework | Customisable Linux-based Amazon Machine Image (AMI) | Python | Microsoft .NET |

and monitoring instances of the AMIs uploaded. Amazon EC2 charges users for the period of time during which the instance is alive, while Amazon S3 charges for any data transfer.

- **Google App Engine**. Google App Engine is a platform for traditional web applications in data centres managed by Google. It allows users to run web applications written using the Python or Java programming language. In addition to the Python standard library, Google App Engine also supports Application Programming Interfaces (APIs) for the data store, Google Accounts, URL fetch, image manipulation, and email services. Current APIs support the following features, including data storage and retrieval from a BigTable non-relational database, making HTTP requests and caching. Google App Engine provides a web-based Administration Console for facilitating users to manage their web applications.
- **Microsoft Windows Azure platform**. Microsoft Windows Azure aims to provide an integrated development, hosting, and control cloud computing environment. Microsoft's Windows Azure platform consists of three components and each of them provides a specific set of services to cloud users, including Windows Azure, SQL Azure, and .NET Services. Windows Azure provides a Windows-based environment for running applications and storing data on servers in data centres, SQL Azure provides data services in the cloud based on SQL Server, and .NET Services offer distributed infrastructure services to cloud-based and local applications. Windows Azure platform can be used both by applications running in the cloud and applications running on local systems. All of the physical resources, VMs and applications in the data centre are monitored by software called the fabric controller.

### *1.1.5 Tools*

Various tools for implementing cloud computing are available in the market [8]. These tools, open source and commercial, provide environments and platforms for developing various cloud services and implementing their own algorithms and mechanisms.

Open source tools, such as Open Nebula, Apache Cloud Stack, Nimbus and Eucalyptus, can be used/accessed free of charge (Table 1.3). Each of these tools has its own features, and shows a different degree of support for security, API, and cloud types.

There are also some commercial tools for cloud computing in the market such as RightScale, Gravitant, VMTurbo and Scalr. These commercial tools are explained in more detail below.

- RightScale: RightScale grid framework can achieve automated management of workflows of messages and jobs. It also provides the mechanism of implementing the elasticity of grid processing solutions. Input queues of the system

**Table 1.3** Open source tools, adapted from [8]

| Tool name | Features | Security | API | Cloud type |
|---|---|---|---|---|
| Open Nebula: It adopts computing, storage, security, monitoring, virtualisation and networking in their data centres | Cloud bursting, on-demand provision of virtual data centres, multiple zones, multi-VM application management | Fine-grained ACLs and user quotas; Integration with LDAP, Active Directory | AWS EC2 and EBS APIs; OGF OCCI APIs | Private |
| Apache Cloud Stack: Easy integration with existing portal and it is fully AJAX-based solution compatible with most of the latest Internet browsers | Powerful API; Multi-role support; On-demand virtual data centre hosting; Dynamic workload management; Broad network virtualisation capabilities | Secure AJAX console access; Secure single sign on; Secure cloud deployments; MPLS support in the cloud | Cloud Stack provides an API that is compatible with AWS EC2 and S3 for organisations to deploy hybrid clouds | Public, hybrid |
| Nimbus: Power and versatility of infrastructure clouds to scientific users; It allows combining Nimbus, OpenStack, Amazon, etc. | Support for proxy credentials for scientific community, batch schedulers, best-effort allocations and others are special targeting features | – | EC2/S3 an API as a compatible IaaS | Private, public |
| Eucalyptus: It helps customers to design and deploy cloud solutions more quickly | Multi-cluster tunnelling and LDAP integration | – | – | Private, hybrid |

are continuously monitored when certain criteria are met. Additional, worker instances are launched to handle the increased processing load.

- Gravitant: Gravitant's cloudMatrix platform is a leading cloud services brokerage and management platform that integrates multiple cloud providers' services (internal or external) into a catalogue and provisioning portal so that enterprises can optimise the consumption of cloud services. Gravitant's cloudMatrix platform enables the core services and features, which can be delivered as packages through a single user interface on myGravitant.com and through a white labelled internal broker platform. Enterprises can deploy these capabilities independently or as an integrated suite based on their cloud service needs.

- VMTurbo: VMTurbo provides a demand-driven cloud and virtualisation control platform for enterprise businesses.
- Scalr: Scalr is suitable for those who look to explore the platform and to build and test their projects on their own. It delivers self-service access to cloud infrastructure and acts as an intermediary management layer between cloud infrastructure and engineering, and provides the ownership of information security back to IT department hands. Scalr enforces cloud infrastructure security such as governance and compliance to create and enforce policies on the basis of budgets, configurations, and user access across entire cloud portfolio. Network policy enforcement allows securing cloud infrastructure by regulating the use of networks. It also enables the delivery of single sign-on across private and public clouds through authentication and authorisation techniques.

### 1.1.6 Challenges

Despite enormous benefits of cloud computing, its adoption is slow due to potential risks and limitations such as data loss, data cleaning, account hijacking, lack of portability/migration from one service provider to another, less reliable, lack of auditability, and less QoS. As a result, there are many challenges in terms of security, interoperability, virtualisation, data leakage, resource sharing, load balancing, multi-tenancy, and SLAs, including those that are concerned with outsourcing data and applications, SLA, extensibility and shared responsibility, cloud interoperability, heterogeneity, multi-tenancy, load balancing, resource scheduling, virtualisation, and privacy and security, etc. The difficulties behind these challenges were identified, and the possible solutions to these challenges were given in [8].

Especially, security is a critical issue in the cloud computing paradigm that can significantly affect the widespread adoption of cloud computing because security is a primary concern for businesses contemplating cloud adoption [16]. For the security issues, the following objectives should be achieved: confidentiality, integrity, availability, authenticity, and accountability. These five objectives represent the basic security requirements. Confidentiality refers to the fact that private or sensitive information should be accessible only to the right people who are authorised; Integrity means protecting against inappropriate information destruction or modification, including ensuring information non-repudiation and authenticity; Availability means ensuring reliable and timely access to and use of information; Authenticity means that a message, transaction, or additional exchange of information is from the source it claims to be from; Accountability means that actions of an entity should be able to be traced uniquely.

Security issues can be divided into three categories [8]: data centre-related security issues, network-related security issues, and other security issues. Data centre-related security issues include those that are concerned with multi-location of service provider, data combination and commingling, restrictions on techniques and logistics, data transfer across gateway, and multi-location of private data. Network-related security issues includes those that are concerned with SQL injection attack, cross-site scripting attack, man-in-middle attack, sniffer attack, reuse IP addresses, security concerns with Hypervisor, DoS attack, cookie poisoning, DDoS attack, and COPTCHA splitting/breaking, Other common security challenges include abuse and nefarious use of cloud computing, insecure application programming interface, malicious insider, shared technology, vulnerability, data loss/leakage, traffic hijacking and account, investigation, and data segregation.

## 1.2 Cloud Manufacturing

Cloud computing aims to realise the idea of offering computing resources as services in a convenient pay-as-you-go manner. Expanding this idea into the manufacturing realm has given rise to the concept of cloud manufacturing [15]. The term and complete concept system of cloud manufacturing were initially introduced by Li et al. [17]. The most prominent and promising feature of cloud manufacturing is the seamless and convenient sharing of a variety of different kinds of distributed manufacturing resources, realising the idea of Manufacturing-as-a-Service (MaaS). Following this concept, companies are provided with the ability to obtain various manufacturing services from the Internet as conveniently as obtaining water and electricity [18]. Cloud manufacturing is a new paradigm that will revolutionise the manufacturing industry [19].

The objective of this section is to provide a comprehensive introduction to cloud manufacturing, including the background for introducing the concept, the concept itself, enabling technologies, research initiatives, applications and challenges. Specifically, Sect. 1.2.1 presents the historical evolution and background of cloud manufacturing. Section 1.2.2 gives a comprehensive introduction to the concept of cloud manufacturing, including its definition, operation principle, resource classification, architecture, service delivery models, deployment models, etc. Section 1.2.3 devotes to the core and supporting technologies that enables the implementation of cloud manufacturing. Section 1.2.4 presents a number of research initiatives of cloud manufacturing around the world. In Sect. 1.2.5, typical applications based on the idea of cloud manufacturing are presented. Finally, Sect. 1.2.6 discusses the challenges of cloud manufacturing to be addressed in the future.

### *1.2.1 Historical Evolution and Background*

Over the last 40 years, many advanced manufacturing paradigms have been proposed, including mass customisation, holonic manufacturing, reconfigurable manufacturing, lean manufacturing, agile manufacturing, networked manufacturing, manufacturing grid, and sustainable manufacturing [20]. The manufacturing focus has shifted from enlarging production scale in the 1960s to cost reduction in the 1970s, from product quality in the 1980s to rapid market response in the 1990s, and lately focusing on service, information and knowledge.

Today, many trends of manufacturing have emerged, including globalisation, individualisation, customisation, deep customer involvement, servitisation, intelligence, etc. In order to adapt to the development trends of the manufacturing industry, companies need to focus on global resource sharing and manufacturing operation collaboration for being agile, cost-effective. In this context, research on collaboration and resource-sharing in all stages of the product lifecycle has received more and more attention. However, there are some major limitations and shortcomings with existing networked manufacturing models and technologies in resolving the collaboration and resource sharing issues. For example, networked concepts such as Internet-based, distributed and manufacturing grid focus on undertaking a single manufacturing task through integration of distributed resources. They do not have centralised operation management of services, freedom to choose different operation modes and embedded access of physical manufacturing equipment, applications and capabilities to the Internet, which are prerequisites for achieving seamless, stable and quality transactions of manufacturing resources. Having little coordination between providers and consumers, these concepts are significantly less effective [17].

Recently, emerged computer, information and especially the Internet technologies, such as cloud computing, IoT, Semantic Web, embedded systems and virtualisation technologies, provide new means for enabling seamless collaboration activities for all the phases of product development. Especially, cloud computing, which delivers computing services over the Internet based on cloud technologies, provides important inspirations for the manufacturing industry, i.e. providing manufacturing resources as services over the Internet, e.g. Design-as-a-Service (DaaS), Machining-as-a-Service (MCaaS), Robot Control as a Service [21], etc.

In this context, cloud manufacturing as a manufacturing paradigm was proposed. Cloud manufacturing promises elasticity, flexibility and adaptability through the on-demand provisioning of manufacturing resources as services, enabling the fundamental and necessary features such as convenient scalability and pay-as-you-go of resources sharing. Cloud manufacturing can effectively address the common challenges that many SME manufacturing companies are facing today such as lack of core technologies, lack of skills of using and managing complex IT systems, lack of opportunities of accessing external resources and capabilities, lack of follow-up services, and more importantly, lack of resource- and capability-sharing mode.

### *1.2.2 Concept*

The concept of cloud manufacturing was first introduced a couple of years ago. Since then, there has been a growing interest in the academic and industrial communities. Thus far, researchers from both academia and industry have purposed a number of definitions of cloud manufacturing from different perspectives and backgrounds. Li et al. [17] defined it as "*a new networked manufacturing paradigm that organises manufacturing resources (i.e. manufacturing cloud) according to customers' requirements for providing on-demand manufacturing services through the Internet and cloud manufacturing platform*". Mirroring NIST's definition of cloud computing, Xu [15] subsequently defined cloud manufacturing as "*a model for enabling ubiquitous, convenient, on-demand network access to a shared pool of configurable manufacturing resources (e.g. manufacturing software tools, manufacturing equipment, and manufacturing capabilities) that can be rapidly provisioned and released with minimal management effort or service provider interaction*". Looking at CPS, Wang et al. [22] defined cloud manufacturing as "*an integrated cyber-physical system that can provide on-demand manufacturing services, digitally and physically, at the best utilisation of manufacturing resources*".

Although there are a number of different definitions, no one has been accepted as the standard one. In spite of this, researchers and members from this particular area, to a large extent, have had a quite clear view on what cloud manufacturing would comprehend and facilitate such as the needs and requirements driving its development and implementation, the services and solutions it would make available and perform, and the concepts and technologies it could build upon are reaching a much higher degree of consensus and agreement [18].

There are overall three categories of participants in a cloud manufacturing system: operator (i.e. cloud provider), resource and service providers, and resource and service consumers (Fig. 1.3) [18].

**Fig. 1.3** Operation principle of cloud manufacturing, adapted from [18]

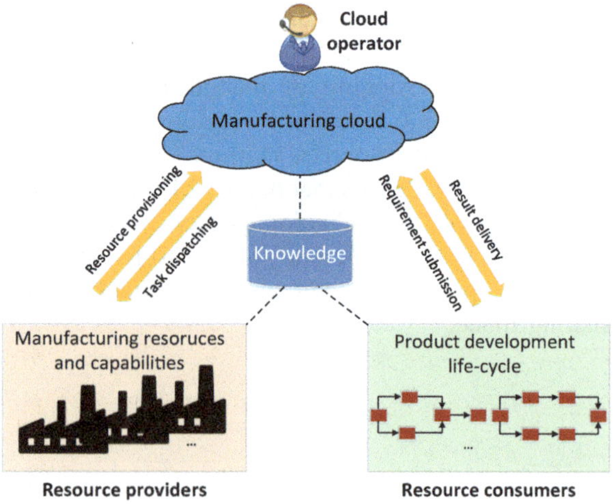

- **Operator (i.e. cloud provider)**. An operator is introduced for managing and operating a cloud manufacturing platform. Introducing the operator is an important feature that differentiates cloud manufacturing from previous manufacturing models. It is also a key means for providing continuous and high-quality services. An operator of a cloud manufacturing platform plays the same role with the cloud provider in cloud computing. Operation and management of the cloud manufacturing platform or system including delivering required support and functions to providers and consumers and maintaining services and technologies required to run the system, as well as finding, combining, controlling and coordinating the required services for fulfilling consumer requirements.
- **Resource and service providers**. Different from the cloud provider of cloud computing (who owns and manages all computing resources necessary for providing all types of computing services), the operator in cloud manufacturing, in general, cannot own all types of manufacturing resources (or does not own any manufacturing resources). As a result, resources in cloud manufacturing come from different providers (e.g. enterprises). In cloud manufacturing, providers from different industries provide their various types of resources or services to a cloud manufacturing platform for the sharing purpose. Hence, manufacturing resources in cloud manufacturing may include all resources encompassed in lifecycles of various types of products. All manufacturing resources provided by different providers exist in the cloud manufacturing platform as services. Depending on resource types and business models, providers may have complete, partial, or no control over resources and services they provide.
- **Resource and service consumers**. The ultimate aim of cloud manufacturing is to provide on-demand manufacturing services to consumers. Cloud manufacturing allows consumers to request services by submitting their requirements to the cloud platform. Consumers are charged on a pay-per-use basis.

It should be noted that the classification of cloud users (including providers and consumers) is based on their functional roles, which may change over time. For example, if an enterprise requests services from a cloud platform, it is a provider. An enterprise may provide some types of resources while request some other types of resources, and in this case, it is a provider and a consumer concurrently. Knowledge plays an important role in cloud manufacturing activities such as perception, connection, virtualisation and encapsulation of manufacturing resources and capabilities, cloud service description, matching, searching, aggregation, and composition, optimal allocation and scheduling of activities and services, etc.

In cloud manufacturing, everything is provided as a service. Although some different classifications of manufacturing resources exist, most agree to the fact that manufacturing resources can be classified into physical manufacturing resources and manufacturing capabilities. Physical resources can be either hard (e.g. manufacturing equipment, computers, networks, servers, materials, logistics facilities, etc.) or soft (e.g. applications, product design and simulation software, analysis tools, models, data, standards, human resources such as personnel of different professions and their knowledge, skills and experience, etc.). Manufacturing capabilities are intangible and dynamic recourses that represent an organisation's capability of undertaking a specific task.

Cloud manufacturing needs a system architecture as a guidance for its implementation. Many research efforts have been made towards the architecture of a cloud manufacturing system, and the proposed cloud manufacturing architectures range from four layers to up to twelve layers. Summarising the architectures proposed, a cloud manufacturing system architecture overall consists of the following layers according to the functions and contents: resource layer, perception layer, virtualisation layer, cloud service layer, application layer, and interface layer as well as other supporting layers, including security layer, knowledge layer, and communication layer (Fig. 1.4) [18].

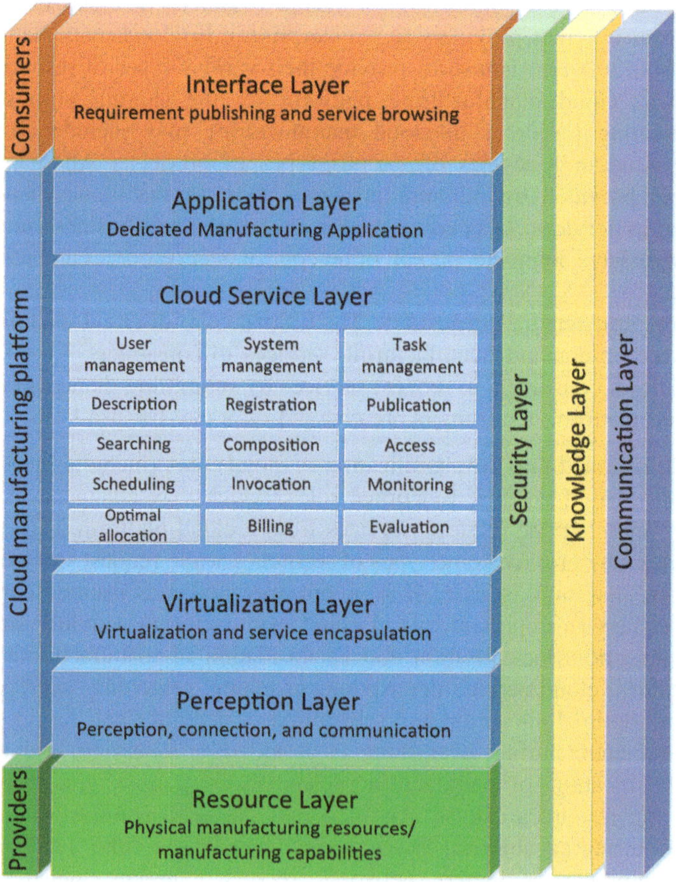


**Fig. 1.4** Cloud manufacturing architecture, adopted from [18]

- **Resource layer**. This layer belongs to the provider domain, encompassing manufacturing resources for the complete product lifecycle from different geographically distributed providers.
- **Perception layer**. This layer is responsible for intelligently sensing manufacturing resources using IoT technologies, enabling them to be connected to the cloud manufacturing platform so as to achieve communication and interactions between a cloud platform and real resources involved such as remote monitoring, prognosis, and control [23, 24].
- **Virtualisation layer**. This layer is responsible for virtualising manufacturing resources and capabilities and encapsulating them into manufacturing cloud services that can be accessed, invoked, and deployed by using virtualisation technologies, service-oriented technologies, and cloud computing technologies. The manufacturing cloud services are classified and aggregated according to specific rules and algorithms, and different kinds of manufacturing clouds can thus be constructed. This layer achieves transformation of manufacturing resources into virtual services, and construction of different manufacturing clouds, thus paving the way for subsequent cloud-based applications in the cloud service layer.
- **Cloud service layer (i.e. core middleware)**. This layer devotes to system, service, resource, and task management, and also supports various service activities and applications such as service description, registration, publication, composition, monitoring, scheduling, and charging.
- **Application layer**. Depending on providers and their offered manufacturing cloud services, dedicated manufacturing application systems such as collaborative design, collaborative manufacturing, collaborative simulation, and collaborative supply chain can be aggregated. Consumers can browse and access these application systems for manual or automatic service configurations. A manufacturing resource provider provides consumers with the ability to select from different possible part properties and predetermined manufacturing constraints (sizes, materials, tolerances, etc.).
- **Interface layer**. As the name implies, this layer serves as an interface between consumers and the cloud platform, providing consumers with an interface for submitting their requirements and browsing available services. The interface supports manual selection and combination of available services, as well as automatic cloud-generated suggested solutions.
- **Knowledge layer**. This layer provides knowledge needed in the different layers above for virtualisation and encapsulation of resources, manufacturing domain knowledge, process knowledge, etc.
- **Security layer**. This layer provides strategies, mechanisms, functions and architecture for cloud manufacturing system security [25].
- **Communication layer**. This layer provides a communication environment for users, operations, resources, services, etc. in the cloud manufacturing system.

Like cloud computing, cloud manufacturing can also be deployed in four models: public, private, community, and hybrid. A public cloud platform is built for sharing manufacturing resources with the general public. It is therefore open for all parties, which means that any enterprise can publish their resources to and request services from the cloud platform. A private cloud is built within an enterprise with the purpose of sharing resources among different subsidiaries of an enterprise. A community cloud platform is built among business partners with shared concerns for better sharing resources and business collaboration. The hybrid cloud, as its name implies, is a combination of public and private clouds.

As in cloud computing, different delivery models of cloud manufacturing can be developed, to support the integration of virtual, intangible and physical resources, i.e. CAD applications and manufacturing capabilities and equipment, as services. Infrastructure, platform and software applications can then be offered as services in cloud manufacturing, all referring to a specific phase of the manufacturing lifecycle, i.e. DaaS, MaaS, etc. [26].

Cloud manufacturing has a number of key characteristics, including Internet of manufacturing resources and ubiquitous sensing, virtual manufacturing society and flexible manufacturing system, service-oriented manufacturing and whole lifecycle capability provisioning, efficient collaboration and seamless integration, knowledge-intensive manufacturing and collective, and social manufacturing innovation [27].

### *1.2.3 Technologies*

The development and implementation of cloud manufacturing is a huge systems engineering. A wide range of technologies are needed for the development, implementation, and operation of cloud manufacturing. In order to be concise, here we focus on the following technologies: cloud computing, IoT, CPS, virtualisation, service-oriented technology, high-performance computing technology, semantic web technology, system management technologies, and big data analytics. Other technologies such as security technology may also be necessary.

- **Cloud computing**. Cloud computing plays a fundamental and key role in the development and implementation of cloud manufacturing. There are two types of cloud computing adoptions in manufacturing: manufacturing with direct adoption of cloud computing technologies and cloud manufacturing—the manufacturing version of cloud computing [15]. In terms of the direct adoption of cloud computing technologies in manufacturing, the key areas are around IT and new business models that cloud computing can readily support, such as pay-as-you-go, the convenience of scaling up and down per demand, and flexibility in deploying and customising solutions. The adoption is typically centred on the BPM applications such as Human Resources (HR), Customer Relationship Management (CRM), and Enterprise Resource Planning (ERP) functions. In cloud manufacturing, the core concept is "Manufacturing-as-a-Service", which is built on and coupled with IaaS, PaaS, and SaaS of cloud computing [26, 28]. Moreover, the business model and related technologies (especially cloud technology) of cloud computing can also provide important references for cloud manufacturing in terms of business model and technology.

- **IoT**. IoT is a core technology for cloud manufacturing. IoT relies on technologies such as radio frequency identification (RFID) tags, sensor and actuator networks, embedded systems and intelligence in smart objects. In cloud manufacturing, IoT plays the role of intelligently sensing and connecting manufacturing resources into cloud manufacturing platform to achieve remote monitoring and control, which paves the way for subsequent virtualisation and servitisation of manufacturing resources. In order to achieve sensing and connection, IoT is responsible for keeping track of resource states and order execution status, and collecting real-time data and information for remote tracking and monitoring purpose. In fact, in the context of cloud manufacturing, IoT helps achieve not only IoT, but also Internet of Services (IoS) and Internet of Users (IoU) [29].
- **CPS**. Essentially, a cloud manufacturing system is a huge CPS with the physical part being physical manufacturing resources on factory floors while the cyber part being the cloud [22]. As a result, CPS can also be regarded as a core technology for cloud manufacturing. Cyber-physical systems are systems that integrate computation and physical processes where embedded computers and networks monitor and control the physical processes with feedback loops where physical processes affect computations. The role of CPS in cloud manufacturing is different from that of IoT in that CPS emphasises the virtual part of a cloud manufacturing system, and focuses more on the interaction and communication aspect between the real and virtual manufacturing resources (i.e. cloud).
- **Service-oriented technology**. SOA refers to systems structured as networks of loosely coupled communicating services and represents an emerging paradigm for integrating heterogeneous systems, platforms, protocols and legacy systems. Service-oriented technology is a technological paradigm that is based on service-orientation paradigm and SOA with the ultimate goal of creating services and assembling them together for large-granularity applications. As a result, service-oriented technology is a key technology in the cloud service layer.
- **Virtualisation**. Virtualisation refers to abstraction of logical resources from their underlying physical characteristics to improve agility, enhance flexibility and reduce cost. Virtualising resources and capabilities requires consideration of resource characteristics and diversity, user requirements and demands as well as performance requirements of resource management. Virtualisation of manufacturing equipment poses a great challenge to the implementation of cloud manufacturing. The critical issues with virtualisation in cloud manufacturing are manufacturing resource modelling. Mapping plays a critical role in the process of virtualisation [15]. Generally, there are three mapping relationships between

resources and services: one-to-one when the functionality or capability of a resource matches one manufacturing requirement, one-to-many for a resource with multiple functions or capabilities which each matches different manufacturing requirements independently, and many-to-one when multiple resources are required to match a manufacturing requirement.

- **Semantic Web**. In cloud manufacturing, there is semantic heterogeneity in business process integration and manufacturing resource and service capabilities. Ontologies provide an effective means describing them in an unambiguous, computer-understandable form. By utilising powerful representation and reasoning abilities of Semantic Web technology, successful matching between requests and services is made possible [30].
- **System management technologies**. System management mainly includes resource and service management, knowledge and data management, task management, and platform management. Effective system management technologies and methods and service management are essential for smooth running and operation of a cloud manufacturing system. Resource and service management activities are encompassed in the entire product lifecycle, and corresponding technologies include those used for resource and service description, publication, discovery, access, virtualisation and encapsulation, composition, integration and scheduling, etc. [31]. In cloud manufacturing, all data, information, models, algorithms, rules, and strategies can be considered as knowledge. Knowledge engineering and management plays a crucial role in all the activities encompassed in the product lifecycle such as resource virtualisation, servitisation, and service composition and scheduling. Cloud manufacturing aggregates numerous manufacturing tasks from different consumers, and thus task management methods and technologies are also needed, including task description, decomposition, classification, and scheduling. Platform management is also part of system management, which focuses on the issues at the platform level, including, for example, platform architecture, security, transaction flows, etc.
- **High-performance computing (HPC)**. Both cloud computing and IoT require the support of HPC solutions. In cloud manufacturing, HPC is inevitable. HPC refers to a broad set of architectures based on multi-processor configurations as a means to enhance performance. It often uses supercomputers and computer clusters to handle multiple tasks at a high speed.
- **Big data analytics**. In cloud manufacturing, huge amounts of data will be generated during the service management and application processes concerning, for example, manufacturing resources, manufacturing services, manufacturing tasks, enterprises and users, as well as the application process. Big data analytics can play a key role in cloud manufacturing. For example, big data about service utilisation can be used for discerning high-value services and reveal cooperative relationship between enterprises [12].

### 1.2.4 Research Initiatives

There are a number of research initiatives on cloud manufacturing with both academic and industrial participants in local, national and international projects of varying sizes and scopes [18]. Some of these initiatives are summarised below.

- **CMfg**. This national Chinese research initiative coordinated by Beihang University is usually thought as the first source of the cloud manufacturing. CMfg presents an application model of cloud manufacturing, describing cloud manufacturing platform activities ranging from user requests to the return of solutions. Also proposed is a cloud manufacturing architecture with the following five layers: (1) physical layer for provider resources and capabilities; (2) virtualised resource layer for virtualising resources and encapsulate them as services; (3) service layer for cloud manufacturing core functions such as service management deployment, registration, searching, matching, composition, scheduling, monitoring, cost and pricing, billing, etc.; (4) application layer for requests within specific manufacturing applications; and (5) user layer with interfaces for both consumers requests and provider input/registration of resources. To demonstrate the feasibility of the CMfg concept, a cloud-based application—cloud simulation—based on the COSIMCSP (Cloud Simulation Platform) has been demonstrated, in which the collaborative work in the multidisciplinary design of a virtual flight vehicle prototype is simulated [17, 26, 32].
- **Cmanufacturing**. A research group at the University of Auckland, New Zealand presented a public cloud infrastructure known as ICMS (Interoperable Cloud-based Manufacturing System) [33]. ICMS has a three-layer architecture: a Smart Cloud Manager, a User Cloud, and a Manufacturing Cloud, which are responsible for assisting and supervising the interaction between consumers and providers, for managing consumers and their requests, and for managing providers and their resources, respectively. It clarifies users into two types: customer users (CUs) and enterprise users (EUs). CUs are defined as customers/organisations requesting a self-contained production task, while EUs are organisations/enterprises seeking additional capabilities and support to fulfil bigger and more demanding production tasks in collaboration with temporary partners and their services. Standard data models for cloud services and relevant features were also developed and described.
- **Cloud-based design and manufacturing (CBDM)**. A research group at Georgia Institute of Technology presented a conceptual reference model called CBDM for their interpretation of cloud manufacturing [34, 35]. CBDM is built on the concept of cloud computing, with manufacturing resources being available as different services. For the implementation of CBDM, they proposed a Distributed Infrastructure with the Centralised Interfacing System model. The Distributed Infrastructure is composed of three groups of assets: human (including consumers, producers, managers), communication (including communication network, network security and two interfaces for communicating with

the human and manufacturing processes' asset groups), and manufacturing process (hardware and software resources). The Centralised Interface System enables the system to function as a whole.
- **Cloud-based Manufacturing-as-a-Service environment**. ManuCloud, a European project funded by the European Commission, has eight consortium members from academy and industry, from four different EU member states (i.e. Austria, Germany, Hungary and the United Kingdom) [36, 37]. The objective of the project is to develop a service-oriented IT environment to support the transition from mass production to personalised, customer-oriented and eco-efficient manufacturing. A conceptual architecture with a front-end system and MaaS infrastructure to support cloud-based manufacturing of customised products has been proposed. The front-end is deployed as part of an integrated web-based portal to support collaborative development, and consists of a Customised Product Advisory System and interfaces for Infrastructure Management. A manufacturing service description language provides a formal description of both production and product-related information, and is used for the integration of the front-end and the MaaS environment.

### *1.2.5 Applications*

Many ongoing applications in different industries or fields that are related to or inspired by the concept of cloud manufacturing have been reported, including automotive industry [38], machine tool industry [39], semiconductor industry [40], etc., and the related areas include waste electrical and electronic equipment recovery/recycling [41, 42], and cloud robotics [43].

### *1.2.6 Challenges*

The ongoing development of cloud manufacturing is facing many challenges in concepts, technologies and standards [18], including (1) how to achieve the integration of various technologies such as cloud computing, IoT, Semantic Web, high-performance computing, embedded systems; (2) how to bring various types of resources and capabilities to the cloud as services, especially the implementation of Hardware-as-a-Service (HaaS) (i.e. knowledge-based resource clouding); and (3) how to achieve overall management and control of clouds (including service composition, collaboration between cloud manufacturing applications, open communication standards, distributed control and coordination of manufacturing equipment, and user interfaces in cloud environments). In order to realise these, a standard or technique for consistently describing equipment and its functionality, behaviour, structure, etc., is required.

## 1.3 Conclusions

This chapter systematically and comprehensively but briefly presented the status and advancement of cloud technologies in cloud computing and cloud manufacturing, which represent the typical and the latest applications of cloud technologies in computing and manufacturing. Cloud computing is an Internet-based computing paradigm for delivering computing resources as service over the Internet. Expanding the idea of offering computing resources as services in cloud computing into the manufacturing field gave rise to the concept of cloud manufacturing. Cloud computing and cloud manufacturing represent the latest status and advancement of cloud technologies in IT and manufacturing industries, respectively. For cloud computing, issues ranging from its evolution path and background to the concept itself, from architecture requirements to platforms, tools, and challenges were presented. The corresponding issues of cloud manufacturing has also been presented and discussed. However, it should be noted that manufacturing resources involved in cloud manufacturing are much more diverse and complicated than those in cloud computing, thus making it more challenging to implement it.

## References

1. L. Kleinrock, A vision for the Internet. ST J. Res. **2**(1), 4–5 (2005)
2. R. Buyya, C.S. Yeo, S. Venugopal, J. Broberg, I. Brandic, Cloud computing and emerging IT platforms: Vision, hype, and reality for delivering computing as the 5th utility. Future Gener. Comput. Syst. **25**(6), 599–616 (2003)
3. F. Berman, G. Fox, A.J. Hey, in *Grid Computing: Making The Global Infrastructure A Reality*, vol. 2, (John Wiley and sons, 2003)
4. I. Foster, Y. Zhao, I. Raicu, S. Lu, Cloud computing and grid computing 360-degree compared. in *Grid Computing Environments Workshop, GCE'08* (2008), pp. 1–10
5. D. Barkai, *Peer-to-Peer Computing: Technologies For Sharing And Collaborating On The Net* (Intel Press, 2001)
6. L.J. Zhang, H. Cai, J. Zhang, *Services Computing* (Springer, 2007)
7. R. Buyya, C.S. Yeo, S. Venugopal, Market-oriented cloud computing: vision, hype, and reality for delivering it services as computing utilities. in *10th IEEE International Conference on High Performance Computing and Communications, HPCC'08* (2008)
8. M.N. Birje, P.S. Challagidad, R.H. Goudar, M.T. Tapale, Cloud computing review: concepts, technology, challenges and security. Int. J. Cloud Comput. **6**(1), 32–57 (2017)
9. L.M. Vaquero, L. Rodero-Merino, J. Caceres, M. Lindner, A break in the clouds: towards a cloud definition. ACM SIGCOMM Comput. Commun. Rev. **39**(1), 50–55 (2008)
10. *P. Mell, T. Grance, The NIST Definition of Cloud Computing* (2011)
11. Q. Zhang, L. Cheng, R. Boutaba, Cloud computing: state-of-the-art and research challenges. J. Internet Serv. Appl. **1**(1), 7–18 (2010)
12. L. Wang, M. Törngren, M. Onori, Current status and advancement of cyber-physical systems in manufacturing. J. Manuf. Syst. **37**, 517–527 (2015)
13. L. Wang, An overview of internet-enabled cloud-based cyber manufacturing. Trans. Inst. Meas. Control **39**(4), 388–397 (2017)
14. B.P. Rimal, A. Jukan, D. Katsaros, Y. Goeleven, Architectural requirements for cloud computing systems: an enterprise cloud approach. J. Grid Comput. **9**(1), 3–26 (2011)

15. X. Xu, From cloud computing to cloud manufacturing. Robot. Comput. Integr. Manuf. **28**(1), 75–86 (2012)
16. R.L. Krutz, R.D. Vines, *Cloud Security: A Comprehensive Guide to Secure Cloud Computing* (Wiley Publishing, 2010)
17. B.H. Li, L. Zhang, S.L. Wang, F. Tao, J.W. Cao, X.D. Jiang, X. Song, X.D. Chai, Cloud manufacturing: a new service-oriented networked manufacturing model. Comput. Integr. Manuf. Syst. **16**(1), 1–7 (2010)
18. G. Adamson, L. Wang, M. Holm, P. Moore, Cloud manufacturing–a critical review of recent development and future trends. Int. J. Comput. Integr. Manuf. **30**(4–5), 347–380 (2017)
19. D. Wu, D.W. Rosen, L. Wang, D. Schaefer, Cloud-based manufacturing: Old wine in new bottles? Procedia CIRP **17**, 94–99 (2014)
20. F. Tao, Y. Cheng, L. Zhang, A.Y.C. Nee, Advanced manufacturing systems: socialization characteristics and trends. J. Intell. Manuf. **28**(5), 1079–1094 (2017)
21. G. Adamson, L. Wang, M. Holm, P. Moore, Adaptive robot control as a service in cloud manufacturing. in *ASME 2015 International Manufacturing Science and Engineering Conference, MSEC 2015* (2015)
22. L. Wang, R. Gao, I. Ragai, An integrated cyber-physical system for cloud manufacturing. in *Proceedings of ASME 2014 International Manufacturing Science and Engineering Conference*, Michigan, USA, 9–13 June 2014
23. L. Wang, Machine availability monitoring and machining process planning towards cloud manufacturing. CIRP J. Manufact. Sci. Technol. **6**(4), 263–273 (2013)
24. R. Gao, L. Wang, R. Teti, D. Dornfeld, S. Kumara, M. Mori, M. Helu, Cloud-enabled prognosis for manufacturing. CIRP Ann. Manuf. Technol. **64**(2), 749–772 (2015)
25. B. Buckholtz, I. Ragai, L. Wang, Remote equipment security in cloud manufacturing systems. Int. J. Manuf. Res. **11**(2), 126–143 (2016)
26. B.H. Li, L. Zhang, L. Ren, X.D. Chai, F. Tao, Y.L. Luo, Y.Z. Wang, C. Yin, G. Huang, X. P. Zhao, Further discussion on cloud manufacturing. Comput Integr. Manuf. Syst. **17**(3), 449–457 (2011)
27. L. Ren, L. Zhang, L. Wang, F. Tao, X. Chai, Cloud manufacturing: key characteristics and applications. Int. J. Comput. Integr. Manuf. **30**(6), 501–515 (2017)
28. Q. Lin, K, Xia, L. Wang, L. Gao, Research progress of cloud manufacturing in China: a literature survey. in *ASME 2013 International Manufacturing Science and Engineering Conference Collocated with the 41st North American Manufacturing Research Conference, MSEC 2013*, (2013)
29. F. Tao, Y. Cheng, L.D. Xu, L. Zhang, B.H. Li, CCIoT-CMfg: cloud computing and internet of things-based cloud manufacturing service system. IEEE Trans. Industr. Inf. **10**(2), 1435–1442 (2014)
30. K. Xia, L. Gao, L. Wang, W. Li, K.M. Chao, A semantic information services framework for sustainable weee management toward cloud-based remanufacturing. J. Manuf. Sci. Eng. **137** (6), 061011 (2015)
31. F. Tao, L. Zhang, Y. Liu, Y. Cheng, L. Wang, X. Xu, Manufacturing service management in cloud manufacturing: overview and future research directions. J. Manuf. Sci. Eng. **137**(4), 040912 (2015)
32. B.H. Li, L. Zhang, L. Ren, X.D. Chai, F. Tao et al., Typical characteristics, technologies and applications of cloud manufacturing. Comput. Integr. Manuf. Syst. **18**(7), 1345–1356 (2012)
33. X.V. Wang, X.W. Xu, An interoperable solution for Cloud manufacturing. Robot. Comput. Integr. Manuf. **29**(4), 232–247 (2013)
34. D. Schaefer, J.L. Thames, R.D. Wellman Jr. D. Wu, S. Yim, D.W. Rosen, Distributed collaborative design and manufacture in the cloud–motivation, infrastructure, and education. in *ASEE 2012 Annual Conference and Exposition*, (University of Bath, Bath, 2012)
35. D. Wu, D.W. Rosen, L. Wang, D. Schaefer, Cloud-based design and manufacturing: a new paradigm in digital manufacturing and design innovation. CAD Comput. Aided Des. **59**, 1–14 (2015)

36. M. Meier, J. Seidelmann, I. Mezgár, ManuCloud: the next-generation manufacturing as a service environment. ERCIM News **83**, 33–34 (2010)
37. M. Wang, J. Zhou, S. Jing, Cloud manufacturing: Needs, concept and architecture. in *Proceedings of the 2012 IEEE 16th International Conference on Computer Supported Cooperative Work in Design, CSCWD 2012* (2012)
38. Z. Jin, Research on solutions of cloud manufacturing in automotive industry. in *Lecture Notes in Electrical Engineering* (2013), pp. 225–234
39. M.H. Hung, Y.C. Lin, H.C. Huang, M.H. Hsieh, H.C. Yang, F.T. Cheng, Development of an advanced manufacturing cloud for machine tool industry based on AVM technology. in *IEEE International Conference on Automation Science and Engineering*, Madison, WI, USA, 2013
40. T. Chen, Strengthening the competitiveness and sustainability of a semiconductor manufacturer with cloud manufacturing. Sustainability (Switzerland) **6**(1), 251–266 (2014)
41. L. Wang, X.V. Wang, L. Gao, J. Váncza, A cloud-based approach for WEEE remanufacturing. CIRP Ann. Manuf. Technol. **63**(1), 409–412 (2014)
42. X.V. Wang, L. Wang, From cloud manufacturing to cloud remanufacturing: a cloud-based approach for WEEE recovery. Manuf. Lett. **2**(4), 91–95 (2014)
43. X.V. Wang, L. Wang, A. Mohammed, M. Givehchi, Ubiquitous manufacturing system based on cloud: a robotics application. Robot. Comput. Integr. Manuf. **45**, 116–125 (2017)

# Chapter 2
# Latest Advancement in CPS and IoT Applications

## 2.1 Introduction

Internet of Things (IoT), as one of the most important new information technologies, has attracted great attention from governments, industries, and academia, and has been widely used in many fields, such as production, healthcare, and logistics. Originated from the radio frequency identification (RFID) systems, the term IoT was first coined by Ashton in MIT Auto-ID Labs in 1999 [1], referring to wireless communication abilities integrated with sensors and computing devices, thus enabling uniquely identifiable *things* to provide data over the Internet with limited or no human interaction. With the new information technologies integrated with IoT, it is hard to define IoT clearly and uniformly, especially its various application backgrounds. For brevity, IoT can be understood from two perspectives, which are "Internet-oriented" and "Things-oriented" [2]. The former can be viewed as the expansion of Internet applications. IP stack that already connects a huge amount of communicating devices, has all the qualities to make IoT a reality, while the latter means that a large number of *things*, which have identities and virtual personalities, form a worldwide network based on standard communication protocols [3]. In general, the architecture of IoT can be divided into four layers, i.e. sensing layer, networking layer, middleware layer, and application layer. The sensing layer is responsible for sensing and capturing the real-time information of resources, devices, and further sharing among the identified units through a constructed wireless network with tags and sensors. The networking layer is to connect all things together to form the physical network of manufacturing systems, and allow things to share the information with other connected things. The middleware layer is to manage, control, and transmit information in real time through a cost-efficient platform integrated by hardware and software functions. The main function of application layer is to integrate the methodologies and functions of the system to achieve IoT-enabled industrial applications (such as remote monitoring of robots, tracking and tracing of manufacturing resources in real time), and IoT-enabled



L. Wang and X. V. Wang, *Cloud-Based Cyber-Physical Systems in Manufacturing*,
https://doi.org/10.1007/978-3-319-67693-7_2

manufacturing systems. Within the powerful functionality, IoT is widely applied in a number of industries [2], and these applications include four main domains: (1) transportation and logistics, including assisted driving, mobile ticketing, monitoring environmental parameters, augmented maps, and so on; (2) healthcare, including tracking, identification and authentication, data collection, sensing, and others; (3) smart environment (home, office, and plant), such as comfortable homes and offices, smart building, smart cities, smart factories, smart museums, and so on; and (4) personal and social domain, including social networking, historical queries, losses, thefts, and so forth.

In the past decades, advancements in Web- and Internet-based systems and applications have opened up the possibility for industries to utilise the cyber workspace to conduct efficient and effective daily collaborations from anywhere in distributed manufacturing environments [4]. Recent advances in manufacturing industry have paved way for a systematic deployment of Cyber-Physical Systems (CPS), within which information from all related perspectives is closely monitored and synchronised between the physical factory floors and the cyber computational space. CPS are engineered systems that are built from and depend upon the seamless integration of computational algorithms and physical components [5]. The term *Cyber-Physical Systems* was first proposed in the US in 2006 [6]. With the wide applications and development of CPS, the definition of CPS is multiple, and not clear and unified. For example, CPS are integrations of computation and physical processes. Embedded computers and networks monitor and control the physical processes, usually with feedback loops where physical processes affect computations and vice versa [7]. In other words, CPS use computations and communication deeply embedded in and interacting with physical processes so as to add new capabilities to physical systems [1]. Unlike traditional embedded systems that are typically standalone, a full-fledged CPS is characterised by a network of interacting elements with physical input and output, resembling the structure of a sensor network. Tremendous progress has been made in advancing CPS technology over the last five years. Certainly, new smart CPS will drive innovation and competition in sectors as diverse as aerospace, automotive, chemical process, civil infrastructure, energy, healthcare, manufacturing, transportation, and so forth. One example of CPS is an intelligent manufacturing line, where a machine can perform a variety of processes by communicating with the components. Ongoing advancement in science and engineering will continue to enhance the link between computational and physical elements, dramatically increasing the adaptability, autonomy, efficiency, functionality, reliability, safety, and usability of CPS [5]. The final aim of CPS is to realise "intelligent monitoring" and "intelligent control" [8, 9]. These are the processes that need to realise real-time information extraction, data analysis, decision making and data transmission. CPS is an emerging discipline and has attracted and engaged many researchers and vendors. For example, many universities and institutes (such as UC Berkeley, Vanderbilt, Memphis, Michigan, Notre Dame, Maryland, and General Motors Research and Development Centre) have joined one research project (http://newsinfo.nd.edu/news/17248-nsf-funds-cyber-physical-systems-project/). The European Union (EU) and other countries, such as China and Korea, also realised the importance and significance of CPS research (http://www.artemis.eu/). In addition, the American Government named CPS as a new development strategy [8]. In conclusions, research and applications of CPS have been active in such areas like transportation, smart home, robotic surgery, aviation, defence, critical infrastructure, etc. [1]. CPS also positively affected manufacturing in form of Cyber-Physical Production Systems (CPPS) in process automation and control [10].

The structure of CPS was outlined in [11], and the IoT, as the Internet layer, networks the "cyber-physical" things for information transfer. IoT can be seen as a bottom-up vision, an enabling technology, which can be used to create a special class of CPS, i.e. systems including the Internet. However, CPS does not necessarily include the Internet. Some visions of the IoT go beyond basic communication, and consider the ability to link "cloud" representations of the real things with additional information such as location, status, and business related data. Therefore, CPS forms the first level and IoT forms the second level of vertical digital integration.

The progress of many research and applications of CPS and IoT is significant, and this chapter systematically illustrates the latest advancements of CPS and IoT, such as in technologies and industrial applications. The remainder of this chapter is therefore organised as follows. The key enabling technologies of CPS and IoT are presented in Sect. 2.2, followed by their key features and characteristics of CPS and IoT in the literature. Advancements of CPS and IoT are provided in Sect. 2.3. Section 2.4 introduces applications of CPS and IoT, before concluding the chapter in Sect. 2.5.

## 2.2 Key Enabling Technologies in CPS and IoT

Along the progress of CPS and IoT research and applications, Wireless Sensor Network (WSN), Cloud technologies, Big Data, and other enabling technologies play an important role to support CPS and IoT. For example, several initiatives cater for the CPS development, such as Advanced Manufacturing Partnership 2.0 [12] and Industrial Internet [13] in USA, Industry 4.0 [14] in Germany, Factories of Future [15] in EU, and even the less-known Japanese "Monozukuri" that stands for Coopetition. Other initiatives on this front include Wise-ShopFloor for web-based sensor-driven e-shop floor [16] and Cyber-Physical European Roadmap and Strategy (CyPhERS) [17]. In addition, IoT, as new emerging technology, is expected to provide promising strategies and solutions to build intelligent and powerful manufacturing systems and industrial applications by using the growing ubiquity of RFID, and wireless sensor devices [18]. According to the International Telecommunication Union (ITU), the key technologies of IoT contain the RFID technology, Electronic Product Code technology, and ZigBee technology. In what follows, a brief account of these technologies, including WSN, cloud technologies, Big Data, RFID technology and Industry 4.0, is provided.

### *2.2.1 Wireless Sensor Network*

Wireless communication and networking is one of the fast-growing research areas. Significant progress has been made in the fields of WSN [19]. WSN is designed particularly for delivering sensor-related data. It consists of a number of sensor nodes working together to monitor a region to obtain data about the environment. The sensor nodes include MEMS components such as sensors, RF components, and actuators, and CMOS building blocks such as interface pads, data fusion circuitry, specialised and general-purpose signal processing engines, and microcontrollers [20]. These sensors are equipped with wireless interfaces with which they can communicate with one another to form a network. Data gathering is the foundation of data processing and transmission. These sensor nodes can sense, measure, and gather information from the environment and, based on some local decision process, they can transmit the sensed data to the user through a communication protocol. A radio is implemented for wireless communication to transfer the data to a base station (e.g. a laptop, a personal handheld device, or an access point to a fixed infrastructure) [21]. Constraints on resources and design for WSN restrict wide application and development with the demands on volume of data collection and complexity of systems. As a result, by integrating WSNs from different domains, CPS represents one of the major driving forces that go beyond the cyber world towards the physical world [7].

### *2.2.2 Could Technologies*

Due to the explosive growth of data volume and real-time service concept, more rapid methods to deal with these data is required. Cloud computing is used to address the problem of calculating speed and volume. Cloud computing refers to 'a large-scale distributed computing paradigm that is driven by economies of scale, in which a pool of abstracted, virtualised, dynamically scalable, managed computing power, storage, platforms, and services are delivered on demand to external customers over the Internet [22]. Cloud computing is considered as a new business paradigm describing supplement, consumption and delivery model for IT services by utility computing based on the Internet [23]. Infrastructure as a Service (IaaS), Platform as a Service (PaaS), and Software as a Service (SaaS) are the basic service models of cloud computing, and they indicate hardware resources, cloud platforms including operating systems, programme execution environments and databases, enabling application developers to develop, test, deploy and run their applications [24]. Cloud computing has changed the way of thinking of both IT service providers and their customers. It offers business and application models that deliver infrastructure, platform, software and applications in forms of services [25]. Figure 2.1 illustrates different levels of services of cloud applications compared

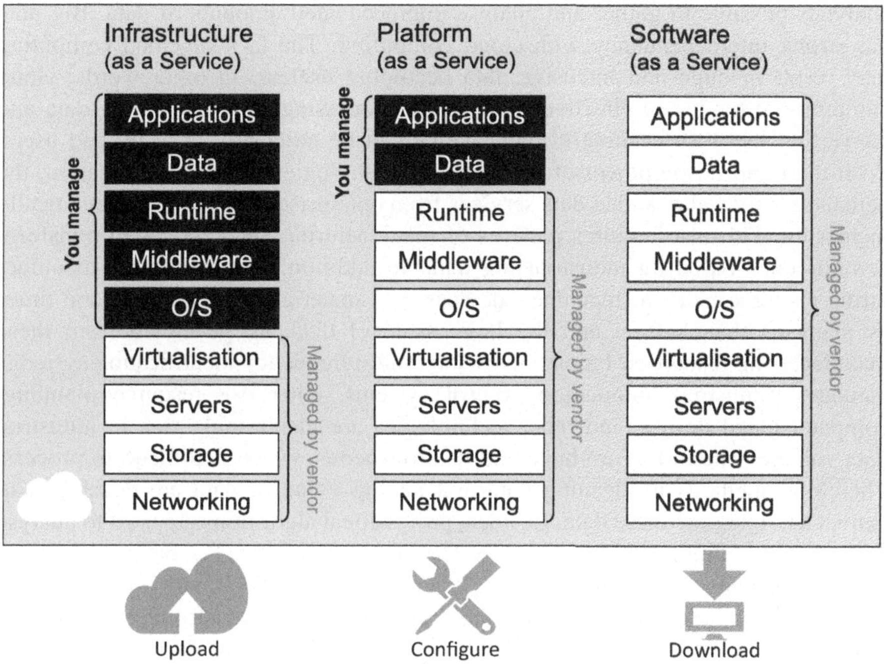


**Fig. 2.1** Different service levels of cloud services

against standalone ones. Inspired by the success of cloud computing, the cloud technology has recently been extended to the manufacturing contexts, leading to the innovation of various cloud manufacturing systems. Cloud manufacturing implies an integrated cyber-physical system that can provide on-demand manufacturing services, digitally and physically, at the best utilisation of manufacturing resources [26, 27]. It aims at offering a shared pool of resources, e.g. manufacturing software, manufacturing facilities, and manufacturing capabilities. However, cloud manufacturing is more than simply deploying manufacturing software applications in the cyber cloud. Besides data storage and virtual machines, the physical resources integrated in the manufacturing cloud must be able to offer adaptive, secure and on-demand manufacturing services, often over the IoT, including work-cells, machine tools, robots, etc.

### 2.2.3 *Big Data*

Big data refers to the analytics based on large data collections. Advancements in computing and memory performance, together with networking have made big data analytics possible to gather and analyse unprecedented amounts of data. Big data has strong interdependency with cloud computing. The lack of cloud computing may result in huge and intensive data becoming useless. In other words, cloud computing provides an effective approach to processing insignificant big data and converting them to meaningful data, which can be efficiently used by end users. With the rapid development of information technologies and industrialisation, the demand for the value-added data services from enterprises and users is dramatically increasing. This also requires patterns of manufacturing and services to transform towards ones based on industrial big data. In addition, the quantity of manufacturing resources such as machines, devices, and materials, etc., have several times as many as than before, and the large scale of data are generated from these manufacturing resources. Industrial sensors, radio frequency identification systems, barcodes, industrial automation control systems, enterprise resource planning, computer aided design, and other technologies are increasingly rich in industrial data volume. These data are huge and heterogeneous, which is difficult to process. Therefore, methods in cleaning data and adding value of data are developed to extract and integrate these data. Besides, professional algorithms are used to analyse these data and find useful information. Therefore, applications of big data in many fields, especially in manufacturing systems, can provide a new paradigm for achieving data-based services and manufacturing pattern transformation. Another important thing in this process is data exchange, including data/information interaction between huge manufacturing resources. The key characteristics of WSN in terms of reliability, flexibility, usability, and security guarantee the stable, high-efficient, and secure data transmissions. CPS and IoT enable further enormous amounts of data related to physical systems to be made available for analysis. Big data is relevant to non-technical systems and IT systems, but becomes even more interesting when applied in the context of CPS due to the implications of physicality in terms of capabilities, technical risks and costs.

### *2.2.4 Industry 4.0*

The term Industry 4.0 was manifested for the first time at the Hannover Fair with the presentation of the "Industry 4.0" initiative [28]. Industry 4.0 is a large German initiative [14] that emphasises the extension of traditional manufacturing systems to full integration of physical, embedded and IT systems including the Internet. It highlights three features for implementation: (1) horizontal integration through value networks, (2) end-to-end digital integration of engineering across the entire value chain, and (3) vertical integration and networked manufacturing systems. The implementation recommendations call for actions in eight key areas including standardisation and reference architecture; managing complex systems; safety and security; work organisation; professional training; regulations, and resource efficiency. Germany's implementation of Industry 4.0 has received great attention throughout the world from researchers and government, and therefore many countries, institutes, and enterprises undertake this research and aim at improving their manufacturing chains. Especially in the process of Industry 4.0, due to the tight integration of micro controller and physical devices, machines and tools are becoming more automated and self-sufficient, increasingly replacing manual labour. Industry 4.0 is a representative of the emergency of the fourth Industrial Revolution through the use of CPS and IoT, followed by *digitisation* through the use of electronics and IT. The goal of the Industry 4.0 is the emergence of digital factories that are to be characterised by the five key features: smart networking, mobility, flexibility, integration of customers, and new innovative business models [29].

### 2.2.5 RFID Technology

IoT can be considered as a global network infrastructure composed of numerous connected devices that rely on sensory, communication, networking, and information processing technologies [30]. A foundational technology for IoT is the RFID technology, which allows microchips to transmit identification information to a reader through wireless communication. An RFID system is composed of an RFID device (tag), a tag reader with an antenna and transceiver, and a host system or connection to an enterprise system [31]. Radio frequency tags in the RFID system are used to store information. RFID tags and readers communicate by non-contact sensors, radio waves or microwaves. The key technologies of RFID include high-adaptive wireless communication technology, high confidentiality; low power consumption, high reliability of RFID devices; small volume, high-efficiency antenna technology; low-cost chip and reader. The most prominent advantage of RFID technology is: non-contact reading and writing, distance from a few centimetres to dozens of metres, to recognise high-speed moving objects, strong security, and can identify multiple targets simultaneously. Compared with the concept, component parts of RFID technology, the industrial applications of RFID technology attract more interests, for example, sensing and capturing information of objects, and identification. Taking an RFID-enabled real-time information sensing and capturing system as an example, auto-ID techniques such as RFID tags will be employed to manufacturing resources such as machines, robots, and raw materials, to make the smart manufacturing objects with the capability of identifying the real-time status of the manufacturing things such as operators, materials, locations, WIP items, etc. When a manufacturing thing comes to a sensing area, this event can be sensed by the registered sensor. Through the communication protocol, the sensor can capture and send the data of the coming manufacturing things to the upper processing unit, which is responsible for processing the primitive information to form the value-added information and data storage.

## 2.3 Key Features and Characteristics of CPS and IoT

In this section, the key features and characteristics of CPS and IoT are presented in order to clarify the related advances. It first points out the key characteristics of CPS, and then discusses some features that have been made. Representatives of the industries reported that CPS are indeed not new technologies, but widely existing and manifested by existing industrial manufacturing systems. Lee et al. [32] pointed out that a CPS consists of two main functional components: (1) the advanced connectivity that ensures real-time data acquisition from the physical world and information feedback from the cyber workspace; and (2) intelligent data management, analytics and computational capability that constructs the cyber space. The case can be made to identify and embody CPS, due to the increasing digitalisation and penetration of embedded systems. The increasing connectivity and capabilities of computational systems promote the emergence of new systems with the typical characteristics of CPS, and these characteristics can be concluded as follows: (1) the deployment of CPS in mass-products applications, such as smart-phone enabled services; (2) the chances for and emergence of new cross-domain applications, for example, the intelligent transportation systems; (3) the increasing openness, adaptability, and autonomy.

Wu et al. [33] identified some unique features of CPS applications as follows:

1. Cross-domain sensor sources and data flows: Multiple types of sensors will be adopted at the same time in intelligent CPS applications. Moreover, these cross-domain sensing data will be exchanged over heterogeneous networks.
2. Embedded and mobile sensing capacity: High-degree mobility of sensors based on the mobile devices makes sensors have the capacity of the mobile sensing coverage over time.
3. User contribution and cooperation through give-and-take-like models: Participatory sensing would be common in CPS.
4. Elastic loads requiring cloud-supported storage and computing capability: With the maturity of cloud computing, the pay-as-you-go concept is likely to be adopted in CPS to serve storage, computing, and communication needs.
5. Accumulated intelligence and knowledge via learning and data mining technologies: Under high dynamics and uncertainty of data in CPS, learning and data mining technologies can be used to retrieve useful knowledge. Then, the feedback from users and actuators may help us to accumulate, or even discover unknown knowledge.
6. Rich interactions among many objects and things through the Internet (such as IoT): A lot of sensor–sensor, sensor–actuator, actuator–actuator, actuator–user, user–user, user–object, object–object, object–thing, thing–thing, and thing–user interactions may occur in CPS applications. Such rich and complex interactions demand flexible communication channels, like the Internet, to facilitate our applications.

These characteristics and features of CPS led the CyPhERS (Cyber-Physical European Roadmap and Strategy) project [17] to carry out a characterisation of CPS, attempting to capture the evolving scope of CPS, from traditionally closed systems, with single jurisdiction, limited adaptability and autonomy. Defining such characteristics would be helpful beyond definitions, because definitions of CPS tend to be very general; instead the characterisation helps to identify various types of CPS. The following aspects of CPS have been identified [17]:

- *Deeply embedded versus IT dominated.* Resource-limited and dedicated computer systems represent the traditional embedded systems, which tightly integrates with the physical processes. The increasing connectivity and capabilities of computing systems enable "embedded" versus IT systems for the intersection. This means that the two types of systems are becoming connected and interacted.
- *Single-domain versus cross-domain.* A traditional embedded system generally is represented by a single domain application system. New cost-efficient communication creates the potential for the demands and applications of new services that cut across existing domains, or for building new CPS domains. The smart home and its connection to the electrical grid represent an example of this trend.
- *Open versus closed.* A traditional embedded system represents a system that is not connected to other computing systems. The difficulties of diagnosing, maintaining and upgrading widely deployed embedded systems provide strong driver towards more open systems. Another driver is provided by the ability to provide new collaborative services.
- *Automation levels and types.* Systems with high autonomy can have the capacity of operating without human supervision/intervention. Automation in many fields is used to replace or relieve the workers' load, especially in the dirty, dull, and dangerous working situation [34]. Based on the environmental, resources efficiency and safety considerations, autonomy is widely developed and applied to all kinds of domains (such as intelligent manufacturing systems, and robot-enabled assembly systems).
- *Governance, indicating the entities responsible for dependable and efficient system operation.* The division of responsibility will associate with the system of system nature.
- *Distributed versus centralised control.* Most CPS already constitute distributed computer systems (or are likely to become so) because of the increasing connectivity. This makes control become more or less decentralised. Control in this context refers to the decision making within the distributed systems.
- *Single jurisdiction versus cross-jurisdiction.* This aspect refers to applicable standards and legislation. Generally, the typical challenge faced by many existing CPS is that jurisdiction is more complicated with more open and cross-domain CPS.

- *Adaptability under uncertain conditions*. The context of the typical CPS may be always varying, such as the environmental conditions, system load, and failures. Adaptable CPS has the capacity of dealing with such varying contexts within given bounds, and potentially contributes to reducing maintenance costs and increasing availability.
- *Human in/outside the loop*. Traditional CPS come in two types; those that are more or less fully autonomous (i.e. act independently of humans, but may be triggered by human inputs), and those with a much closer interaction with humans, including shared control.
- *Degree of integration*. Various types of integration are built based on the effective connectivity. A CPS will have a certain degree of horizontal and vertical integration in a certain context and application domain. Horizontal integrationis the integration of services and functions of similar type, and vertical integration refers to integration across system hierarchies.

CPS may pose a different mix of the key features and depend on their utilisation domain. CPS consider the computational decisional components that use the shared knowledge and information from physical processes to provide intelligence, responsiveness, and adaptation. In conclusion, the differentiating factor among all areas, is not the distinct characteristics but which of them they employ (depending on the scenario) and at which degree [35]. Similarly, IoT focuses mostly on the interaction and integration part while cooperation is optional. According to the definition and applications of IoT, characteristics of IoT are different because of specific-domain applications with typical unique features. As discussed in the previous section, IoT can provide technical support and new opportunities for innovative applications in many fields. Some applications strictly belong to a specific domain and exhibit characteristics peculiar of that domain. Conversely, others applications exhibit characteristics cross-cutting multiple domains. As a result, it is unsuitable or hard to identify characteristics of IoT, which is common for applications in all the fields. Under the context of this chapter, the key IoT features in manufacturing systems are introduced as follows, in contrast to the current manufacturing systems [36].

- Introduce an easy-to-use and easy-to-deploy architecture and solution for implementing smart manufacturing in the whole manufacturing systems using the IoT.
- Design the smart framework and models for improving the intelligence of the bottom-level manufacturing resources such as smart stations and smart vehicles because they are the key to the intelligent manufacturing system.
- Develop a new decision strategy and method for real-time information based production scheduling and internal logistics optimisation, which can be directly applied to the manufacturing system, for example, shop floor, for improving the efficiency.
- Present a critical event-based real-time key production performances analysis model, so as to actively identify real-time production exceptions.

These key IoT features provide the technical support and basis for addressing the "4Cs" (Connection, Communication, Computing, and Control) of resources and devices for the following different applications in manufacturing, and achieving the connection of objects.

## 2.4 Applications of CPS and IoT

CPS has achieved varying applications in different sectors, and they include highly reliable medical devices and systems, traffic control and safety systems, advanced automotive systems, and systems for process control, environmental control, energy conservation, instrumentation, critical infrastructure control, distributed robotics, smart structures, manufacturing, and defence [1, 6, 37]. Widely recognised and accepted attributes of a CPS, such as timeliness, distributed, reliability, fault-tolerance, security, scalability and autonomous, are also identified. In addition, the concept and technologies of IoT can be extended into many fields and various application backgrounds to achieve connection, communication, and interaction of the physical things in a constructed Internet-of-Manufacturing-Things environment, such as manufacturing systems, logistics, intelligent transportation, and supply chain management, etc. Moreover, in a constructed sensing environment, real-time data of things, such as manufacturing resources, can be sensed and captured by the registered sensors. Based on a carefully chosen communication protocol, these accurate, timely, consistent, and value-added data can be transmitted to the upper data processing units, and further shared among manufacturing managers and suppliers. Real-time monitoring, tracking, and tracing of manufacturing resources and devices through the entire manufacturing chain can be achieved. In this chapter, by taking the main research topics into considerations, the main focus is given to applications of CPS and IoT in manufacturing systems and services. Consequently, typical and representative cases are introduced in the following sections, including service oriented architecture, could manufacturing, an IoT-enabled manufacturing system, and CPS in the cloud environment. These examples of manufacturing systems and services can reflect many of the CPS and IoT characteristics and features described in Sect. 2.4.

### *2.4.1 Service Oriented Architecture*

Service-oriented architecture (SOA) is one of the key technologies for information communication [38], and plays an important role in connection between the cyber world and physical world. The advantages of SOA are essential for manufacturing systems, and they are comprehensive, distributed, transparent, and secure [39]. These advantages provide technical and strategic support for enterprises to manage their workflows, optimise their manufacturing chains, select the most optimal suppliers, share information, monitor the performance of devices, and guarantee the information security. They cater for the requirements and demands of a growing number of industrial systems, including integration flexibility and the ability for processes to be composed. CPS has the capability of the integration of the computational and physical worlds. Under the constructed infrastructure service platform, sensors registered/installed in the workspaces of factories sense and capture the physical information of manufacturing resources around them. After capturing these data, instead of directly entering optimisation stage, adaptive control mechanisms are entailed to achieve the added value and management of real-time information [40, 41]. These data are transformed to the upper control systems integrating computation, sensors and actuators in devices, which will then give instructs to actuators for execution. Finally, manufacturing systems supply the support of M2M (including man-to-man, man-to-machine, and machine-to-machine) communication, break the limitation of source constraints, and achieve resource sharing and distributed computing.

In this section, a representative application of SOA is introduced and it is a Ford Motor Company application to the Valencia assembly plant (Ford Transit models) [1], which is one of the first service-oriented architectures effectively deployed in industry and is still used at present. Developed by a system integrator (IntRoSys SA), this approach left all current product lifecycles (PLCs) as they were. According to the architecture shown in Fig. 2.2, agent technology (having Product, Knowledge Manager, Machine agents) is the key enabling technology encapsulated as "wrappers" to the PLCs.

As shown in Fig. 2.2, a Product Agent (PA), a Knowledge Manager Agent (KMA), and a Machine Agent (MA) form the basis of agent enabling technology for the service-oriented architecture of an IntRoSys Multi-Agent System. The PA is responsible for formulating and dispatching the workflow, and the KMA is to perform a check of the physical feasibility of the proposed workflows. The MA translates the workflows into specific machine instructions. The procedure of these agents can be described as follows. The PA receives the order and identifies the Atomic Skills required. The following step is a match of such required Atomic Skills with a database of workflows already executed in the past. If the order can be executed with an existing workflow, then such a workflow is dispatched to the MA, else the PA elaborates a new workflow that is sent for a feasibility check to the

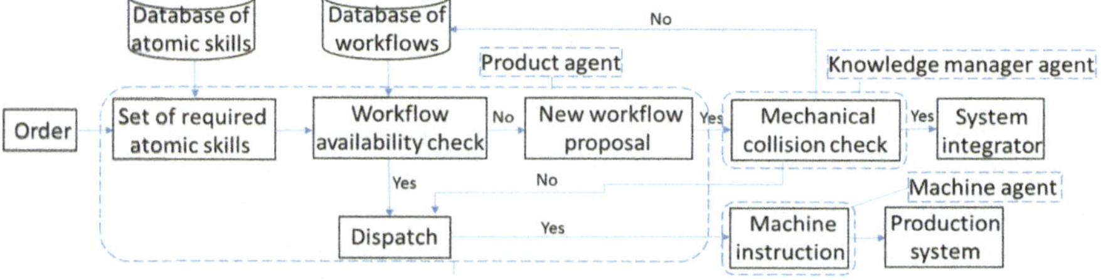


**Fig. 2.2** Service-oriented architecture of IntRoSys multi-agent system

KMA. If the KMA does not detect any problems, the examined workflow is sent back to the PA that dispatches it to the MA. The newly found feasible workflow is also included in the database of existing workflows. Vice versa, in the case of problems with the proposed workflow, the MA warns the System Integrator for the necessity of a human intervention to sort out the related order. Finally, all the dispatched workflows are processed by the MA that sends the necessary machine instructions to the production system. This service-oriented agent architecture is open in the sense that if a new atomic skill (for example a new process for a new variant, or a safety routine) is required, it can be integrated in the system without modifying the existing code but simply by coding it independently and eventually adding it to the related database.

### 2.4.2 Could Manufacturing

Cloud manufacturing is emerging as a new manufacturing paradigm as well as an integrated technology, which is promising in transforming today's manufacturing towards service-oriented, highly collaborative and innovative manufacturing in the future. Combining recently emerged technologies, such as IoT, Cloud Computing, Semantic Web, service-oriented technologies, virtualisation, and advanced high-performance computing technologies, with advanced manufacturing models and information technologies [42], Cloud Manufacturing is a new manufacturing paradigm built on resource sharing, supporting and driving the flexible usage of globally distributed, scalable, sustainable, service-oriented manufacturing systems and resources [43]. Cloud manufacturing is also a smart networked manufacturing model that embraces cloud computing, aiming at meeting growing demands for higher product individualisation, broader global cooperation, knowledge-intensive innovation and increased agility in market response. In cloud manufacturing, customers can conveniently obtain on-demand services supporting the entire lifecycle of a product through network access to a shared pool where distributed manufacturing resources are virtualised and under unified management in a configurable and optimised manner [24]. Xu [44] defined could manufacturing as a model for enabling ubiquitous, convenient and on-demand network access to a shared pool of configurable manufacturing resources (e.g. manufacturing software tools, manufacturing equipment, and manufacturing capabilities), which can be rapidly provisioned and released with minimal management effort or service provider interaction. Over the past years, there have been different definitions of cloud manufacturing. In general, cloud manufacturing is a system that can provide manufacturing services digitally and physically to best utilise manufacturing resources [27]. In essence, it must connect to the real manufacturing equipment to form a CPS [37]. Along this direction, an integrated CPS for cloud manufacturing is presented hereafter, aiming for improved remote accessibility and controllability of factory equipment, such as CNC machines and robots, by combining 3D models, sensor data and camera images in real-time. It is realised by significantly reducing network traffic over the Internet, and building cloud-based services of monitoring, process planning, machining and assembly in a decentralised environment.

Cloud-DPP (cloud-based function-block enabled adaptive distributed process planning) is a joint research effort between KTH and Sandvik, Sweden, aiming for cloud-based distributed and adaptive process planning in a shared cyber workspace. Figure 2.3 depicts the architecture of distributed process planning (DPP), real-time process monitoring, dynamic resource scheduling, and remote device control of CPS. Based on the real-time information of machines and their status, DPP handles adaptive decision making of process planning. It is also possible for the Cloud-DPP to generate machining process plans adaptively to changes through well-informed decision making [45]. This is done by linking sensors embedded/attached to each machine to a manufacturing cloud in the cyber workspace, and delivering process plans in function blocks to the machine controller on the physical shop floor for execution. By properly dividing process planning tasks and assigning them to the cloud and function blocks, adaptive process planning and machining become possible. Cloud-DPP elaborates a two-layer distributed adaptive process planning based on function-block technology and cloud concept.

Cloud-DPP service obtains data from the monitoring service (machine tool ID, current status, and available time windows) and the feature list of a new part to be machined. The service uses a three-layer structure comprising supervisory planning (SP), execution control (EC), and operation planning (OP) [46, 47] (Fig. 2.4). At first, after having received input from the higher-level production planning, SP generates a generic process plan through generic setup planning and sequencing.

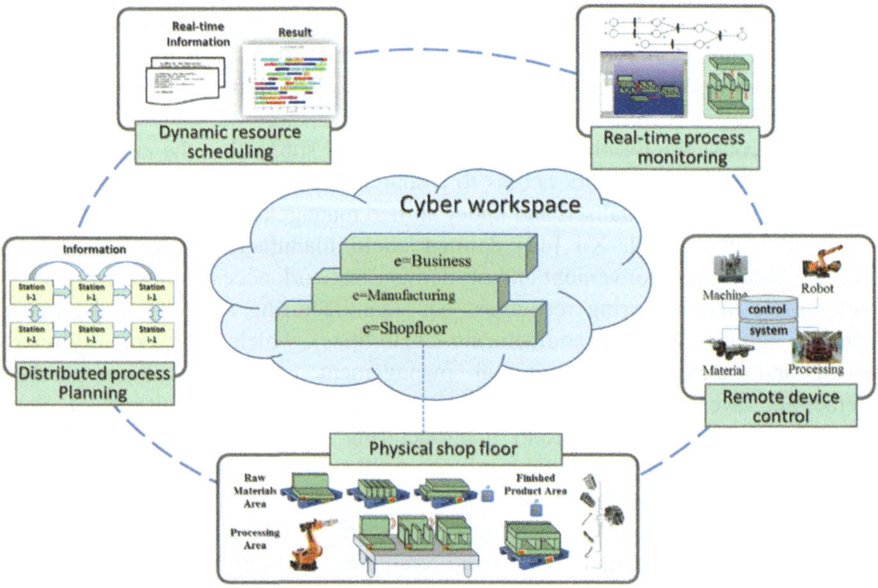


**Fig. 2.3** Could-DPP in a cyber-physical system

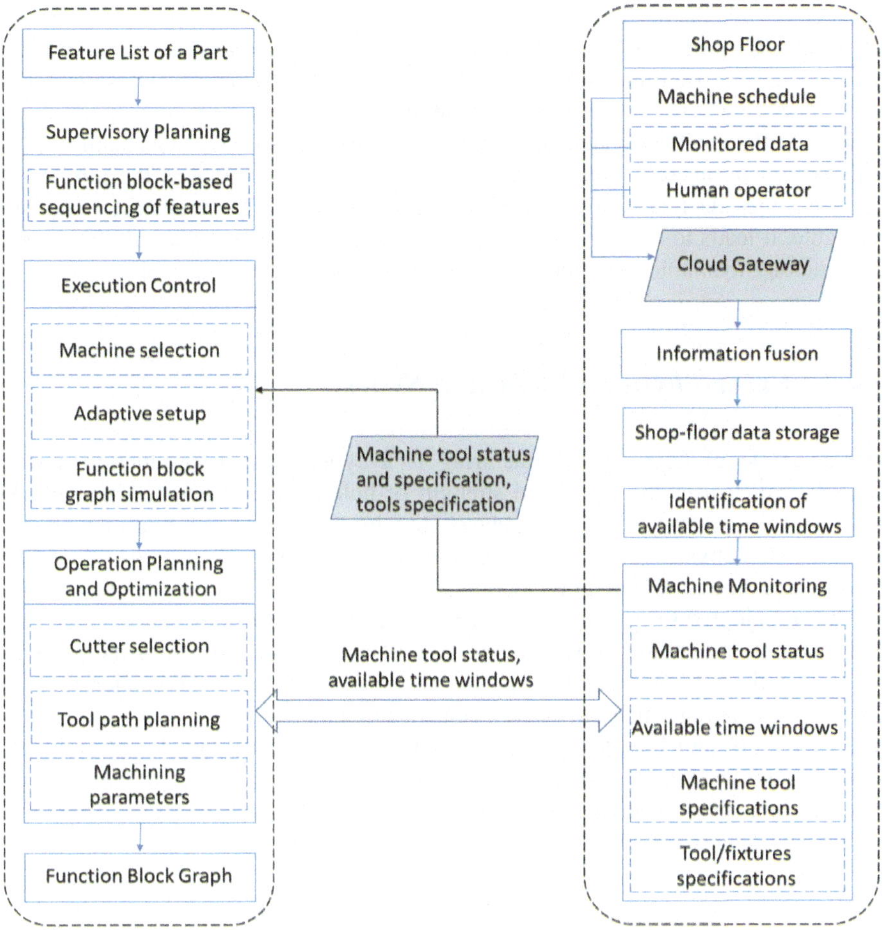


**Fig. 2.4** Workflow of Could-DPP

A generic process plan is a nested directed graph in which the nodes are the sequenced setups with machining features grouped in each setup, and the edges represent precedency relations among the nodes. A generic setup is a group of machining features with the same tool access direction that is possible to cut at least on a 3-axis machine. Later, it would be possible to assign any generic setup to any 3-, 4-, or 5-axis machine independently or merge it with other generic setups if the machine can handle them in a single setup. The precedence relations are decided by sequencing algorithms that infer necessary and preferred precedence among the machining features and setups by means of various rules addressing different aspects such as datum references, tolerances, and machinability. For each machining feature in the generic process plan, one machining feature function block instance is created. A generic process plan is completely resource-independent and the function blocks at this stage of their lifecycle are called meta function blocks. Then, the EC performs machine selection, adaptive machine-specific setup planning, and job dispatching. During this phase, in case the machine tool has more than three machining axes, the generated generic setups should be further merged and adapted to the available machine configuration due to higher tool reachability [46]. This will also ensure a higher efficiency of machining process as the number of setup changes decreases. When multiple machines of different characteristics are available, it leads to an optimisation problem in which the best way of distributing tasks among available machines based on defined criteria is targeted.

### 2.4.3 Cyber-Physical Production Systems

One example of CPS in manufacturing is a cyber-physical production system (CPPS), which uses the combined strength of holons, agents and function blocks [48]. In this context, a CPS is represented by a holarchy of multiple holons. A logical and physical parts that mimic the cyber and physical entities of the CPPS constitute a holon. When implementing this holonic CPPS, agents and function blocks are adopted to realise the two aspects of a holon for information processing and materials processing, respectively. Holons and agents have attracted growing interests in the field of manufacturing control because of the typical challenges from continuous changing manufacturing demands and patterns such as decentralisation, frequently shifting technologies, and various market perturbations. Function blocks as machine-level execution and control standards, are regarded a suitable approach to modelling distributed manufacturing systems and fit well with the concepts of holons and agents [49, 50]. Integration of holons, agents and function blocks can represent and model a CPPS, which is intelligent and adaptive, and can cope with challenges faced by manufacturing systems in the future. In order to understand the holonic CPPS and its potential in manufacturing, in the following we briefly introduce holons, agents and function blocks in terms characteristic.

Agent, as the core of agent technology, is defined as a software entity situated in some environment, that is capable of autonomous action in this environment in order to meet its design objectives. Another definition of agent is as follows: an agent is an autonomous component that represents physical or logical objects in the system, capable of acting in order to achieve its goals, and being able to interact with other agents, when it does not possess knowledge and skills to reach alone its objectives. This means that an agent should have the capacity of interaction without the direct instruction of higher authorities or intervention of humans, but instead in negotiation with other agents if necessary. A network of agents can be used to build a multi-agent system (MAS), which can be best characterised as a software technology having the capacity of modelling and implementing the individual and social behaviour in distributed systems. Six main characteristics of agents are concluded as follows: (1) reactive: agents should be able to sense their surrounding environment and react to the changes that occur; (2) proactive: agents should be capable of achieving their assigned goal; (3) autonomous: agents should have enough knowledge and authority to operate and act on their own without direct instructions or intervention from humans or other agents; (4) cooperative: agents can interact with other agents if necessary to achieve the global objective of the system; (5) adaptive: agents can learn from their previous behaviours and can apply their experiences to future scenarios; and (6) mobile: agents can move through a network [48].

Holonic manufacturing system (HMS) is defined as a group of holons that integrate the whole process of manufacturing activities from order booking through design, production and marketing to realise the agile manufacturing [51]. Where holon is an autonomous and cooperative building block of a manufacturing system for transforming, transporting, storing and/or validating information and physical objects. A holon consists of an information processing part and often a physical processing part, and also be a part of another holon. The characteristics of holon are autonomous, cooperative, open, proactive, reactive, learner, and recursive. Some of them have the similar attributes and capacities with that of agents. We briefly introduce some of them, for example, autonomy indicates the capability of the building block to create and control the execution of its own plans and strategies, and cooperation is defined as a process whereby a set of building blocks develops mutually acceptable plans and executes these plans. In addition, the holarchy denotes a system of holons that can cooperate to achieve a goal or objective, and defines the basic rules for the cooperation of the holons and thereby limits their autonomy. A holon can dynamically belong to multiple holarchies.

Function blocks can be applied in distributed environments and can be dispatched to machine tool controllers. In fact, each function block is a control software unit that is embedded with necessary information and algorithms that are required for performing a task at the controller level. An internal finite state machine would be responsible for controlling the different states and transition of sub-tasks within the function block. Function blocks have the capacity of producing different outputs using an identical input through changing the internal states of the function block to automatically expand their application. Function blocks can be classified into two main categories, namely basic function blocks and composite function blocks. The basic function blocks use their internal data and algorithms to perform a task, and a composite function block only relies on its containing basic function blocks' behaviours, but is not embedded with internal information. The event input of the function block triggers the execution of tasks. An execution control chart is responsible for denoting the execution control states, transitions and actions of that function block instance. Finally, when the task is finished, the function block's event output will send triggering signal to the next function block.

To make a holonic CPPS clear, the overall architecture of a robot holon that can be implemented by the multi-agent technology and function blocks is presented shown in Fig. 2.5. It is composed of a layered structure from the bottom to the top, including the physical part layer, the control system layer, and the planning, scheduling and execution control layer. The physical part layer contains the actual physical equipment on the shop floor, such as sensors, robots, and actuators, and is

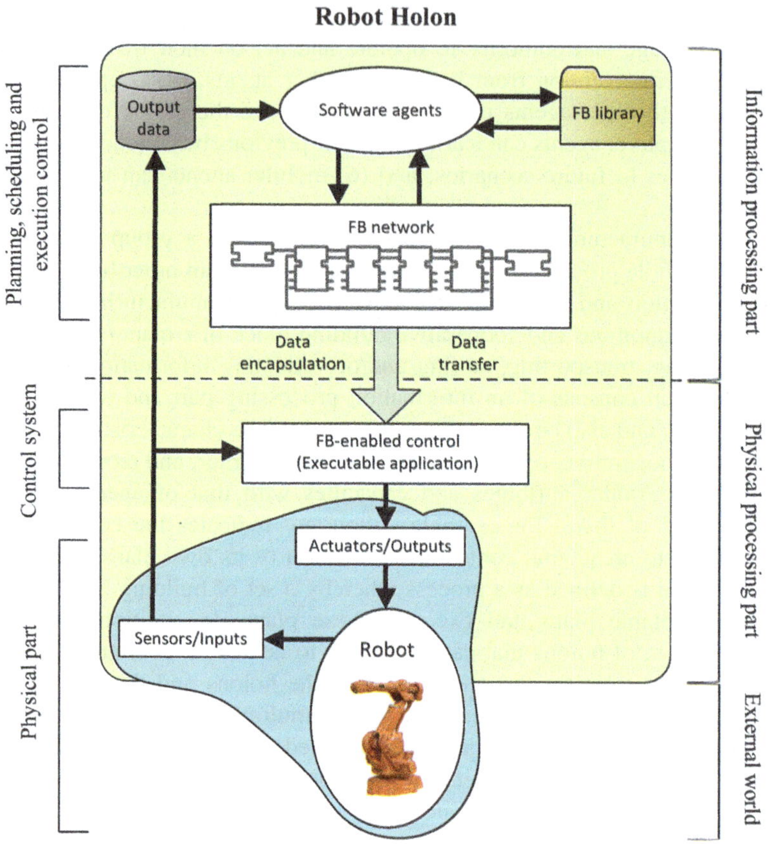


**Fig. 2.5** MAS and function blocks embedded in a holon

connected to the control system layer. In the control system layer, function block-enabled control system completes the work of translating tasks to native machine control codes that are transmitted into the physical part layer, then, machines in the physical part layer execute operation instructs. The planning, scheduling and execution control layer is responsible for planning and scheduling, decision making, developing control instructions and transferring them to the control layer where these instructions can be executed as runtime codes. The software agents would be responsible for high-level decision-making and coordination, and function blocks at the lower controller level would be responsible for execution of processes on the machine and quick fault recovery in case of disturbances on the shop floor. However, data will be processed by the software agents in complex cases because of the lack of the capacity of fault recovery for function blocks. The decision-making agent, and communication/collaboration agent are integrated into the software agents at the information processing part of the holon, where the decision-making agent achieves communication and interaction with other decision-making agents embedded in other individual holons within a holarchy and humans through the communication agent. As mentioned before, a CPS considers two different layers of "cyber" and "physical". The layered structure of the robot holon shown in Fig. 2.5 matches the structure of CPS and allows for distributed control. The information processing and physical processing of a holon can be connected through a wide area network (WAN) that can be geographically distributed. These separated parts of the holons can replace the traditional constitutive elements of the CPS; for example, the logical parts of the holons as decision-making units in the cyber part of the CPS, and the physical parts of the holons as execution units in the physical part of the CPS. Agents in the logical part of a holon for information processing can sense their surrounding environment and the conditions of their representative physical part on the shop floor. Due to limited awareness of a single agent, they need to build the intercommunication with other agents to obtain knowledge of the surrounding environment. As a result, agents of different holons can create cooperative groups (within a holarchy) for improving the quality of decision making. Humans as another element of a CPS can monitor the status and actions of physical entities through visual interfaces and can also enter the decision-making process when needed via communication agents.

Finally, we introduce a hypothetic manufacturing CPS that consists of two conveyors, two robots, and one machine, and produces a single type of product, and this can show the future potential of CPS in manufacturing. The raw material is fed through the first conveyor and is then transferred to the machine by a robot. Once the machining process is finished, the second robot would unload the finished part from the machine and place it on the second conveyor.

For the sake of modelling manufacturing CPS, the following assumptions are made: there is no buffer limit for the products after the second conveyor and before the first conveyor, and different robots are used for loading and unloading. One holarchy of eight defined holons is used to represent the hypothetic CPS: (1) a raw material holon, (2) a product holon, (3) an order holon, (4) conveyor holon-1, (5) conveyor holon-2, (6) robot holon-1, (7) robot holon-2, and (8) a machine holon. By applying the design concepts [48], CPS can be modelled. Different sets of connections among these holons represent the possible inter-holon communications for dynamic planning, scheduling and execution that can be implemented by agents and function blocks. In the case of tool breakage occurring while machine holon is in the middle of machining process. In order to stopping the spread of damage, the hard real-time control system will halt the process in real time and retract the tool away from the part. Considering the current conditions, function blocks will then adjust the process plan using the remaining available tools. The information of the new condition (i.e. tool breakage) is also sent to the information processing part of the machine holon. The information processing part will then communicate with the conveyor and robot holons, and updates them about the new condition so that they can make appropriate decisions such as delaying their processes or slowing down. To find an alternative process plan for the machine holon, its current machining state, geometry, etc., will be updated through building the communication between the machine holon and in-process part. For example, a second machine (machine holon-2) exists on the shopfloor in parallel to machine 1, and is currently off. It is assumed that machine holon-1 has found the best alternative process plan with its remaining tool sets, however, the cycle time using this plan would be more than the previous approach. As a result, machine holon-1 communicates with order holon and finds out that current production rate will not satisfy the demand in the remaining timeframe. Furthermore, the replacement tool is not currently available in the inventory and would not become available in time. Machine holon-1 will communicate with machine holon-2, to see if it can help with the production of parts in order to satisfy the demand in time. Machine holon-2 will then communicate with the part holon, raw material holon and order holon to get information on the part and its specifications, and to see if it can fabricate the part. The process plan is then generated, encapsulated in function blocks and dispatched to the machine 2 controller for execution. The presented scenario is a simple case that may frequently occur on the shop floor. A CPS modelled through holons, function blocks and multi-agent systems are capable of automatically processing these data and adjusting the system to the new conditions. Humans can also monitor and control the processes through designed user interfaces.

Different sets of connections among these holons (shown as dashed arrows in Fig. 2.6) indicate the possible inter-holon communications for dynamic planning, scheduling and execution that can be implemented by agents and function blocks.

### 2.4.4 IoT-Enabled Manufacturing System

In manufacturing, IoT technology is rapidly developing under the support of RFID, sensors, smart technology, and nanotechnology, and it is expected to promote interconnection of anything. Furthermore, IoT is helpful to construct a platform for sharing and interconnecting all kinds of manufacturing resources. Applying the generalised IoT into manufacturing industry can be used to address the connection, communication, computing, and control of manufacturing resources. Coupled with the rapid development of embedded systems and technologies, it provides enabling technologies for realising intelligent embedding of physical terminal equipment and the interconnection of M2M in manufacturing.

This section presents an IoT-enabled manufacturing system (IoTMS), and the focus is placed on sensing and capturing real-time information of manufacturing resources, real-time system monitoring, and the optimisation of manufacturing systems. IoTMS denotes a multisource real-time data-driven manufacturing system, covering monitoring, decision making and management from the production orders assigned to the required work-in-progress or finished products [36]. Hardware such as RFID devices, sensors, manufacturing resources, etc., are used to build an IoT-enabled sensing environment through the configuration of RFID devices, and sensors, and further sensing and capturing real-time information of manufacturing resources. Software integrates algorithms, modelling methods, and communication technologies, and therefore, has the capacity of analysing, computing, reasoning,

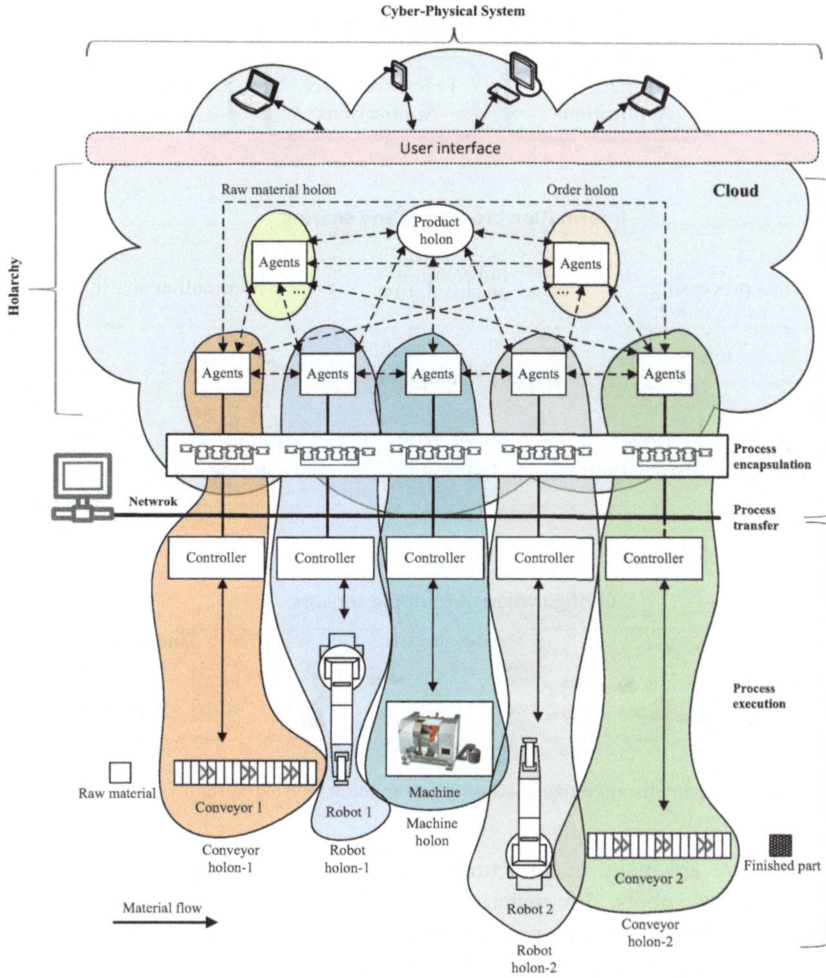


**Fig. 2.6** A hypothetic CPPS modelled using holons, agents and function blocks

controlling, and decision making. As one of the key technologies of the IoTMS, a real-time and multisource manufacturing information sensing system is responsible for sensing and capturing real-time information of manufacturing resources in the IoTMS, and this is the foundation of IoTMS. The architecture of real-time and multisource manufacturing information sensing system is designed as Fig. 2.7. This architecture is composed of four layers from the bottom to the top, namely the configuration of multiple sensors, sensors management, multisource manufacturing information processing and sharing, and management systems.

The configuration of multiple sensors, as shown in the bottom of Fig. 2.7, is to construct an IoT-enabled smart sensing environment for the physical manufacturing

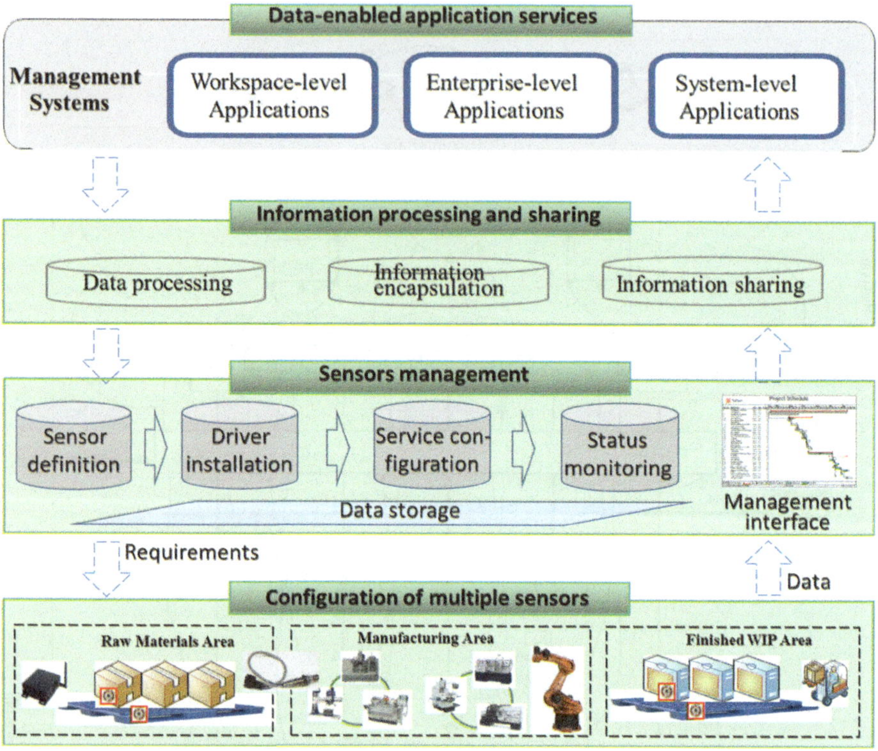


**Fig. 2.7** A real-time multisource manufacturing information sensing system [36]

resources. There are many manufacturing resources in the shop floor/workspace, such as machines, robots, raw materials, devices, and work-in-process, the status information of these manufacturing resources are constantly changing along the changes of manufacturing processes. As a result, huge amount of data related to these manufacturing resources are created in real-time. However, these data are heterogeneous because of the different attributes of manufacturing resources. To select fitting type of sensors for the configuration of intelligent sensing environment, the manufacturing information that is required to be collected is analysed and determined, and then, the needed sensors can be selected and configured in this module. This is based on the consideration of cost in the construction of intelligent sensing environment. For example, users in a manufacturing system show more interest in data related to machines, process quality, production objects, worker, and manufacturing environment. The information of different sensor, e.g. type, hardware interface, cost, target manufacturing resource, function, and reliability, would be analysed, and the fitting sensors are selected to build a smart sensing environment in the physical layer of manufacturing systems.

The sensors management module is responsible for managing sensors and monitoring the working status of the sensors. First, the sensors deployed in the manufacturing systems are registered into the platform of this module. The parameter information of the sensors is input in the registering system, including type of the sensor, its frequency (UHF or HF), interface (USB or COM), intercommunication protocol, data type, record means, reading method, etc. This step is to ensure the data collection in real time after sensors installation. Then, right sensors are installed in the right locations. In addition, the sensor drivers are installed to ensure the sensors operate reliably. However, due to the fact that the heterogeneous sensors have various embedded software, communication protocol, and access right, and so forth, the standard interface and drive library are used to address the heterogeneity of sensors. For example, the standard interfaces are used to drive the sensor, and the system downloads the third-party driver from the Internet according to sensor type, brand, and version, and then install it in the system and update the driver library with the latest edition. The service-oriented architecture is developed to integrate the sensors with different working mode into a uniform pattern under the same platform. In this architecture, the heterogeneous sensors can be published, searched, and invoked through the Internet. To invoke and manage all the sensors registered in the manufacturing system, each sensor will be designed and assigned a single service address and service ID, and all of the service addresses and ID are encapsulated into the standard web service. As a result, the heterogeneous sensors can be managed effectively, and the multisource and heterogeneous manufacturing data can be captured easily. Finally, the operating status of sensors is monitored by this module in real time, Further, the exceptions of sensors can also be sensed and handled to guarantee a stable condition of the sensor network.

The multisource manufacturing information processing and sharing module consists of information processing, information encapsulation, and information sharing. Huge amounts of real-time manufacturing data sensed and captured by the sensors are chaotic, and insignificant. As a result, the information processing is necessary and important to generate value-added information. On one hand, the value-added sensor data are the real interest of the managers/users; on the other hand, the processing of information added value can filter out the primitive and meaningless data and reduce the size of data. Besides, the data storage space can be saved. The information sensed by sensors originally is defined as the primitive information, and the value-added information is defined as the key information. After achieving the added value of information, information encapsulation is used to encapsulate value-added manufacturing data into a standard information template. After the encapsulation, the value-added manufacturing information can be stored under a standard form, easily accessible by different managers. Manufacturing information sharing module is responsible for sharing the valuable information among managers and users. Information sharing relies on the information communication technologies. Therefore, in the designed "Push model" and "Get model" communication methods, users are required to register their information first. Then the real-time and multisource manufacturing data can be published, and users can subscribe, search, and invoke the data they need. For example, in the Push model, the users submit their basic information, such as the user name, web address, and requirements, etc. Once submitted information is captured in the system, value-added real-time information meeting the needs of the users will be sent to the web address through wireless communication such as Wi-Fi.

Enterprise information system is responsible for providing information application services, which consist of application in the workshops, and enterprises, as well as among the big enterprise systems. The first is to provide the data service for the access, identification, and control of the physical manufacturing execution process from materials and semifinished products to the final products. The data identified and acquired from the IoT-enabled workshop manufacturing layer are materials, product, and production related the workshop information system. The second is to provide data service of integrating the production-related information, the product-related information, and other business management information, as well as the integration of the IoT-based workshop and other enterprise information subsystems. The third is to share the manufacturing data, manufacturing resources, and manufacturing services with other enterprise systems. This can achieve the optimal collaboration of manufacturing resources, and data sharing services, and dynamic optimisation of enterprise information.

### *2.4.5 CPS in Cloud Environment*

Over the last decades, Industry 4.0, initiated in Germany, is to promote the effective use of the latest information and communications technologies in real industrial applications towards smart factories of the future [52]. CPS, as the core technologies of Industry 4.0, is usually connected through the Internet, and recently applied the concept of the cloud to form cyber manufacturing [4]. Using the power of cloud computing and facilitated by the real-time connectivity to physical machines and robots (IoT), cyber manufacturing is able to realise true-sense CPS with various functions in one closed loop. In essence, cloud-based cyber manufacturing is an integrated CPS that can provide on-demand manufacturing services digitally and physically to best utilise distributed manufacturing resources and capabilities from anywhere [27]. The architecture of cloud-based cyber manufacturing system, shown in Fig. 2.8, adopts a three-tier view-control-model (VCM) design pattern and a segmented (public vs. private) network structure to address the requirement of efficient and secure data communication between cyber systems and physical systems in the cloud environment.

The Application tier is the application server in the cloud, which handles major security concerns; for example, session control, user registration, sensor data collection and distribution, and physical systems manipulation. This tier connects with the physical factory network, which contains the manufacturing resources and devices of factories, such as machines, and robots. *Signal Collector*, a server-side module, is responsible for collecting sensor data from networked physical machines or robots. After capturing the sensor data, *Signal Publisher* receive these data, and publishes and transmits the sensor data to the registered users, and uses the popular

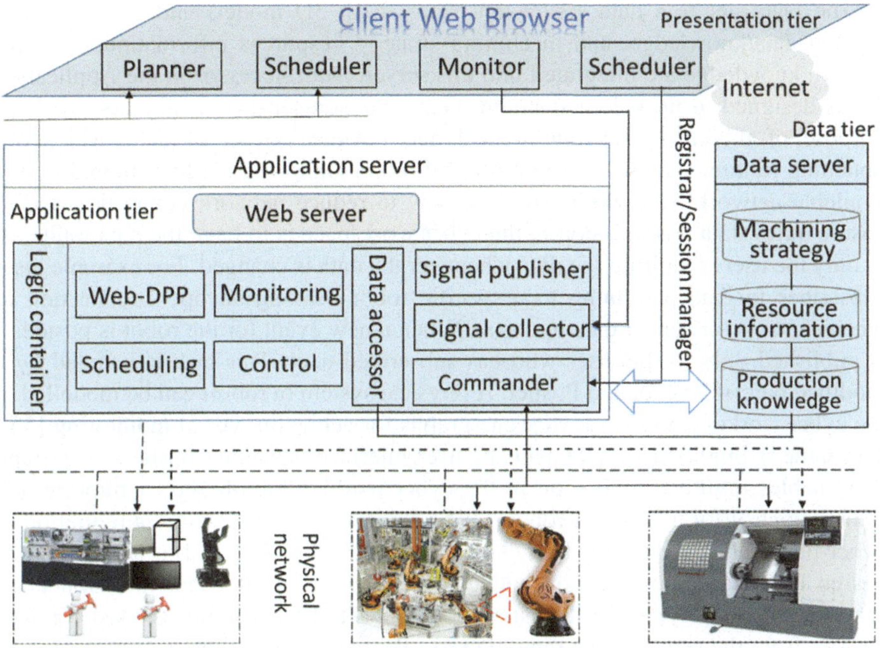


**Fig. 2.8** Architecture of a cloud-based cyber manufacturing system

publish-subscribe design pattern to distribute the sensor data at the right time to the right users. A *Session Manager* is designed to search for issues such as user authentication, information synchronisation, session control and data logging. All the initial transactions must pass this module for access rights authorisation. The *Registrar* is designed to maintain a list of subscribers (users) together with the requested sensor data. For the sake of security, the *Commander* that functions like a proxy, through a distributed device network with tight security constraints, is used to control a physical machine or robot [53]. The *Signal Publisher* uses a Pushlet [54] for implementation.

The Presentation tier is essentially a browser-based user interface where planner and scheduler can operate. Users still have the flexibility of monitoring machining and assembly operations from different perspectives, even though, the real-time sensor signals control the behaviours of the 3D models, e.g. selecting different 3D models and viewpoint control and so forth through the *Cyber Viewer*. An authorised user may submit a control command from anywhere to the cloud application server through the *Controller*. The *Commander* at the server-side then takes over the control rights for real machine/robot manipulations. The *Monitor* provides the operator with runtime status of the real equipment. For troubleshooting purpose, a user-side *ChatRoom* (not shown in Fig. 2.8) is designed for synchronised messaging among connected users.

The Data tier is a data server that stores both 3D models and relevant engineering data/knowledge, and machining strategy, resources information, and production knowledge are integrated into this server. *Data Accessor* in the Application tier is designed to provide a standard means for non-sensory data access, and the purpose of this is to separate logical and physical views of data. Moreover, obtaining runtime status of a robot for real-time monitoring is often limited by the available network bandwidth. The best way to reduce network congestion and to ensure quick data transmission in the cyber workspace is to have the data multicast to only the users requiring the data whenever the data is changed. For example, user subscribes to data pertaining to a specific robot, leaving an open connection to receive events (or sensor data updates). When a new event for the robot is posted, it is published only to this user who has subscribed to it. This task is handled by a modified pair of Pushlet and Postlet. A physical system (a robot) can be modelled in the cyber workspace using Java scene graphs for achieving visual monitoring [55]. The same is applied to implement the user interface, specified in the scene graph that enables intuitive navigation in the cyber world. Cloud-based monitoring and remote control for a physical robot can be achieved in the system configuration in cyber-physical environment. TCP (Transmission Control Protocol) is adopted in the design for data communication between the robot controller and the application server, whereas (Hypertext Transfer Protocol: HTTP) streaming is used for data sharing from the server to the remote users. These control and transfer protocol can provide hardware and software protection for robots. Using this design, the CPS allows a remote user to monitor the motions of all joints and to control the robot for remote assembly operations. For the cloud-based remote monitoring and control mentioned above, an operator mainly focuses on what is going on at the robot side, once separated from a robot. This means that motion monitoring must be presented intuitively to guide the operator for remote control. This is then facilitated by condition monitoring in terms of vibration, force, temperature, and so forth, on how well it is doing. While camera-based approach is common and instrumental for motion monitoring, its bandwidth consumption can easily create a bottleneck for cloud-based real-time applications, which is the main concern of this application. To address this problem, a virtual 3D model entirely driven by real sensor data is used for cloud-based monitoring. The testing results of a mini-cell assembly case study reveal that a roundtrip latency of 30 ms is achieved using this approach, which is fast enough to be considered as real-time at the system level. This delay depends more on the network speed than the CPU speed.

## 2.5 Conclusions

This chapter introduces the latest advancement of Cyber-Physical Systems (CPS) and Internet of Things (IoT) from multiple aspects. Firstly, a brief introduction is presented to better understand CPS and IoT. Wireless sensor network, cloud technologies, big data, Industry 4.0, and RFID technology, as the key enabling technologies related CPS and IoT, are also introduced. Then, key features, and characteristics of CPS and IoT, especially in real applications and projects, are explained. Finally, the typical and representative application examples of CPS and IoT are outlined to highlight the latest advancements.

## References


1. L. Wang, M. Törngren, M. Onori, Current status and advancement of cyber-physical systems in manufacturing. J. Manufact. Syst. **37**, 517–527 (2015)
2. L. Atzori, A. Lera, G. Morabito, Internet of things: a survey. Comput. Netw. **54**(15), 2787–2805 (2010)
3. B. Sterling, Shaping things, vol. 39 (2005, October), p. 152
4. L. Wang, Cyber manufacturing: research and applications, in *TMCE* (2014), pp. 39–49
5. Q. Chang, R. Gao, Y. Lei, L. Wang, C. Wu, Cyber-physical systems in manufacturing and service systems. Math. Prob. Eng. **2015**, 1–2 (2015)
6. E.A. Lee, *Cyber-physical systems—are computing foundations adequate?* vol. 1 (2006, October), pp. 1–9
7. E.A. Lee, Cyber physical systems: design challenges. in *11th IEEE International Symposium on Object Oriented Real-Time Distributed Computing* (2008), pp. 363–369
8. J. Wan, H. Yan, H. Suo, F. Li, Advances in cyber-physical systems research. KSII Trans. Internet Inform. Syst. **5**(11), 1891–1908 (2011)
9. J. Wan, D. Zhang, S. Zhao, L. Yang, J. Lloret, Context-aware vehicular cyber-physical systems with cloud support: architecture, challenges, and solutions. IEEE Commun. Mag. **52** (8), 106–113 (2014)
10. L. Monostori, Cyber-physical production systems: roots, expectations and R&D challenges. Procedia CIRP **17**, 9–13 (2014)
11. M. Mikusz, Towards an understanding of cyber-physical systems as industrial software-product-service systems. Procedia CIRP **16**, 385–389 (2014)
12. AMP (2014), http://www.manufacturing.gov/amp.html
13. IIC (2014), http://www.industrialinternetconsortium.org/
14. H. Kagermann, W. Wahlster, J. Helbig, Recommendations for implementing the strategic initiative INDUSTRIE 4.0. in *Final Report of the Industrie 4.0 Working Group* (2013), http://www.acatech.de/fileadmin/user_upload/Baumstruktur_nach_Website/Acatech/root/de/Material_fuer_Sonderseiten/Industrie_4.0/Final_report__Industrie_4.0_accessible.pdf
15. Horizon 2020 (2014), http://ec.europa.eu/programmes/horizon2020/
16. L. Wang, Wise-ShopFloor: an integrated approach for web-based collaborative manufacturing. IEEE Trans. Syst. Man Cybern. C Appl. Rev. **38**(4), 562–573 (2008)
17. CyPhERS, *Cyber-Physical European Roadmap and Strategy* (2013), http://www.cyphers.eu/
18. L.D. Xu, W. He, S. Li, Internet of things in industries: a survey. IEEE Trans. Industr. Inf. **10** (4), 2233–2243 (2014)
19. I.F. Akyildiz, W. Su, Y. Sankarasubramaniam, E. Cayirci, A survey on sensor networks. IEEE Commun. Mag. **40**(8), 102–105 (2002)
20. G.J. Pottie, Wireless sensor networks. *International Telecoms Week* (1998), pp. 22–26
21. J. Yick, B. Mukherjee, D. Ghosal, Wireless sensor network survey. Comput. Netw. **52**(12), 2292–2330 (2008)
22. I. Foster, Y. Zhao, I. Raicu, S. Lu, Cloud Computing and Grid Computing 360-degree compared. in *Grid Computing Environments Workshop* (2008)
23. M.A. Vouk, Cloud computing–Issues, research and implementations. J. Comput. Info. Technol. **16**(4), 31–40 (2008)

24. L. Ren, L. Zhang, L. Wang, F. Tao, X. Chai, Cloud manufacturing: key characteristics and applications. Int. J. Comput. Integr. Manuf. **30**(6), 501–515 (2017)
25. P. Mell, T. Grance, The NIST Final Version of NIST Cloud Computing Definition Published. *NIST Special Publication*, 145 (2011), p. 7
26. L. Wang, A. Mohammed, M. Onori, Remote robotic assembly guided by 3D models linking to a real robot. CIRP Ann. Manufact. Technol. **63**(1), 1–4 (2014)
27. L. Wang, X.V. Wang, L. Gao, J. Váncza, A cloud-based approach for WEEE remanufacturing. CIRP Ann. Manufact. Technol. **63**(1), 409–412 (2014)
28. Industry 4.0, http://blog.bosch-si.com/industry-4-0-germany-takes-first-steps-toward-the-next-industrial-revolution/
29. N. Jazdi, Cyber physical systems in the context of Industry 4.0. in *2014 IEEE Automation, Quality and Testing, Robotics* (2014), pp. 2–4
30. L. Tan, Future internet: The Internet of Things. in *3rd International Conference on Advanced Computer Theory and Engineering* (2010), V5-376–V5-380
31. C.M. Roberts, Radio frequency identification (RFID). Comput. Secur. **25**(1), 18–26 (2006)
32. J. Lee, B. Bagheri, H.A. Kao, A Cyber-Physical Systems architecture for Industry 4.0-based manufacturing systems. Manufact. Lett. **3**, 18–23 (2015)
33. F.-J. Wu, Y.-F. Kao, Y.-C. Tseng, From wireless sensor networks towards cyber physical systems. Pervasive Mob. Comput. **7**(4), 397–413 (2011)
34. L. Takayama, C. Nass, W. Ju, Beyond dirty, dangerous and dull: what everyday people think robots should do. in *International Conference on Human Robot* (2008), pp. 25–32
35. P. Leitão, A.W. Colombo, S. Karnouskos, Industrial automation based on cyber-physical systems technologies: prototype implementations and challenges. Comput. Ind. **81**, 11–25 (2016)
36. Y.F. Zhang, F. Tao, *Optimization of Manufacturing Systems Using the Internet of Things.* (Elsevier, 2016). (ISBN 978-0-12-809910-0)
37. L. Wang, R. Gao, I. Ragai, An integrated cyber-physical system for cloud manufacturing. in *ASME International Manufacturing Science and Engineering Conference*, MSEC2014–4171 (2014)
38. F. Chen, P. Deng, J. Wan, D. Zhang, A.V. Vasilakos, X. Rong, Data mining for the internet of things: literature review and challenges. Int. J. Distrib. Sens. Netw. **2015**, 1–14 (2015)
39. X. Yue, H. Cai, H. Yan, C. Zou, K. Zhou, Cloud-assisted industrial cyber-physical systems: an insight. Microprocess. Microsyst. **39**(8), 1262–1270 (2015)
40. T. Qu, S.P. Lei, Z.Z. Wang, D.X. Nie, X. Chen, G.Q. Huang, IoT-based real-time production logistics synchronization system under smart cloud manufacturing. Int. J. Adv. Manuf. Technol. **84**(1–4), 147–164 (2016)
41. T. Qu, H.D. Yang, G.Q. Huang, Y.F. Zhang, H. Luo, W. Qin, A case of implementing RFID-based real-time shop-floor material management for household electrical appliance manufacturers. J. Intell. Manuf. **23**(6), 2343–2356 (2012)
42. B.H. Li, L. Zhang, L. Ren, X.D. Chai, F. Tao et al., Typical characteristics, technologies and applications of cloud manufacturing. Comput. Integr. Manuf. Syst. **18**(7), 1345–1356 (2012)
43. G. Adamson, L. Wang, M. Holm, P. Moore, Cloud manufacturing—a critical review of recent development and future trends. Int. J. Comput. Integr. Manuf. **30**(4–5), 347–380 (2017)
44. X. Xu, From cloud computing to cloud manufacturing. Robot. Comput. Integr. Manuf. **28**(1), 75–86 (2012)
45. L. Wang, Machine availability monitoring and machining process planning towards Cloud manufacturing. CIRP J. Manufact. Sci. Technol. **6**(4), 263–273 (2013)
46. M. Givehchi, A. Haghighi, L. Wang, Cloud-DPP for distributed process planning of mill-turn machining operations. Robot. Comput. Integr. Manuf. **47**, 76–84 (2017)
47. D. Mourtzis, E. Vlachou, N. Xanthopoulos, M. Givehchi, L. Wang, Cloud-based adaptive process planning considering availability and capabilities of machine tools. J. Manuf. Syst. **39**, 1–8 (2016)
48. L. Wang, A. Haghighi, Combined strength of holons, agents and function blocks in cyber-physical systems. J. Manuf. Syst. **40**, 25–34 (2016)

49. L. Wang, G. Adamson, M. Holm, P. Moore, A review of function blocks for process planning and control of manufacturing equipment. J. Manuf. Syst. **31**(3), 269–279 (2012)
50. L. Wang, An overview of function block enabled adaptive process planning for machining. J. Manuf. Syst. **35**, 10–25 (2015)
51. P. Leitão, Agent-based distributed manufacturing control: A state-of-the-art survey. Eng. Appl. Artif. Intell. **22**(7), 979–991 (2009)
52. S. Wang, J. Wan, D. Zhang, D. Li, C. Zhang, Towards smart factory for Industry 4.0: a self-organized multi-agent system with big data based feedback and coordination. Comput. Netw. **101**(4), 158–168 (2016)
53. Y. Xu, R. Song, L. Korba, L. Wang, W. Shen, S. Lang, Distributed device networks with security constraints. IEEE Trans. Industr. Inf. **1**(4), 217–225 (2005)
54. J. Van den Broecke, Pushlets, Part 1: Send events from servlets to DHTML client browsers. in *JavaWorld* (2000)
55. L. Wang, B. Wong, W. Shen, S. Lang, A java 3D-enabled cyber workspace. Commun. ACM **45**(11), 45–49 (2002)

# Chapter 3
# Challenges in Cybersecurity

## 3.1 Introduction

Manufacturing technology changes with the needs of consumers. For example, the concept of Cloud Manufacturing (CM) was largely created due to the globalisation of the world economy. Reference [1] notes that the 20th century had a "rapidly accelerating rate of technological development, including vast improvements in manufacturing abilities and global markets." Following this trend, manufacturing in the 21st century has shown the need to be versatile and scalable [2]. The authors also mentioned that in order for manufacturing to be versatile, the method of controlling machines must also be versatile; they currently are not. Though complexity is increasing, low cost manufacturing is still desired to maintain profits. Many of the key technologies have already been developed to make CM a dependable configuration in today's manufacturing industry.

While CM has the potential to solve issues in manufacturing, it has its concerns. Among these is cybersecurity, which is collecting considerable attention because it is a major reason that users or consumers are not confident in adopting CM [3]. More specifically, cybersecurity measures for remotely controlling machines and the information sent to them is limited [1, 2]. Reference [4] defines this problem as communication security; when using the Internet to communicate, information must be maintained and managed from risk. Essentially, CM uses resources that can be transmitted anywhere from a central location to a device with Internet access. Cybersecurity is a topic growing in popularity, but it still lacks adequate information available.

A major concern of implementing CM systems is the assurance that proprietary information about the intellectual property owned by the organisation or information about the company's operations is available only to the authorised individuals. Any cybernetic system must accommodate privacy of the individuals and organisations involved. Privacy is intended to ensure data security and limit the data that



L. Wang and X. V. Wang, *Cloud-Based Cyber-Physical Systems in Manufacturing*,
https://doi.org/10.1007/978-3-319-67693-7_3

any user can access. Protecting privacy during operations ensures the safety of physical machines and their operators in the cloud manufacturing context. Information discussing the safety of equipment operators is limited. Cybersecurity is a must to protect both the privacy and safety in any cybernetic system. It is also intended to protect the software and machines in a CM system.

The purpose of this chapter is to provide an overview of cybersecurity measures for the protection of data being sent to physical machines in a CM system; the cybersecurity measures will also ensure the protection of the physical machines. While this chapter does not present new data, it does gather information about the emerging topics of remote equipment security and allows the authors to provide input on the issue. The remaining sections of the chapter are revealed as follows. Section 3.2 introduces how Internet of Things (IoT) enables remote machine control; Sect. 3.3 discusses remote equipment control and its applications in CM; Sect. 3.4 covers cybersecurity concerns in remote equipment control and provides security methods for ensuring data safety in CM; Sect. 3.5 discusses the future outlook of remote equipment control and its security; and Sect. 3.6 brings the chapter to a conclusion.

## 3.2 Internet of Things

A brief account of IoT is provided to understand its importance to remote equipment control. It is understood that CM integrated technologies such as Cloud Computing and IoT to provide more competitive and efficient manufacturing. IoT allows distributed manufacturing resources to connect to one another virtually, thus allowing them to become virtual resources that can be offered as services [3].

Sensors, such as RFID (Radio Frequency Identification) tags that are constantly transmitting signals will be used to connect *things* with one another [5]. Sensors are able to provide real-time machine information such as pressure, force, and temperature in order to allow better machine management and be able to send the remote operator an alarm should a problem arise; the problem can then quickly be corrected [6]. Using sensors and RFIDs for example, IoT allows humans and machines to be connected throughout the manufacturing process [7]. For example, these sensors will allow a user at any location to view the operating conditions of a machine, how many machines are being utilised, and also what quality the machines are producing [7]. These sensors can additionally be used for intelligent identification, locating, and tracking parts and processes [6]. Figure 3.1 demonstrates the role of IoT in providing machining data and feedback to the remote operator in a CM environment. As wireless sensors become more advanced, communication from human to machine and machine to machine will become more efficient and occur "from anywhere at any time" [7]. Hitachi for example, currently has the smallest RFID in the world, measuring only 0.15 mm$\times$ 0.15 mm $\times$ 7.5 μm. Reference [8] discusses additional information on IoT and its implementation in CM.

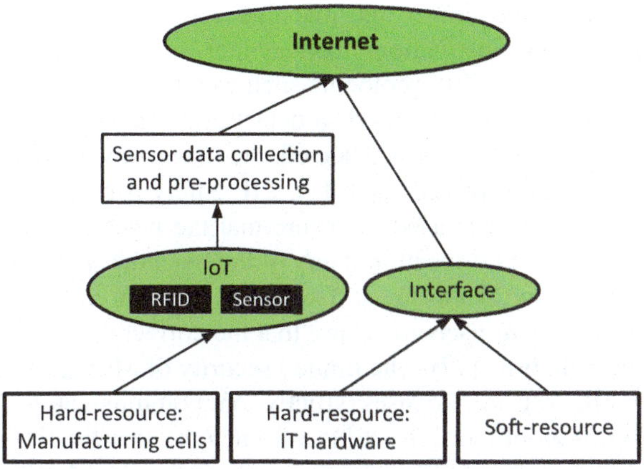


**Fig. 3.1** Role of IoT in cloud manufacturing. Adapted from [6]

IoT is key to improving automation in manufacturing and implementing CM [7, 9]. Reference [10] states that in order for IoT to be successful, the connections must be omnipresent in all devices. Reference [11] concluded from a National Compliance Management Services (NCMS) report that "there is a consensus that linking factory devices to the Internet will become the backbone technology for future manufacturing". The application and cybersecurity measures of connecting the Internet to industrial machines will be discussed in Sects. 3.3 and 3.4. However, as IoT grows and expands, cybercrime is expected to grow as well [5, 12]. References [10, 13] foreshadow that security risks will be far greater with IoT than with the current Internet.

## 3.3 Remote Equipment Control

One of the main applications of IoT is the ability to remotely control and monitor machinery. Since machine operators will be physically separated from the machine in a decentralised environment, what is happening on the machine is of primary concern to the operator [11]. Using sensors, as described in the previous section, events such as a machine tool moving can be tracked and transported over the Internet and be viewed on a screen in order to assist the operator in making decisions [7]. "In all kinds of manufacturing resources, the machining equipment is one of the most important resources [14]." Remote equipment control is key in situations where machines are too remote to reach or that have extreme hazards for human safety [11, 15]. A rescue operation is another application for remote equipment control [15]. For example, situations such as searching the ocean floor for a wrecked plane or ship is a case for remote controlled equipment. Furthermore, remote equipment control could be used to economically assemble components [15].

One method to assist in remotely operating machinery is to utilise Machine Control as a Service (MCaaS). MCaaS physically separates the machine location from the location of the machine operator [1, 2]. MCaaS mainly facilitates communicating instructions to machinery and ensuring the instructions can be safely completed. Safety protocols such as CIP (Common Industrial Protocol) Safety and PROFI Safe can be used to detect errors in communication such as missing or old data; when an error is detected, the machine will follow its programmed emergency behaviour to protect itself [1]. CIP Safety is used to communicate information from machinery; it is used to ensure that the machines are functioning safely [16]. This can include notification if any of the machine safety mechanisms such as guards and control hardware are at fault [16]. PROFI Safe is a method of communication between two users to ensure that the correct data is delivered to the right location at the right time [17]. The limited security of MCaaS is discussed in detail in Sect. 3.4. While MCaaS focuses strictly on communicating safe machining instructions, Wise-ShopFloor [18–20] is a framework that focuses on monitoring machinery and limiting who has the ability to remotely control machinery.

Wise-ShopFloor is proposed for remotely monitoring and controlling machinery in real time. Its architecture can be seen in Fig. 3.2 [21]. Reference [20] states that real-time monitoring is impractical due to "real-time constraints". This is because most machines today do not have the capabilities built-into transmit and receive data. There are existing systems to monitor machines, however, they can only monitor machines online; they must do all other process planning offline. Reference [6] refers to this as OnCloud and OffCloud. OnCloud refers to services provided through the CM platform, such as monitoring machinery, while OffCloud refers to activities that must be performed by a human such as material logistics [6]. Many tasks require a combination of OnCloud and OffCloud activities [6]. Existing systems use cameras to monitor equipment. To achieve real-time monitoring, Wise-ShopFloor uses sensors rather than video streams to track the machine's motion. When comparing the number of bytes used in a 640 × 800 image for

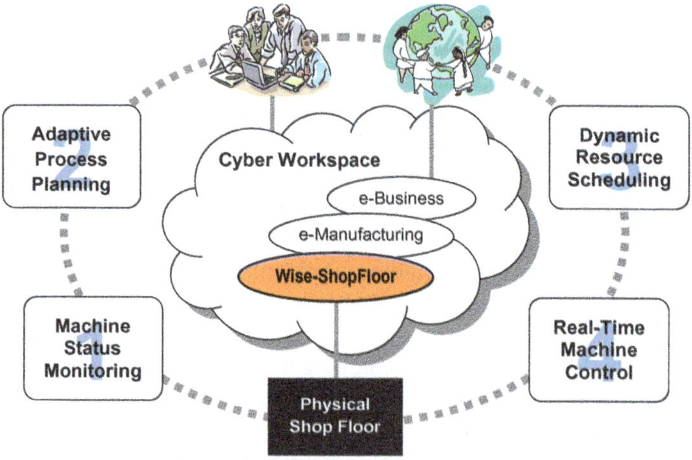


**Fig. 3.2** Architecture of wise-shopfloor [21]

monitoring a machine to that used for a scene shown in Wise-ShopFloor using sensors, the sensors require only 0.02% the number of bytes as video does [19]. Therefore, the sensors are able to monitor a machine with little time delay, unlike the delay occurring using video [19]. The sensors send data to a 3D model in JAVA which visually allows the user to look at the machine's movements [19, 20]. Using sensors, information such as temperature and machine vibration can be collected and displayed on the 3D model in JAVA through the use of colours [20]. To reduce bandwidth traffic and ensure quick data transfer, it is also suggested to relay machine information to only the required users when an operating condition has been changed [11]. Wise-ShopFloor aims to be able to control machine operations from anywhere at any time [1].

## 3.4 Security Concerns and Methods

### *3.4.1 Security Concerns*

Security in CM is a major concern [3, 6, 22]. It is causing the cloud to grow at a much slower rate [23–27]. Security risk is increased because the cloud requires two-way communication between a customer and service provider [28]. The increasing collaboration in CM among companies is also causing security threats with the amount of data interacting in the cloud [3]. Security related to remote equipment control is focused on ensuring only those with proper access rights have the ability to remotely operate machinery. For that reason, this section will mainly focus on *control* who has the ability to remotely operate equipment and focus less on CM security in general.

Risks in security are known to exist. Reference [29] categorises cloud security risks into three categories: provider related vulnerabilities, availability, and third-party data control. Security in regards to equipment control is covered in the last category: third-party data control. Unfortunately, security risks with third-party data control are the least known due to a lack of knowledge on how the third-party stores data. The third-party has control over the data being stored and this causes concern over who accesses the data [3, 29]. In an environment with multiple users, access security measures must be created to ensure data is protected [27]. This should include user authentication and authorisation [30]. Reference [30] feels that firewalls will not ensure data access protection. Authentication is the process of proving a user's identity; once the user's identity is proven, authorisation determines what privileges a user has and what actions they can perform [3]. Some factors that can affect the effectiveness of these access controls include turnover of employment in the third-party and also the changing roles of users within a company [3]. However, Reference [3] suggests auditing the authorisation records, authentication process, and activities of users to ensure security is sufficient to prevent and be able to detect data breaches.

The Cloud Security Alliance (CSA) defined seven security issues that they see in the cloud [31]. Some of the issues mentioned were malicious insiders, insecure interfaces and APIs (Application Programming Interfaces), and account or service hijacking. Companies are aware of the threats from malicious insiders [31]. Because of their inside access, malicious insiders could control sensitive data with little chance of being caught [31]. Some ways for a company to handle malicious insiders are to separate system privileges, log server accessing, and implement two-factor authentication [32]. The next issue mentioned, security of the API and interfaces, directly controls the security of the cloud service because this is where the machines are monitored and managed [31]. Reference [33] discussed how George Wrenn, the Security Solutions Director at Unisys, feels that user authorisation might not be enough to prevent unauthorised access to the API. To combat this risk, service providers must use strong access controls with encrypted transmission [31]. The last issue mentioned, account or service hijacking occurs when a user's login information is stolen [33]. Using this information, attackers can alter data, direct clients to illegitimate sites, and send false information [31]. George Wrenn states that ways to combat this are to employ cryptographic authentication systems and to only authorise access using devices serving the company's interest [33]. Devices considered to be serving the company's interest could include workstation computers or other company-owner equipment. Two-factor authentication and monitoring worker activity are other solutions [31]. Additional security risks of using the cloud can be found in [34, 35].

### *3.4.2 Security Methods and Architecture*

Security protocol varies depending on the amount of cloud services being provided. "Just as capabilities are inherited, so are the information security issues and risks" [25, 36]. However, security is still a topic of debate with its capabilities in the cloud still unclear [37]. Reference [38] discusses the requirements for a CM platform. Data and communication security are key in providing cloud applications; security aspects for infrastructure, communication, and application levels are described [38]. For infrastructure security, a firewall can be used to separate cloud data and allow selective access from authorised users. Tracking which users access certain data can also be used to prevent and detect data hacks. In communication security, JAVA Messaging Service (JMS) is recommended to be used. This will allow security to be set which can limit items such as message sizes and user interfaces. The JMS will also prevent data access from unauthorised users. Security in applications will be provided through setting access rights and encrypting data. Accessing data will have set limits on what can be seen and also be protected with passwords. Reference [39] mentions that limited access, data privacy, and accountability with data access also need to be addressed. Security is a critical aspect in this platform because it is key to gaining cloud acceptance in industry.

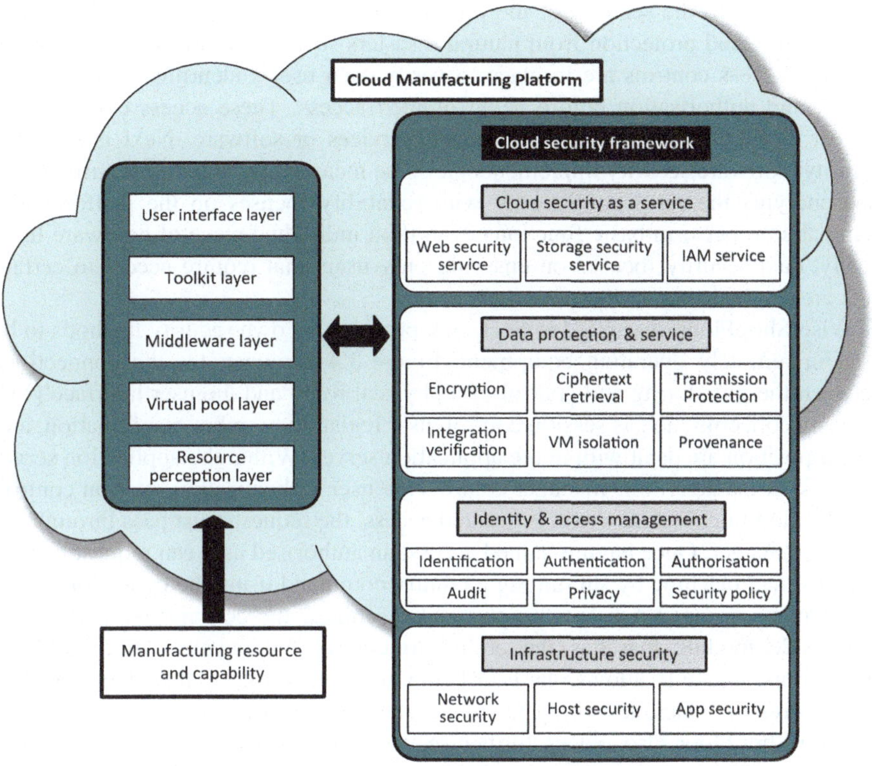


**Fig. 3.3** Cloud manufacturing security framework. Adapted from [3]

To address security concerns, reference [3] presents a four-layer architecture as shown in Fig. 3.3, being infrastructure security, identity and access management, data protection and security, and cloud security as a service [3]. The infrastructure security focuses on network security [3]. The identity and access management maintains data privacy and protection by verifying users. This layer ensures that only those who have rights to perform a certain action can do so; this maintains data privacy. This is especially important when multiple users can have access to a resource; it can also detect false identities and security threats [3]. The data protection and security ensures data is encrypted or secured in other ways; this is because service providers may be lacking in data security. The cloud security as a service provides the security to ensure privacy and data security against unwanted users. Security is key in preventing cloud failure [3].

Another security platform for manufacturing is Virtual Fort Knox. As the name implies, security is the main focus of this platform [40]. The security is divided into five sections: physical, network, software, reliability, and data security. Physical security refers to the security of the physical servers from unwanted access, hardware failure, and protection from natural disasters such as earthquakes. In network security, access controls are in place to verify each user's identification, authentication, and authorisation before being granted access. These access controls will protect the system from intruders altering services or software. Next is software security; software security implements the same measures as network security but it also encrypts the services. Security with reliability focuses on the platform still being able to perform basic functions even if an individual piece of hardware fails. Lastly, data security focuses on ensuring only users that require access to certain data are granted access.

Wise-ShopFloor, discussed in Sect. 3.3, proposes its own security methods to be used in remotely operating equipment. Figure 3.4 demonstrates the connections between the monitoring and control of a physical robot and the user interface [11]. Security concerns such as session control, user registration, robot manipulation, and sensor position are dealt with in the application server. Within the application server is a session manager; this manager controls the user authentication, session control, and data logging. Before a user is granted access, the request must pass through the session manager. Once having gained access, an authorised user can request control of a physical machine by submitting a control command using the cyber controller. To limit the security risks of the physical machines, the commander is the only server-side module that has the ability to control the physical machine [18]. Wise-ShopFloor only allows one user being able to control the machine at a time; this ensures the machine is only receiving one set of instructions.

Wise-ShopFloor also places further security measures that are used for communication between the users and the robots as shown in Fig. 3.5 [11]. Transmission Control Protocol (TCP) is used for communication between the machine and application server and Hypertext Transfer Protocol (HTTP) is used for sharing data with the users. TCP can only occur on one server and between two users at a time [41]. Data is sent in packet sizes that are controlled by the connection speed; when a packet is sent, the sender waits to send another packet until the receiver acknowledges it received the first packet. One security method built in is that if the receiver detects that information was manipulated in sending, it ignores the sent packet and waits for the sender to resend the packet again. One disadvantage with TCP is that the packets it sends may not be sent in the correct order. One current use of TCP is in electronic mail applications [41]. TCP is only a physical security, however HTTP utilises a firewall and is more protected over the Internet. The security design of Wise-ShopFloor allows the system to monitor the motions of the physical machine and also control machine movement [11]. Reference [15] sees Wise-ShopFloor as being proven through case studies to be a viable and effective system for manufacturing floors.

Most security measures for MCaaS, discussed in Sect. 3.3, focus on ensuring that the data is transmitted safely, not security on controlling who accesses the data. However, security risk can still be reduced in MCaaS because it only allows users

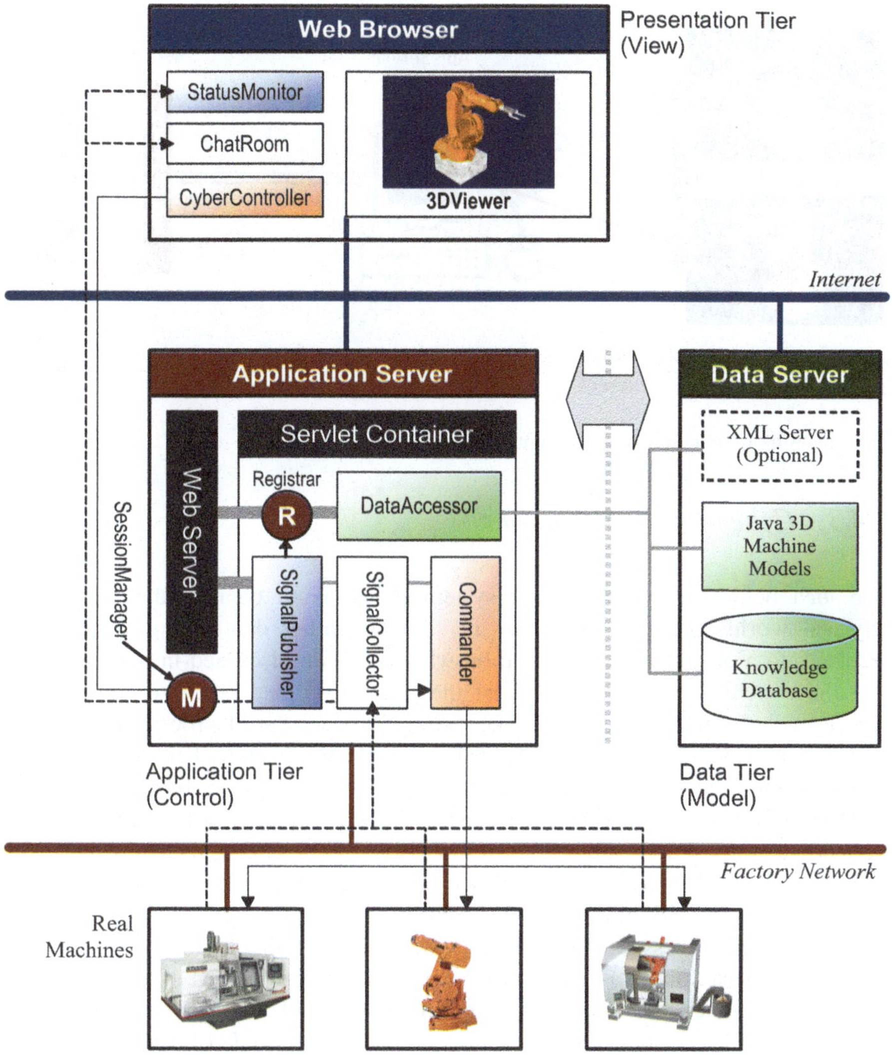


**Fig. 3.4** System for remote monitoring and control [11]

with granted access to control the machine rather than the traditional manufacturing facility where any worker can access a machine [1]. The authors suggest also using firewalls, virus scanners, and other existing IT security measures to protect the process information [1]. Reference [21] suggests security measures such as virtual LANs and packet filters as examples to ensure a security. However, to create an effective MCaaS, further control engineering is still needed [1].

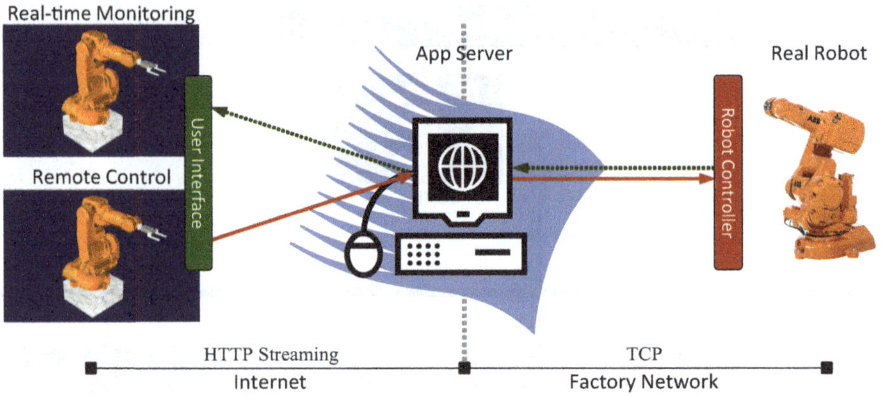


**Fig. 3.5** Robot communication configuration [11]

### *3.4.3 Cyber-Physical Systems*

The idea behind a cyber-physical system (CPS) is to connect the virtual and physical worlds using sensors that are able to gather data and provide useful feedback to a physical entity. Based on what has been discussed in Sects. 3.2 and 3.3, the conclusion can be made that the technology and processes involved in remotely controlling equipment make the equipment control process a CPS. The sensor data gathered from the physical machine is transmitted through IoT and allows a remote operator to monitor and control the equipment based on the data received. Therefore, the security of controlling who can access the remotely controlled machines directly affects the security of the CPS. One of the challenges with cyber-physical systems is security. However, research regarding security is limited. A CPS allows the communication between machines and humans. These systems can gather and process data and use interfaces to communicate with humans [42]. CPSs are expected to be a major contributor to the design of CM systems. As CPS research continues, communication between physical machines, sensors, and engineering software will improve and meet the levels needed for secure autonomous manufacturing in the future [7].

Several security protocols have been suggested to ensure data security in a CPS. For example, Reference [43] suggests a convenient, low cost method of authentication to be used with IoT. The method requires no passwords or codes; it uses telebiometric authentication. A biometric is a trait unique to a person such as their fingerprint, voice, or face. The authentication works by using biometric sensors that are registered to a user; these sensors are connected to telecommunication networks. Therefore, as a user requests access to a resource on IoT, the sensors will provide the authentication for the user to be granted access without having to enter a password [43].

## 3.5 Future Outlook

A number of web-based systems such as WebCADET, CyberCut, and NegotiationLens have been developed to assist in the manufacturing environment [15]. However, the systems are either for only monitoring machining or only providing offline simulations of machining [15]. In addition, these systems cannot communicate in real-time and often require special software to utilise them; this limits their convenience of use [15]. Reference [11] confirms that when these systems are considered for machine control and real-time monitoring, their applications are not practical. The need for real-time monitoring is achieved through frameworks such as Wise-ShopFloor being researched; it can be assumed that additional frameworks will be developed to meeting this need and to enable remote machine operation.

Currently, equipment security through IoT is based on byte-code verification, user permission, and security policies [11]. Byte-code verification is used when new instructions are uploaded to a virtual machine; each byte-code is inspected and verified that it can be performed without being damaging to the machine [44]. User permission limits the number of tasks and files that a user can access [44]. However, future considerations for security measures include data encryption, digital rights management, confidentiality agreements, and equipment/operator protection [11]. When a user has been authorised, the data they were trying to access becomes visible to them in the system; however, with encryption, once the user is authorised, the data is visible, but not able to be understood without the necessary key [44]. This can ensure only those with permitted access can read the data to ensure confidential data can be kept safe. Reference [3] warns that with limited support for a standard method of identifying users, customers of cloud services may be required to develop their own security controls. Other future security methods to control user access to equipment have been discussed throughout the paper. For example, telebiometric authentication and Wise-ShopFloor's multiple security methods were discussed in Sect. 3.4.

On the other hand, Reference [35] states that new cloud security measures are not necessarily needed because security measures that already exist can be used. Once information is into the cloud and is simply being stored in one place, it can be protected using current security measures because the data is not moving [35, 45]. Current security measures that could be used are firewalls, endpoint security, and network intrusion prevention systems because they can be adjusted to protect a virtual server instead of protecting physical servers as they currently do [35]. Reference [46] discusses how a firewall is used to maintain security and accessibility control by managing the incoming and outgoing traffic. The firewall secures the user, data, and hardware in the manufacturing system. It also verifies user identity to provide protection. By verifying the user's identity, the safety of the software and data are guaranteed. The user can only control a machine once they successfully make it through the firewall and the request is delivered to the Interface

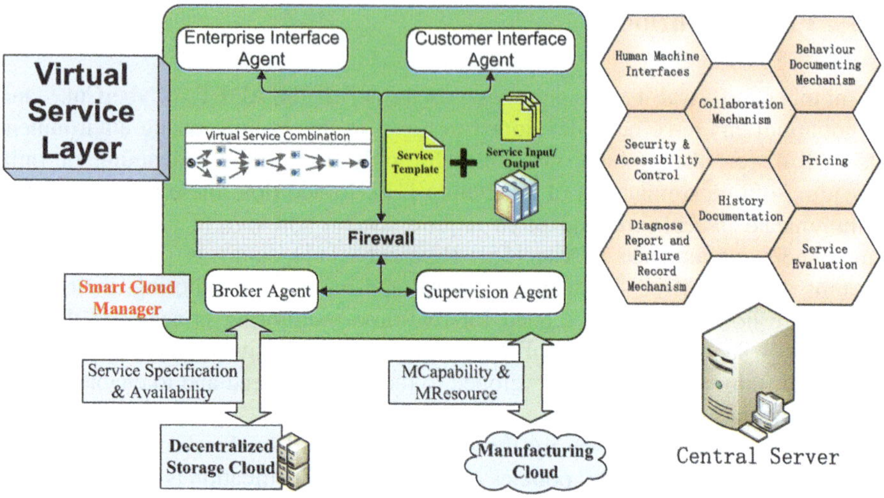


**Fig. 3.6** Virtual service layer [4]

Agent, as can be seen in Fig. 3.6 [46]. Furthermore, the firewall limits the user to only be able to access what they have the rights to access. To protect the physical machines, there are agreements which ensure that only one user can have access to a machine at a time. Thus, the firewall helps solve the problem suggested by Reference [3] of providing security when there are multiple users that can access a resource.

Reference [22] states that in the future, a mature CM platform should only require the inputs and service requirements from the user. The user should have enough confidence in the cloud where they will not need to monitor or provide any technical support during the manufacturing operations [22]. The cloud will be able to accomplish the task and provide the user with their end product [22]. The users will see the cloud as a "black box" only showing them the input and output of the manufacturing process [22].

## 3.6 Conclusions

Issues such as ensuring the correct user is accessing the given data can be reduced by implementing access controls that limit the user's interface. Table 3.1 outlines security measures that can be taken to improve remote equipment security. It is acknowledged that controlling network devices, in this case, machines, is a concern that is not yet solved [15]. One method of controlling machines is to employ MCaaS which will allow the scalability of manufacturing equipment to meet the varying needs of modern manufacturing; however, there is still research needed to

**Table 3.1** Possible security measures

| Threat | Equipment damage |
|---|---|
| Security method | Byte-code verification [11, 44] |
| Benefit | Inspects instructions to protect machine from performing destructive actions [44] |
| Issue | Verification can be easily disabled with a command [44] |
| Threat | Equipment control and monitoring |
| Security method | Wise-ShopFloor [18–20] |
| Benefit | Collect various types of machine data in near real-time [20], multiple stages to verify users and set access controls [18], only one user at a time can control machine movement [18] |
| Issue | Must install sensors on machine to track position [20] |
| Threat | Communication data tampering |
| Security method | Transmission Control Protocol (TCP) [41] |
| Benefit | If system detects tampering in data being sent, data is ignored and new machine instructions are sent until they are correct [41], sends data in small file amounts instead of all at once [41] |
| Issue | Limited by internet connection speed [41], data is sent and received in random file order [41], only a physical security [41], users must be on same server [41] |
| Threat | Data access |
| Security method | Encryption [3, 11, 31, 38, 40, 44] |
| Benefit | Ensures data safety against unauthorised users, need a key to understand data [7], serves as protection in case providers are lacking in security [3] |
| Issue | None mentioned |
| Threat | Equipment data tampering |
| Security method | Hypertext Transfer Protocol (HTTP) [16, 41] |
| Benefit | Same as TCP, uses a firewall to safely transfer data over the Internet [16] |
| Issue | Limited by the Internet connection speed [41], data is sent and received in random file order [41], users must be on the same server [41] |
| Threat | Malicious insiders |
| Security method | User authentication [30, 31, 40, 45] |
| Benefit | Detects false users [3], can track data being accessed [31], keeps data private [3, 40] |
| Issue | User login information can be passed out to unauthorised users |
| Threat | Insecure interface |
| Security method | Access controls [27, 31, 33, 40, 44, 45] |
| Benefit | Protects machines with multiple users [27, 44], restricts unwanted access [3, 44] |
| Issue | Might not prevent data from being accessed by unauthorised users [33] |
| Threat | Account hijacking |
| Security method | Monitor activity [31] |
| Benefit | Tracks what data is being accessed [38] |
| Issue | Does not prevent it, only detects it [33] |

(continued)

**Table 3.1** (continued)

| | |
|---|---|
| Threat | Equipment damage |
| Threat | Unwanted data access |
| Security method | Firewall [1, 38, 46] |
| Benefit | Limits accessible data and verifies users [38, 46], monitors user access [46], protects machines from multiple users [46] |
| Issue | Cannot guarantee protection [27] |
| Threat | Communication security |
| Security method | JAVA messaging service [38] |
| Benefit | Limits message sizes and sets access controls [38], removes unwanted users [38], encrypts data [38] |
| Issue | None mentioned |
| Threat | Data tampering |
| Security method | Digital rights management [47] |
| Benefit | Prevents data from being saved or altered unless the provider permits it [47], may require special software to view data [47] |
| Issue | None mentioned |
| Threat | Equipment control |
| Security method | MCaaS [1, 2] |
| Benefit | Physically separates machine from controls [1, 2], ensures data is not tampered as it is sent to machines [1] |
| Issue | Does not limit user access [1] |
| Threat | Platform security |
| Security method | Virtual Fort Knox [40] |
| Benefit | Secures physical and virtual knowledge from accidents and unwanted access [40], verifies each user [40], encrypts data [40] |
| Issue | Requires users to have substantial in-house IT infrastructure [40], data must be stored by company, not stored in the cloud [40] |

secure the data being sent to the machines. Using remote equipment security methods such as mentioned in Wise-ShopFloor and Virtual Fort Knox, users will be verified before being allowed to access data. For example, Reference [4] states that it is important to implement "trust mechanisms" in all levels of CM to ensure data is safe. The aim of this chapter is to provide an overview of current security measures being considered to ensure the protection of data being sent to physical machines in a CM system. While this chapter does not present new data, it does gather information about the emerging topics of remote equipment security and allows the authors to provide input on the issue. The authors outline existing security issues and solutions in Table 3.1. This chapter also provides the reader a broad understanding of current and future outlook of related topics. These topics are still in debate and more research needs to continue into the future.

## References


1. A. Verl, A. Lechler, S. Wesner, A. Kirstadter, J. Schlechtendahl, L. Schubert, S. Meier, An approach for a cloud-based machine tool control. Procedia CIRP **7**, 682–687 (2013)
2. J. Schlechtendahl, F. Kretschmer, A. Lechler, A. Verl, Communication mechanisms for cloud based machine controls. Procedia CIRP **17**, 830–834 (2014)
3. Y. Wang, S. Ma, L. Ren, A security framework for cloud manufacturing. ASME 2014 international manufacturing science and engineering conference, Detroit, MI, USA, 9–13 June 2014. doi:10.1115/MSEC2014-4082
4. Y. Yang, R.X. Gao, Z. Fan, J. Wang, L. Wang, Cloud-based prognosis: perspective and challenge. ASME 2014 international manufacturing science and engineering conference, Detroit, MI, USA, 9–13 June 2014
5. D. Shin, A socio-technical framework for internet-of-things design: a human-centered design for the internet of things. Telemat. Inform. **31**(4), 519–531 (2014)
6. L. Ren, L. Zhang, C. Zhao, X. Chai, Cloud manufacturing platform: operating paradigm, functional requirements, and architecture design. ASME 2013 international manufacturing science and engineering conference, Madison, WI, USA, 10–14 June 2013
7. L. Wang, M. Givehchi, G. Adamson, M. Holm, A sensor-driven 3D model-based approach to remote real-time monitoring. CIRP Ann. Manuf. Technol. **26**, 493–496 (2011)
8. B. Buckholtz, I. Ragai, L. Wang, Cloud manufacturing: current trends and future implementations. ASME J. Manuf. Sci. Eng. (2015)
9. G. Colistra, V. Pilloni, L. Atzori, The problem of task allocation in the internet of things and the consensus-based approach. Comput. Netw. **73**, 98–111 (2014)
10. J. Gubbi, R. Buyya, S. Marusic, M. Palaniswami, Internet of things (Iot): a vision, architectural elements, and future directions. Future Gener. Comput. Syst. **29**(7), 1645–1660 (2013)
11. D. Wu, D.W. Rosen, L. Wang, D. Schaefer, Cloud-based design and manufacturing: a new paradigm in digital manufacturing and design innovation. Comput. Aided Des. **59**, 1–14 (2015)
12. R. Roman, P. Najera, J. Lopez, Securing the internet of things. Computer **44**(9), 51–58 (2011)
13. E. Oriwoh, P. Sant, G. Epiphaniou, Guidelines for internet of things deployment approaches —the thing commandments. Procedia Comput. Sci. **21**, 122–131 (2013)
14. S. Wang, L. Guo, L. Kang, C.S. Li, X. Li, Y.M. Stephane, Research on selection strategy of machining equipment in cloud manufacturing. Int. J. Adv. Manuf. Technol **71**, 1549–1563 (2014)
15. D. Zhang, L. Wang, Z. Gao, An integrated approach for remote manipulation of a high-performance reconfigurable parallel kinematic machine. J. Manuf. Syst. **29**, 164–172 (2010)
16. S. Radeka, S. Karelb, The common industrial protocol in machine safety. 25th DAAAM international symposium on intelligent manufacturing and automation, Vienna, Austria, 23–30 November, 2014. doi:10.1016/j.proeng.2015.01.431
17. Profibus, PROFIsafe system description—safety technology and application. Profibus, (2010). http://www.profibus.com/
18. L. Wang, W. Shen, S. Lang, Wise-ShopFloor: a web-based and sensor-driven e-shop floor. Trans. ASME J. Comput. Inf. Sci. Eng. **4**(1), 56–60 (2004)
19. X. Xu, From cloud computing to cloud manufacturing. Robot. Comput. Integr. Manuf. **28**(1), 75–86 (2012)
20. L. Wang, Wise-ShopFloor: an integrated approach for web-based collaborative manufacturing. Systems, Man, and Cybernetics, Part C: Applications and Reviews, IEEE, pp. 562–573
21. M. Holm, M. Givehchi, A. Mohammed, L. Wang, Web based monitoring and control of distant robotic operations. ASME international 2012 manufacturing science and engineering conference, Notre Dame, IN, 2012

22. X.V. Wang, X. Xu, Cloud manufacturing in support of sustainability. ASME 2014 international manufacturing science and engineering conference, Detroit, MI, USA, 9–13 June 2014
23. O. Kuyoro'Shade, I. Frank, A. Oludele, Cloud computing security issues and challenges. Int. J. Comput. Netw. **3**(5), 247–255 (2011)
24. L. Almutair, S. Zaghloul, A new virtualization-based security architecture in a cloud computing environment. 3rd international conference on digital information processing and communication, Dubai, UAE, 30 Jan–1 Feb 2013, pp. 676–686
25. S. Subashini, V. Kavitha, A survey on security issues in service delivery models of cloud computing. J. Netw. Comput. Appl. **34**(1), 1–11 (2011)
26. J. Che, Y. Duan, T. Zhang, J. Fan, Study on the security models and strategies of cloud computing. Procedia Eng. **23**, 586–593 (2011)
27. F. Shahzad, State-of-the-art survey on cloud computing security Challenges, approaches and solutions. Procedia Comput. Sci. **37**, 357–362 (2014)
28. L. Rabai, B. Arfa, M. Jouini, A.B. Aissa, A. Mili, A cybersecurity model in cloud computing environments. J. King Saud Univers. Comput. Inf. Sci. **25**(1), 63–75 (2013)
29. R. Chow, P. Golle, M. Jakobsson, E. Shi, J. Staddon, R. Masuoka, J. Molina, Controlling data in the cloud: outsourcing computation without outsourcing control. ACM workshop on Cloud Computing Security (CCSW), Chicago, IL, 2009, pp. 85–90
30. A.K. Talukder, L. Zimmerman, H.A. Prahalad, Cloud economics: principles, costs, and benefits, in *Cloud Computing: Principles, Systems, and Applications*, 1st ed., (Springer, London, 2010)
31. Cloud Security Alliance, *Top Threats to Cloud Computing* (2010). http://www.cloudsecurityalliance.org
32. J. Spring, Monitoring cloud computing by layer, part 1. Secur. Priv. IEEE **9**(2), 66–68 (2011). doi:10.1109/MSP.2011.33
33. T. Khorshed, S. Ali, S. Wasimi, A survey on gaps, threat remediation challenges and some thoughts for proactive attack detection in cloud computing. Future Gener. Comput. Syst. **28** (6), 833–851 (2012)
34. J. Chaudhary, H. Patel, D. Patel, S. Patel, Tradeoffs between performance and security of cryptographic primitives used in storage as a service for cloud computing, *CUBE* international information technology conference, New York, NY, 2012, pp. 557–560
35. C. Vecchiola, T. Selvi, R. Buyya, *Cloud platforms in industry* (McGraw-Hill, Mastering Cloud Computing, 2013)
36. Cloud Security Alliance, Security Guidance For Critical Areas Of Focus In Cloud Computing, vol. 2.1, 2009. http://www.cloudsecurityalliance.org
37. M.G. Avram, Advantages and challenges of adopting cloud computing from an enterprise perspective. Procedia Technol. **12**, 529–534 (2014)
38. D. Stock, M. Stöhr, U. Rauschecker, T. Bauernhansl, Cloud-based platform to facilitate access to manufacturing IT. 8th international conference on digital enterprise technology, Stuttgart, Germany, 2014. doi:10.1016/j.procir.2014.10.045
39. A. Kang, J.H. Park, L. Barolli, H.Y. Jeong, CMMI security model for cloud manufacturing system's network. Eighth international conference on broadband, wireless computing, communication and applications, Compiegne, France, 28–30 Oct, 2013
40. P. Holtewert, R. Wutzke, J. Seidelmann, T. Bauernhansl, Virtual fort knox federative, secure and cloud-based platform for manufacturing. Forty sixth CIRP conference on manufacturing systems, 2013, pp. 527–532. doi:10.1016/j.procir.2013.06.027
41. P. Loshin, Transmission control protocol In *TCP/IP Clearly Explained*, 4th edn 2003, pp. 351–382. doi:10.1016/B978-155860782-8/50021-X
42. C. Gurkok, Securing cloud computing systems, in *Network and System Security*, ch. 4 (2013)
43. X.V. Wang, X.W. Xu, Virtualize manufacturing capabilities in the cloud: requirements and architecture. ASME 2013 international manufacturing science and engineering conference, Madison, WI, USA, 10–14 June 2013

44. C.S. Horstmann, G. Cornell, Core Java, vol. II–Advanced Features, 8th edn, Prentice Hall, 2008. ISBN 978-0-13-235479-0
45. L. Monostori, Cyber-physical production systems: roots, expectations and R&D challenges. Var. Manag. Manuf **17**, 9–13 (2014)
46. P.H. Griffin, Telebiometric authentication objects. Procedia Comput. Sci. **36**, 393–400 (2014)
47. A.M. Bahaa-Eldin, M.A.A. Sobh, A comprehensive software copy protection and digital rights management platform. Ain Shams Eng. J. **5**(3), 703–720 (2014)

# Part II
# Cloud-Based Monitoring, Planning and Control in CPS

# Chapter 4
# Machine Availability Monitoring and Process Planning

## 4.1 Introduction

Today, the global market is characterised by turbulent demands for highly customised products. Customers are increasingly demanding for higher quality products at low cost with quick delivery, and for shorter times between successive product generations. Cooperation among different companies becomes product-specific, customer-centric and dynamic. Manufacturing jobs are diversified and urgent. Moreover, outsourcing, joint ventures, and cross-border collaborations have led to a shop-floor environment geographically distributed across corporate and national boundaries. Moreover, the uncertainties of today's machining operations make this distributed environment further complicated. Companies and decision systems must be more flexible and adaptive to unplanned deviations on turbulent shop floors where metal-cutting processes should be adjusted dynamically to the changes. It is evident that factories of the future must contain smart decision modules that can fine-tune runtime operations adaptively to achieve specified production objectives. However, today's manufacturing systems still exhibit various limitations, especially in flexibility and adaptability.

On the other hand, modern manufacturing industries have shown clear trends in recent years—away from long standing and well-established products and relevant production that have been stable over many years, away from comprehensive trusts that may cover all the processes of production, and also away from the single economic consideration of production; instead, companies increasingly focus on their core manufacturing competencies, develop and produce adaptive and customised products, enter more often into alliances for manufacturing and resource optimisation, and integrate environmental and social responsibilities into their operations. This trend will lead to an Internet- and Web-based service-oriented Cloud manufacturing in the future to overcome today's limitations in rigid system structure, standalone software usage, centralised resource utilisation, unidirectional information flow and offline decision making.

 
L. Wang and X. V. Wang, *Cloud-Based Cyber-Physical Systems in Manufacturing*,
https://doi.org/10.1007/978-3-319-67693-7_4

As one of the core components of a manufacturing system, computer-aided process planning (CAPP) is desired to be responsive and adaptive to the changes in production capacity and functionality. Unfortunately, conventional CAPP systems are neither flexible nor adaptive, if applied directly to dynamic operations. Quite often, a process plan generated in advance is found unfeasible or even unusable to targeted resources, resulting in wasted time and effort spent in early process planning—a productivity drop when idle machines must wait for re-planning the remaining operations. Within the context, an adaptive approach is considered suitable and is thus introduced in this chapter for dealing with the dynamic situation, e.g. job-shop machining.

Targeting cloud manufacturing, the aim of this chapter is to present an Internet- and Web-based service-oriented system for machine availability monitoring and process planning. Particularly, this chapter introduces a tiered system architecture and introduces an event-driven approach using IEC 61499 function blocks. By connecting to Wise-ShopFloor framework, it enables real-time machine availability monitoring and machining status monitoring during metal cutting, locally or remotely. The closed-loop information flow makes process planning and monitoring feasible services for the cloud manufacturing.

The remainder of the chapter is organised as follows. Section 4.2 reviews the state of the art of the relevant research works. Section 4.3 introduces a new Web-DPP concept, which is extended to system architecture design in Sect. 4.4. System analysis of the Web-DPP is reported in Sect. 4.5 in form of IDEF0. The system is implemented and outlined in Sect. 4.6. In Sect. 4.7, a sample part machining is chosen to demonstrate and validate the capability of the prototype system in terms of process planning and machine availability monitoring. Finally, scientific contributions and future directions are summarised in Sect. 4.8.

## 4.2 Literature Review

The concept of cloud manufacturing is based on cloud computing, e.g. Software as a Service (SaaS), Platform as a Service (PaaS) and Infrastructure as a Service (IaaS). It is a new-generation service-oriented approach to supporting multiple companies to deploy and manage services for manufacturing operations over the Internet. Cloud manufacturing would provide cost-effective, flexible and scalable solutions to companies by sharing complex manufacturing software tools as services with lower support and maintenance costs. The development of cloud manufacturing includes design of four layers [1]:

- Manufacturing resource layer, such as manufacturing equipment, sensors, servers, etc.;
- Manufacturing virtual service layer, in which manufacturing resources are identified, virtualised and packaged as services. Identification and

communication technologies have been researched, including wireless sensor network, RFID, Internet of Things, MTConnect [2], etc.;

- Global service layer, which relies on a suite of cloud computing technologies such as PaaS to take global service computing and supporting for various demands and requirements;
- Application layer, which is the interface for users to invoke services for applications. Wise-ShopFloor [3] is such an example.

Although cloud computing has been active during the past decade, there have been no research projects on cloud development for process planning at machine and execution control level, which is an imperative research area for developing a comprehensive cloud manufacturing environment for factories of the future.

Process planning is commonly known as a bridge between design and manufacturing of a mechanical product. The tasks involved are generally preparatory, including process sequencing, resource (machine and cutter) selection, cutting parameter assignment, tool path planning, operation optimisation, and numerical control code generation. Process planning also relates to setup and fixture planning, closely. Computer-aided process planning (CAPP) dates back to 1960s when computers were first introduced to the field. Two decades later, more than 156 CAPP systems were reported in the literature [4]. Since its beginning, CAPP research has continuously attracted a large amount of interest over the past four decades. To date, a huge volume of literature has been published on this very subject. Among many others, previous research studies on CAPP include machining feature-driven approach [5], object-oriented approach [6], Petri net-based approach [7], neural network-based approach [8–10], genetic algorithm-based approach [11], constraint-based approach [12], multi-agent bidding-based approach [13], and knowledge-based approach [14, 15]. In terms of specific application domains, the reported approaches, together with their variants, have been applied to process sequencing [16], cutter selection [17], cutting parameter selection [18], tool path planning [19], and setup planning [20], etc. Today, machining feature-based approaches combined with artificial intelligence-based methods are the popular choices for process planners. Although the existing approaches can address the core decision-making problems involved in process planning, they are often *centralised* in decision making, *static* in system structure, and *time consuming* in computation, with many unrealistic assumptions on the availability of resources and production environment.

In terms of process planning methodologies, research efforts have recently been shifted to distributed process planning [21, 22], reconfigurable process planning [23], integrated planning and scheduling [24], and energy-efficient process planning based on the capacity profile of machine tools [25, 26]. Despite the naming differences, their common objective is to generate robust, flexible, precise yet adaptable process plans, effectively. Nevertheless, process planning research is facing new challenges today owing to the dynamic market and business globalisation in much more decentralised manufacturing environment than ever before. It demands for a new way of thinking in process planning that is collaborative among engineering teams and adaptive to environmental changes on manufacturing shop floors. On this front, the latest ICT technologies including the Internet, Web, Java, and XML etc. are popularly used for collaborative process planning to support a networked manufacturing environment. Within this context, a process planning system must be able to accommodate the variation and distribution of manufacturing resources and materials processing tasks, in collaboration with different process planners. In other words, it deals with how to support collaborative process planning among the planners at different places, and how to improve instantaneous communication among each other. In the work by Xu et al. [27], they put forward an idea that used computer screen-sharing technique to support a multi-user co-operation. This approach overcomes the limitation on processing resources and knowledge in the traditional narrow-sense process planning, and improves planners' collaborative work. In the same year, Java language was adopted to transfer a CAPP system to Web-based environment so that its functions and operations can be distributed to various computer systems to reduce the computational load on a single computer [28]. The distributed computing environment is based on J2EE, enabling the manufacturing processes to be planned effectively over the Internet. In the research by Qiu et al. [29], a distributed multi-user environment over the Internet was suggested. It was implemented as a web-based system by combining an external authoring interface and Java. This system allows users carrying out manufacturability evaluation based on a predefined process plan. Another Internet-enabled system was reported in [30] for setup planning in machining operations using Java and Web technologies, where XML was used to transfer data and information between various manufacturing systems. Agent technology was also popularly used in collaborative process planning in recent years. A multi-agent system for distributed process planning was presented in [31]. Three autonomous agents, (i.e. Global Manager Agent, Design Agent, and Optimisation Agent) are capable of communicating with each other through XML. Hence, it enables process planning in a distributed e-manufacturing environment.

Another dimension of process planning is the adaptability of a process plan to unforeseen changes on manufacturing shop floors. Here, the dynamic changes are dubbed uncertainty, such as frequent production change, job delay, missing or broken cutters, unavailable fixtures/machines, rush orders from clients, and even the short notice of sick leave of a chief operator. Such dynamic characteristics of manufacturing shop floors pose an unprecedented challenge to CAPP systems. In this situation, a process plan generated in advance is often found unfeasible and non-applicable due to the dynamic changes. Subjecting to re-planning, the process plan may jeopardise production by putting those available machines on hold. In order to address this problem effectively, Wang et al. [32, 33] proposed to using enriched machining features and function blocks to handle the uncertainty. The goal is to generate detailed and adaptive operation plans at runtime by CNC controllers to best utilise the capability of the available machines. However, monitoring the availability of distributed manufacturing resources, yet in real time, is missing from the CAPP systems reported in the literature. The absence of up-to-date information on machine availability leads directly to the problem that a generated process plan may be inapplicable to the anticipated machine (non-availability of the machine) and unfit to an alternative machine. This is also due to the disconnection between process planning and resource scheduling systems. The latter has led to the research topic of process planning and scheduling integration. On this front, an integrated CAPP system (ICAPPS) was developed [34], aiming for single-piece, small-lot and make-to-order production. On the basis of the integrated model, the functions of ICAPPS were implemented on design layer, part planning layer, shop planning layer and scheduling layer. Another integrated process planning and scheduling tool was reported in [35], using the integrated definition (IDEF) methodology. An activity model was used to develop their system that allows a user to plan both the process and the production at the same time. Sormaz and Khoshnevis [36] discussed a method for generating alternative process plans that takes production schedules into due consideration.

With mounting environmental concerns, remanufacturing from end-of-life product components has gained more attention in recent years. In addition to the traditional objectives such as time, cost and quality of machining, economic viability and energy consumption are also taken into consideration during process planning. One example along this direction is the process planning for IT-equipment remanufacturing by Kernbaum et al. [37], which derived a process planning method based on the description of the market situation and the involved actors for remanufacturing processes in a given facility. For more comprehensive review of process planning activities during the past decade, readers are referred to [38] for details.

From the literature survey, it is found that the reported process planning approaches and systems are mostly limited to problems of a static nature, with decisions made well in advance of their actual use. Their adaptability to unforeseeable changes on shop floors, however, remains limited and insufficient. The CAPP software tools available today are centralised in decision making, static in system structure, and offline in data processing. Due to the lack of actual conditions on the shop floors, it is difficult for a centralised offline system to make adaptive decisions. In the case of machining, any number of possible process plans may exist depending on the actual machine chosen. Even if the same machine tool is considered, it is likely that there is more than one way of doing the job. Therefore, planning with alternatives is likely a practical approach, which is sometimes called *non-linear process planning* (NLPP). NLPP is further complicated by issues such as possible alternative feature-based interpretations of the same part; hence different final process plans can be generated for the same part. NLPP deserves further research as it opens avenues for scheduling with alternative routings, which is in fact one of the key issues addressed in this chapter.

Addressing the aforementioned shop-floor uncertainty, this chapter introduces an approach for distributed process planning [39] over the Web. In this chapter, the latest development is presented towards the implementation of a prototype system, the ultimate goal of which is to improve system performance when planning machining operations on a shop floor with frequent unplanned changes, with high adaptability, by integrating real-time monitoring information of machine availability into service-oriented process planning.

## 4.3 Concept of Distributed Process Planning

Figure 4.1 depicts the relationship among real-time machine availability monitoring, dynamic resource scheduling, distributed process planning (DPP), integrated process simulation, and remote machine control in a shared cyber workspace, where DPP as one of cloud services handles adaptive decision making of process planning based on real-time process status, machine availability, and dynamic resource scheduling. As shown in Fig. 4.1, the five modules close the loop of information flow to address the uncertainty or changes of machines and machining processes on shop floors. Based on the real-time information of machines and their availability, it is possible for DPP to generate process plans adaptively to the changes through well-informed decision making.

As mentioned earlier, machining process planning is the task that transforms the design information of a product into relevant machining operations, and determines at the same time an optimal or near-optimal sequence of the machining operations. A process plan generally consists of two types of information: *generic* ones (machining method, machining sequence and machining strategy) and *machine-specific* ones (cutter data, cutting parameters, and tool paths). A two-tier system architecture is, therefore, considered suitable to separate the generic data from those machine-specific ones in DPP. This concept is illustrated in Fig. 4.2.

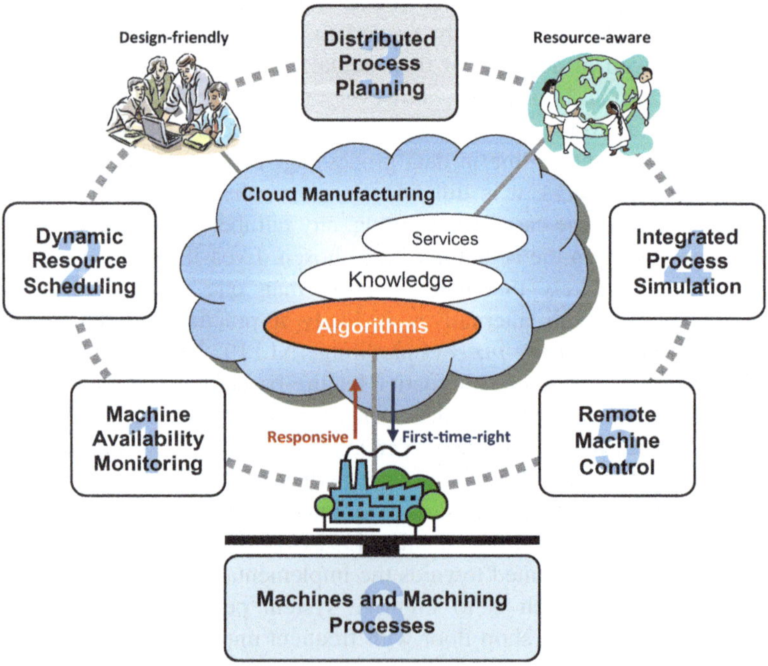


**Fig. 4.1** Distributed process planning in cloud manufacturing

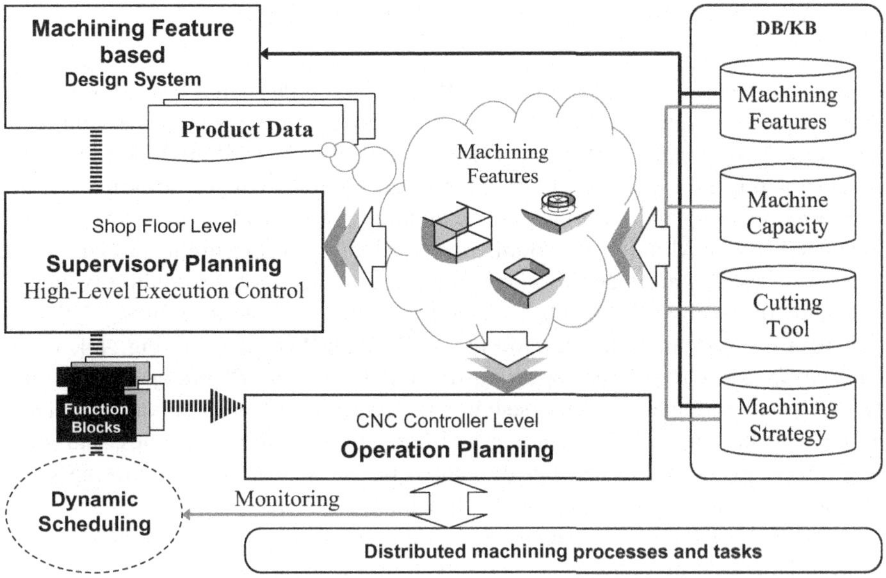


**Fig. 4.2** Staged decision making in DPP

According to the DPP concept, machining features and function blocks are two enabling technologies. They carry the machining technological information and pass through various system modules until final machining. Machining features, such as hole, pocket and step shown in Fig. 4.2, are first defined, created and stored in a machining feature library for ready use by a machining feature-based CAD system (for non-feature-based CAD systems, a third-party utility tool may be used for features extraction and recognition). In the DPP, the *divide-and-conquer* strategy is adopted. The tasks of process planning are first divided into two groups and then accomplished at two different stages (both in time and in location): factory-level *supervisory planning* (SP) and machine-level *operation planning* (OP). The SP module handles product data parsing, machining feature decomposition, initial setup planning, machining process sequencing, fixture and machine selection, etc. The OP module, on the other hand, focuses on the detailed operation procedures for each machining operations, including cutter selection, cutting parameters assignment, tool path planning, and numerical control code generation. Between SP and OP, dynamic scheduling functions can be integrated by event-driven function blocks. Owing to the two-tier system structure, the decision-making in DPP becomes distributed in terms of timing (SP *in advance* vs. OP *at runtime*) and location (SP in *one* computer vs. OP in *many* controllers). The objective of the decisions separation is to make the high-level SP plans generic and portable to alternative machines in case of the non-availability of a given machine. In other words, since the final OP plans are generated adaptively at runtime by machine controllers, there is no need to prepare for redundant process plans, resulting in significantly reduced re-planning effort and machine idle time.

## 4.4 Architecture Design of a Web-Based DPP

As an extension to DPP, a Web-based DPP (or Web-DPP) is not limited to process planning. It also handles machining job dispatching and oversees job execution in machines. Such functionalities are designed into the Wise-ShopFloor framework [3] as shown in Fig. 4.3. Web-DPP in the Logic Container shares information with other modules, e.g. Monitoring and Scheduling, for adaptive decision making. Facilitated by the Monitoring module, the availability of machining resources and their current status are made available for dynamic resource scheduling, which in turn helps the Web-DPP for job dispatching to the available machines.

A detailed system architecture design of Web-DPP is shown in Fig. 4.4, where *Supervisory Planning*, *Execution Control* and *Operation Planning* are the three major system modules. In this design, the Execution Control module is placed in-between the Supervisory Planning and Operation Planning modules, and looks after jobs dispatching (in the unit of setups) based on real-time monitoring information, availability of machines and scheduling decisions.

Web-DPP assumes that machining features are already made available in the product data–they are either generated directly by using a feature-based design tool or recognised by a third-party machining feature recognition solution.

Setup planning is generally considered as a part of process planning. During SP, a generic setup plan is generated by grouping machining features according to their tool access directions (TAD). Since 3-axis machines are most common on

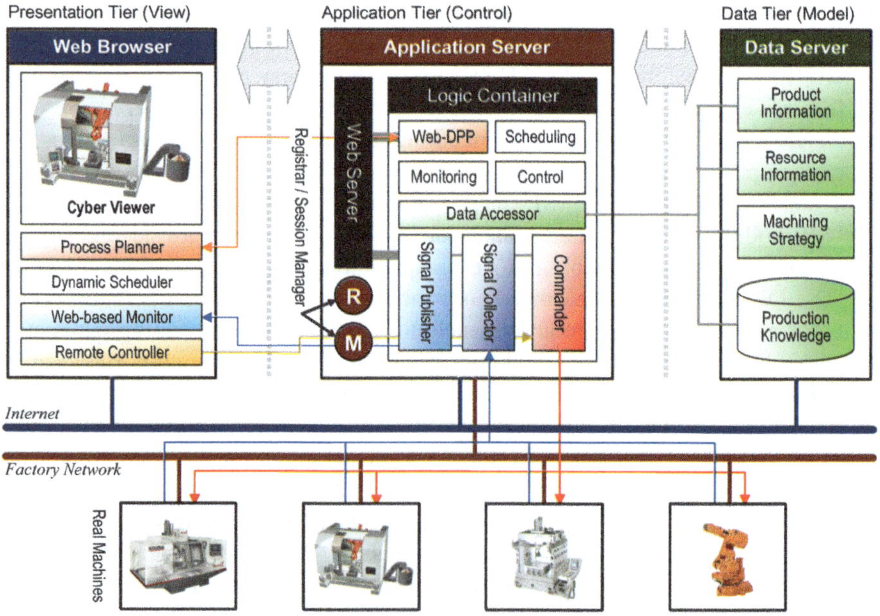


**Fig. 4.3** Web-DPP as part of Wise-ShopFloor framework

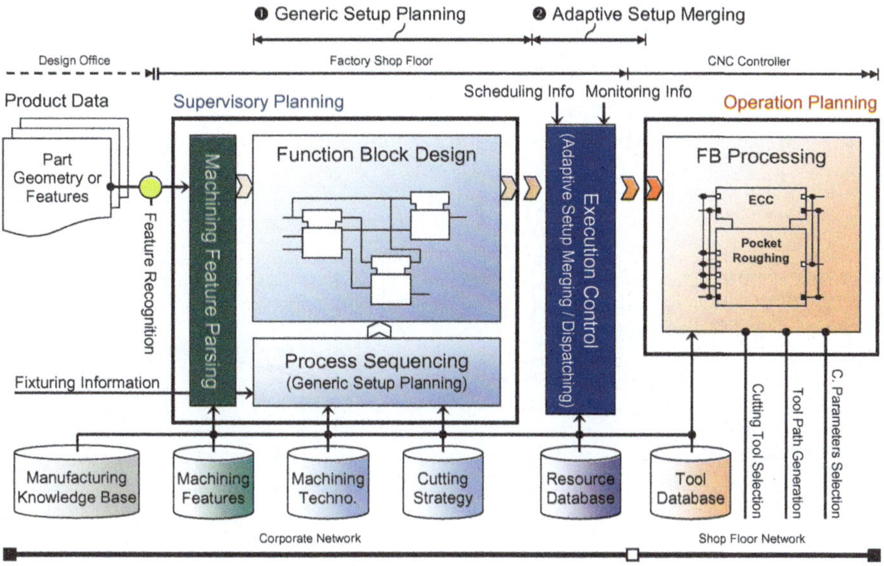


**Fig. 4.4** System architecture of Web-DPP

machining shop floors and constitute the basic configuration of machining resources, a generic setup plan at this stage is made for 3-axis machines only. Although a generic setup plan is applicable to other types of machines, setup merging for 4- or 5-axis machines is needed to best utilise the capability of the higher-end machines. This is performed before job dispatching by the Execution Control. Setup merging is beyond the scope of this chapter and will be reported separately.

## 4.5 Functional Analysis of Web-DPP

IDEF0 is adopted for analysing Web-DPP functionalities. The three core modules of the Web-DPP in Fig. 4.4 are modelled in IDEF0 in Fig. 4.5, together with the relationship and data/information flow among the three modules, where M1–M5 represent human, computer, network, security and machine, respectively.

Different from conventional process planning approaches, function blocks of varying types are applied in Web-DPP. Meta function blocks (MFBs) are the output of SP. They are used to encapsulate machining sequences (of both setups and machining features). An MFB only contains generic process information of a product. In other words, it serves as a high-level process template, with suggested cutter types (e.g. drill, square end mill, etc.) and tool-path patterns (e.g. zigzag, spiral-out, etc.), for subsequent machining tasks. (Readers are referred to [33] for more details on how to design function blocks.) Execution function blocks (EFBs)

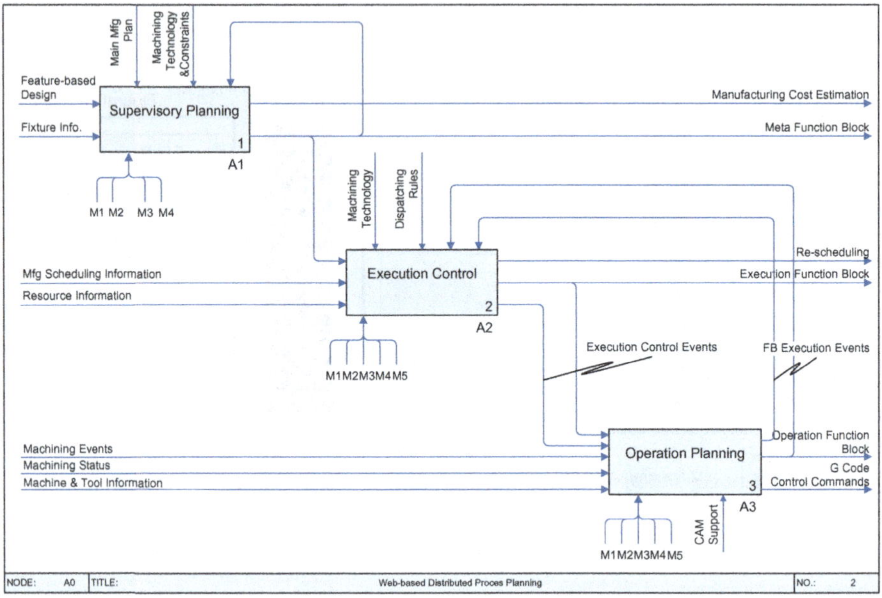


**Fig. 4.5** IDEF0 model of Web-DPP

are those function blocks that are ready to be dispatched to a chosen machine. An EFB is created by instantiating an appropriate MFB when associated with a specific machining operation like drilling a hole. Each job with multiple operations corresponds to its own set of EFBs, so that a monitoring function can be integrated with the set of EFBs. The structure of an operation function block (OFB) is similar to that of an EFB. However, an OFB specifies and completes an EFB with machine-specific details about a machining operation. Moreover, during OP, it is possible to update and override the values of variables of an EFB, so as to make it optimal and adaptable to actual situations during machining operations. The two different terms of EFB and OFB are used to distinguish a given function block. This is because they are two separate entities with different level of details, fulfilling different level of execution, residing in different resources, and moreover, they may be deployed to physically distributed machine controllers.

In summary, a function block contains a set of predefined functions/algorithms that can be triggered by a known event arriving at the function block. By executing its associated algorithm, a planning decision can be made at runtime to process a machining feature. The process of information enrichment from machining features to function blocks together with their relationship is depicted in Fig. 4.6.

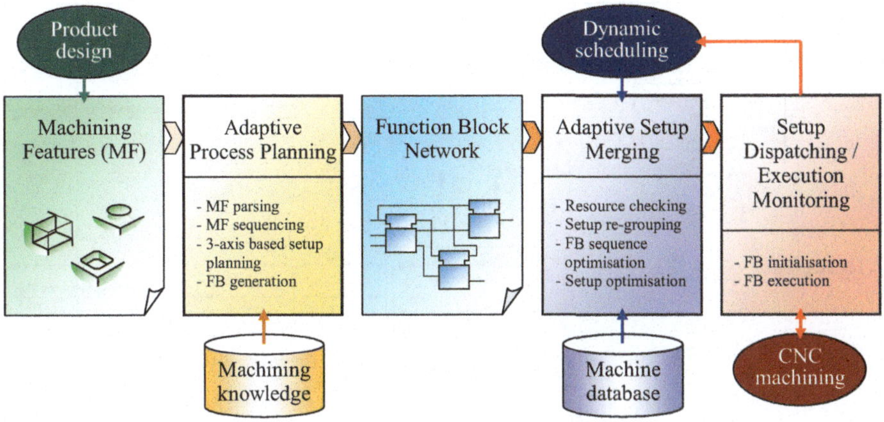


**Fig. 4.6** Information evolution from machining features to function blocks

### *4.5.1 Supervisory Planning*

Incoming from a higher-level production planning system, a main manufacturing plan triggers the SP to make a generic machining sequence plan of a given product. In Web-DPP, machining feature-based reasoning is adopted in the SP by considering the most common machining resources (e.g. 3-axis machine tools, standard jigs and fixtures, and popular cutters, etc.), established machining technology and known manufacturing constraints. A so-generated machining sequence plan is then passed to a function block designer and packed as a network of MFBs (Fig. 4.4). Details of the internal structure and data flow of SP are illustrated in Fig. 4.7. Within the SP, the function block designer is used to: (1) define new function block types, (2) specify task-specific algorithms for each defined function block type, and (3) map machining features to MFBs according to the generated machining sequences.

### *4.5.2 Execution Control*

The execution control module receives scheduling information and monitoring events, making itself an important integration point of actions and decision making of the entire Web-DPP system. The functionalities of the execution control module include setup merging (on a 4- or 5-axis machine, if chosen), job (EFB) dispatching, and execution monitoring of an OFB, as shown in Fig. 4.8. Within the context, the job execution monitoring is facilitated by triggering a function block-embedded algorithm that can collect the current machining status (including machining feature ID, machine tool ID, current cutting conditions, job completion rate, etc.) and send them back to the execution control module. The real-time job

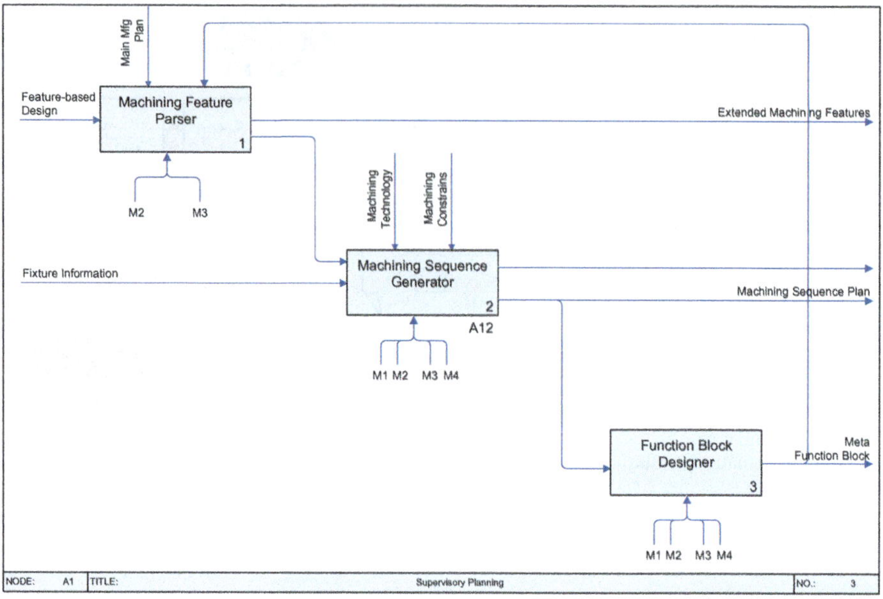


**Fig. 4.7** IDEF0 model of supervisory planning

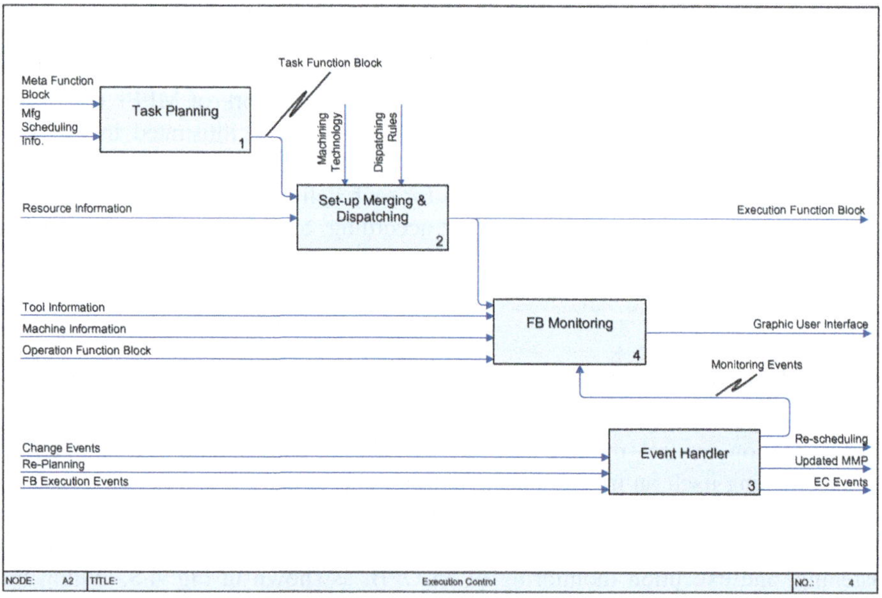


**Fig. 4.8** IDEF0 model of execution control

execution information is crucial for dynamic resource scheduling and job dispatching of the next batch of machining jobs between machines according to the availability of the machines on the shop floor.

### *4.5.3 Operation Planning*

Inside of a machine controller, OP functions as a real-time execution module of OFBs. In other words, OP is runnable in a machine controller. The OP module not only specifies and optimises the process plan received from the SP, including cutter selection, machining sequence optimisation, machining parameters assignment, and tool path planning, but also sequentially executes the OFBs in an explanation engine (the Executor in Fig. 4.9). According to this design, the operation planning process in a machine controller can be truly adaptive, meaning it can dynamically modify the process plan depending on the dynamics of the actual machining process. At the current stage, since most commercial CNC controllers are of closed nature and do not recognise instructions other than G codes, the proposed function blocks cannot be executed directly by the existing commercial controllers yet. Alternatively, our implementation and testing were carried out in an open-architecture CNC controller. In order to utilise the legacy machine tools already installed in industry, conventional G codes can be generated by OFBs as an option. Details of the OP are illustrated in Fig. 4.9.

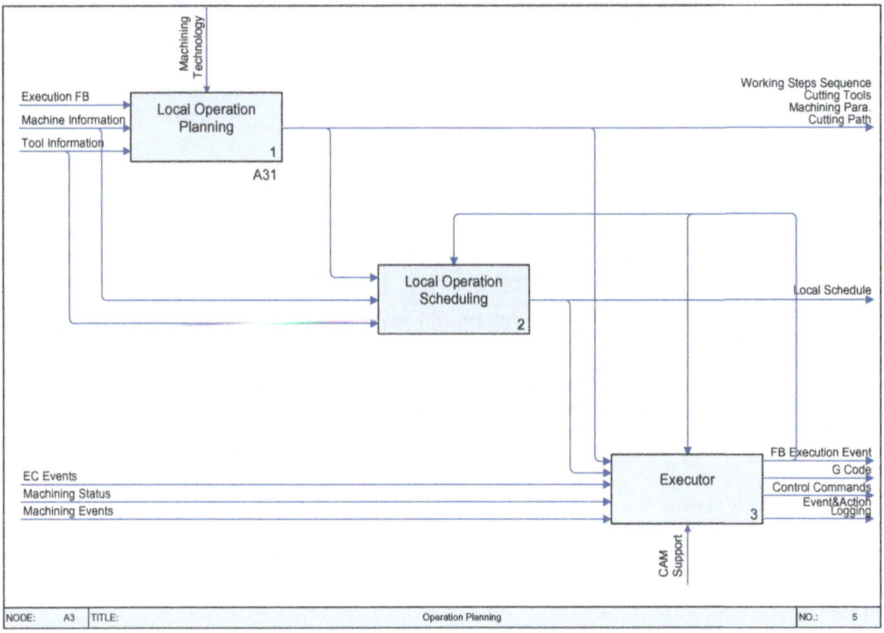


**Fig. 4.9** IDEF0 model of operation planning

## 4.6 Web-DPP Prototype Implementing

Being a part of the Wise-ShopFloor framework, Web-DPP adopts the same browser/server architecture and view-control-model (VCM) design pattern for prototype implementation. It has been designed with the built-in secure session control and data protection as shown in Fig. 4.3. In order to meet the user requirements of rich visual data sharing and satisfy the real-time constraint of data transmission over the Internet, the following solutions are implemented in the Web-DPP prototype:

- Use interactive 3D models (scene graph-based Java 3D) for visualisation;
- Provide process planning functions as services via web-based graphical user interface;
- Deploy major services (process planning and control logics) in a secure application server.

The Web-DPP prototype is implemented according to the package diagram depicted in Fig. 4.10. The eight core modules are clustered into supervisory planning, execution control and operation planning, to deliver the designed functionalities as illustrated in Fig. 4.5. These system modules are accessible via the dedicated web user interfaces. Figure 4.11 reveals one snapshot of the Web-DPP user interface implemented in Java applet. The prismatic sample part shown in Fig. 4.11 is used in the case study in Sect. 4.7 to showcase the capability and validate the feasibility of the two-tier DPP concept.

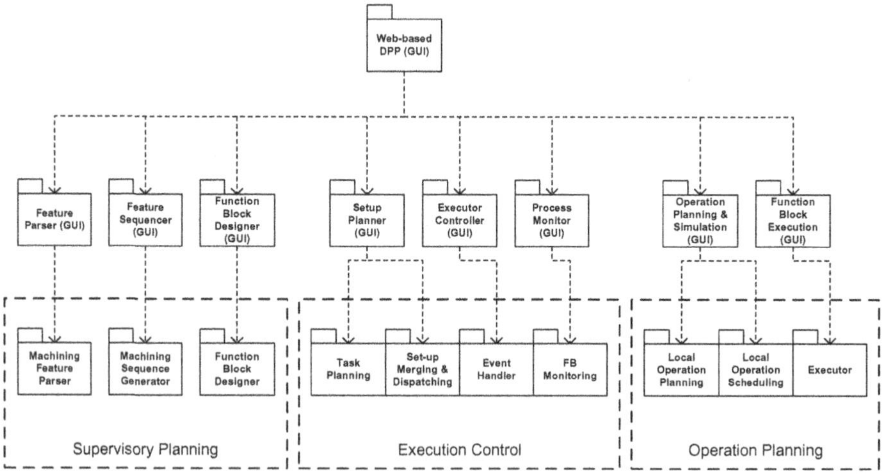


**Fig. 4.10** A package diagram for Web-DPP implementation

## 4.7 A Case Study

The prismatic part shown in Fig. 4.11 consists of 14 machining features. Each machining feature is a basic geometrical shape such as step, slot, pocket or hole that can be readily machined by using a standard cutter. After applying the five defined feature-based geometry-reasoning rules reported in [32], the 14 machining features are grouped into two setups, each with a primary machining sequence that must be followed. The reasoning at this stage only considers datum references and manufacturing constraints. Other non-critical machining sequence (neither datum reference nor manufacturing constraint involved) remains in parallel. For example, the four holes F11–F14 in Fig. 4.11, their machining sequence is not linear at this stage and will be determined later by the controller-level operation planning.

For the sake of brevity, only Setup 1 of the sample part is mapped to a single composite function block (CFB) as shown in Fig. 4.12, consisting of seven basic function blocks (BFBs). Each BFB represents one type of machining feature, and each CFB forms one setup for dispatching to one machine. After the mapping process, the required machining sequence of the sample part is represented by the event flow among the BFBs. Note that the same BFB can be called more than once to fabricate the machining features of the same type, e.g. the four holes on the top surface of the sample part. In Web-DPP, a CFB (setup) is the basic units for job dispatching to the available machines. Here, machine selection is dealt with by a separate scheduling utility of Wise-ShopFloor. Upon dispatching to a chosen machine, detailed operation planning takes place for each machining feature to specify machine-specific operations, including cutter and cutting parameter

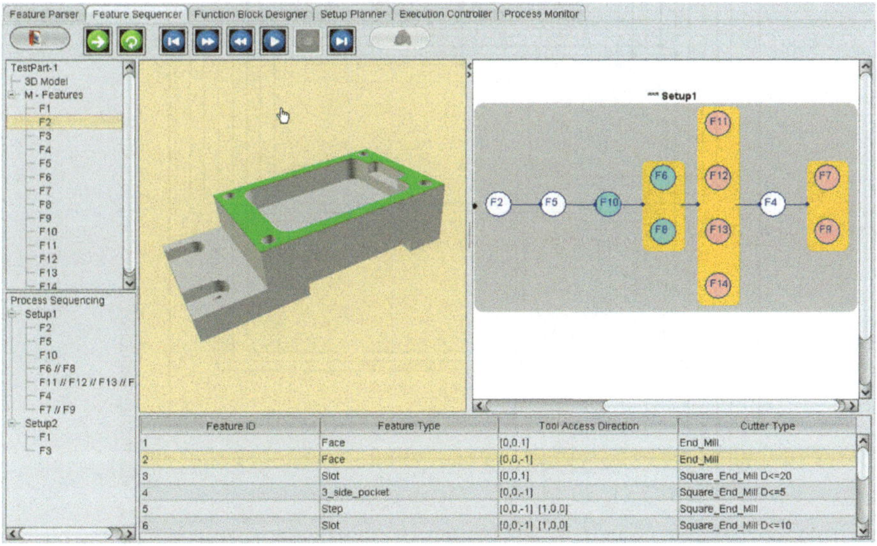


**Fig. 4.11** Setup grouping and process sequencing

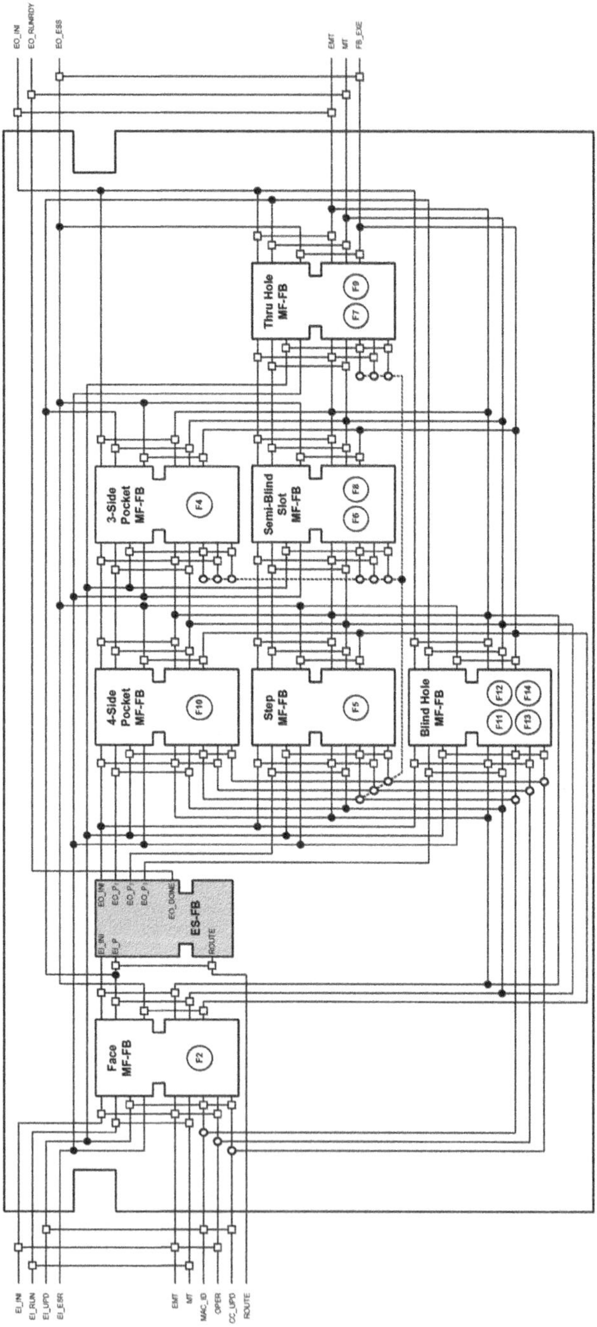


**Fig. 4.12** Machining sequence embedded in a CFB

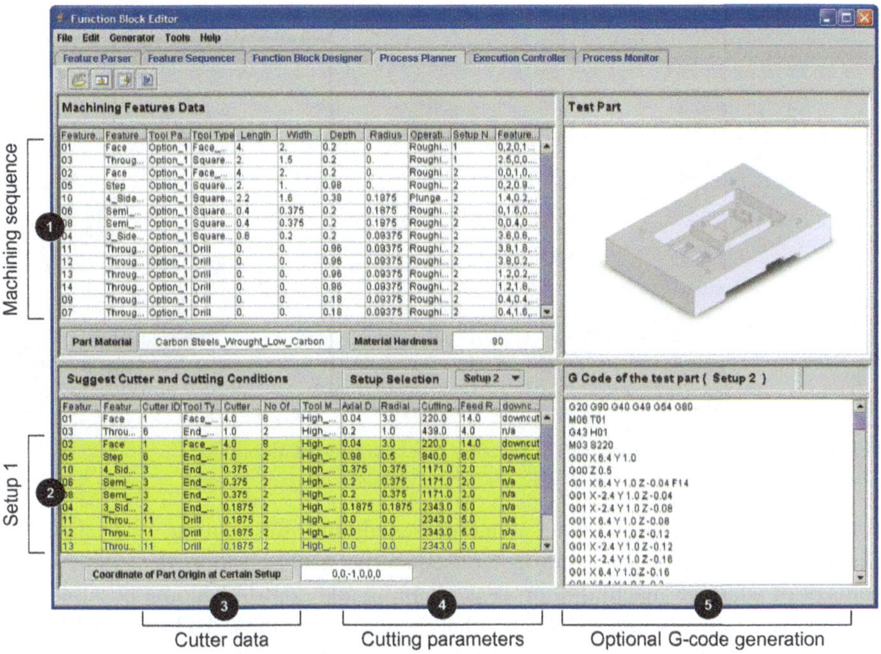


**Fig. 4.13** Operation planning by function blocks

selection, and tool-path planning. Due to the limited access to existing controllers with closed architecture, the operation planning in its current implementation is performed in a front-end computer of the machine.

A pictorial view on how OP works during operation planning is illustrated in Fig. 4.13. The final machining sequence ❶, setup formation ❷, cutter data ❸, cutting parameters ❹, and optional G-code ❺ for legacy machines, are all derived by the embedded algorithms of the function blocks; whereas each line of ❸ and ❹ is used to cut one corresponding machining feature. The ultimate goal in the future is to execute the function blocks directly when they can be recognised by CNC controllers, instead of generating optional G-code at the last minute. It would give controllers more flexibility for adaptive machining.

As shown in Figs. 4.2 and 4.8, execution monitoring is one of the designed features of Web-DPP, which is realised by triggering one algorithm embedded in a function block in running state. This situation is further visualised in Fig. 4.14, where the cutting conditions of a 3-side pocket are displayed together with the current cutter location (*x*, *y*, *z*) and the job completion rate (running progress) of 64%. The added feature of status monitoring provides a process planner or a production manager with a holistic view of the entire shop floor, if every machine is networked and monitored. According to the job completion rate of a machining job on a given machine, it is possible to predict the availability of the machine in the near future. The execution monitoring information retrieved in real-time can thus

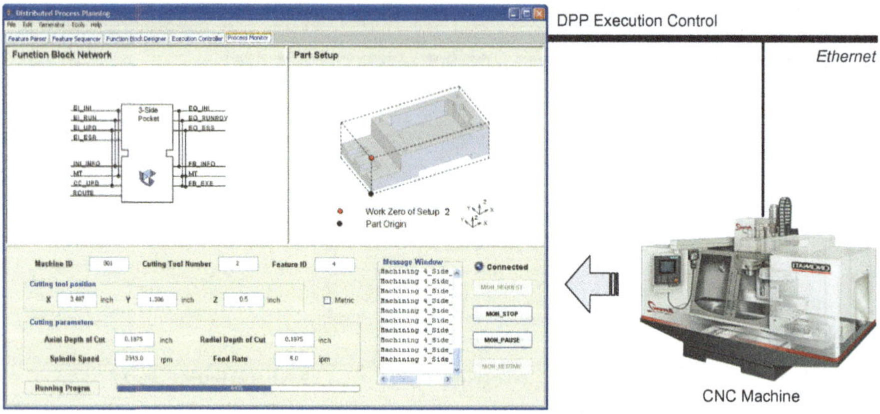


**Fig. 4.14** Execution monitoring by a function block-embedded algorithm

contribute to resource planning, dynamic scheduling, job routing, line balancing, and shop floor execution control. Consequently, a closed-loop decision making concept of adaptive machining process planning and control become possible. The sample part (slightly altered from what is shown in Figs. 4.11 and 4.14, with surrounding supporting materials for quick fixturing) is processed (in Fig. 4.13) and machined on a 5-axis legacy machine tool as shown in Fig. 4.15. The G-code used for machining the part is generated at runtime by function blocks.

**Fig. 4.15** Function block-enabled machining

## 4.8 Conclusions

Targeting the future manufacturing shifting towards cyber-physical manufacturing, this chapter introduces a web-based and service-oriented approach for distributed machining process planning in a decentralised and dynamic manufacturing environment, particularly for SMEs of job-shop machining operations with uncertainty. The advantages of this approach include network-wide accessibility and adaptive decision-making capability to process planning with unpredictable shop-floor changes. This is facilitated by a web-based user interface and a real-time execution monitoring service of machine availability. A Web-DPP prototype has been designed and implemented as web services, which was extended from a Wise-ShopFloor framework to separate generic information from machine-specific ones. The advantages of this work can be summarised as:

- Two-tier system architecture for distributed decision making;
- Machining feature-based geometry reasoning for machining sequence planning;
- Design of function blocks for controller-level operation planning;
- Algorithm-based process execution and machine availability monitoring; and
- Closed-loop information flow for scheduling and job dispatching via real-time monitoring.

The presented Web-DPP prototype runs inside a standard web browser, whereas the decision modules reside in one or more application servers, constituting a part of the cloud manufacturing services. As a result of cloud manufacturing, no dedicated software is needed to be installed in local computers at client side. The limitation of this prototype in its current implementation is the inability of dealing with complex products with freeform surfaces. The future work should be aligned to explore along the direction to cover more product variety and complexity. At the same time, more research efforts need to focus on functionality enhancement, innovative processing algorithm development, and testing using real-world cases via open-architecture CNC controllers in dynamic shop-floor environment. Moreover, integration with a third-party scheduling system, a more sophisticated feature-parsing system and a function block compliant CNC controller also deserves to be investigated, the results of which will be of great interest to peer researchers and practitioners.

It is envisioned that cloud manufacturing will reorganise the manufacturing practices of today by means of cloud services, where resources (software and hardware) can be shared cost-effectively by many. It is particularly useful and beneficial to SMEs who do not have the luxury of resources that are expensive for hosting and maintenance. Web-DPP intends to share knowledge and solutions in machining process planning with SMEs based on a *pay-as-you-go* or *pay-per-use* business model. Clients willing to use the service would open their machining resources for availability monitoring and thus adaptive machining by function blocks will become a reality in the future.

## References


1. X. Xu, From cloud computing to cloud manufacturing. Rob. Comput. Integr. Manuf. **28**(1), 75–86 (2012)
2. A. Vijayaraghavan, W. Sobel, A. Fox, D. Dornfeld, P. Warndorf, Improving machine tool interoperability using standardized interface protocols: MTConnect, in *Proceedings of 2008 International Symposium on Flexible Automation* (Atlanta, USA, 2008)
3. L. Wang, Wise-ShopFloor: an integrated approach for web-based collaborative manufacturing. IEEE Trans. Syst. Man Cybern. Part C Appl. Rev. **38**(4), 562–573 (2008)
4. L. Alting, H. Zhang, Computer aided process planning: the state-of-the-art survey. Int. J. Prod. Res. **27**(4), 553–585 (1989)
5. L. Wang, D.H. Norrie, Process planning and control in a holonic manufacturing environment. J. Appl. Syst. Stud. **2**(1), 106–126 (2001)
6. Y. Zhang, S.C. Feng, X. Wang, W. Tian, R. Wu, Object oriented manufacturing resource modeling for adaptive process planning. Int. J. Prod. Res. **37**(18), 4179–4195 (1999)
7. P. Xirouchakis, D. Kiritsis, J.-G. Persson, A petrinet technique for process planning cost estimation. Ann. CIRP. **47**(1), 427–430 (1998)
8. M.W. Park, H.M. Rho, B.T. Park, Generation of modified cutting condition using neural network for an operation planning system. Ann. CIRP. **45**(1), 475–478 (1996)
9. C.R. Devireddy, K. Ghosh, Feature-based modeling and neural network-based CAPP for integrated manufacturing. Int. J. Comput. Integr. Manuf. **12**(1), 61–74 (1999)
10. L. Monostori, Z.J. Viharos, S. Markos, Satisfying various requirements in different levels and stages of machining using one general ANN-based process model. J. Mater. Process. Technol. **107**, 228–235 (2000)
11. F. Zhang, Y.F. Zhang, A.Y.C. Nee, Using genetic algorithms in process planning for job shop machining. IEEE Trans. Evol. Comput. **1**(4), 278–289 (1997)
12. A. Márkus, J. Váncza, A. Kovács, Constraint-based process planning in sheet metal bending. Ann. CIRP. **51**(1), 425–428 (2002)
13. P. Gu, S. Balasubramanian, D.H. Norrie, Bidding-based process planning and scheduling in a multi-agent system. Comput. Ind. Eng. **32**(2), 477–496 (1997)
14. D. Lutters, T.C. Wijnker, H.J.J. Kals, Information management in process planning. Ann. CIRP. **48**(1), 385–388 (1999)
15. B. Denkena, M. Shpitalni, P. Kowalski, G. Molcho, Y. Zipori, Knowledge management in process planning. Annals of the CIRP. **56**(1), 175–180 (2007)
16. S.H. Yeo, B.K.A. Ngoi, H. Chen, Process sequence optimization based on a new cost-tolerance model. J. Intell. Manuf. **9**, 29–37 (1998)
17. T. Lim, J. Corney, J.M. Ritchie, D.E.R. Clark, Optimizing tool selection. Int. J. Prod. Res. **39** (6), 1239–1256 (2001)
18. B. Arezoo, K. Ridgway, A.M.A. Al-Ahmari, Selection of cutting tools and conditions of machining operations using an expert system. Comput. Ind. **42**, 43–58 (2000)
19. R.M. Boogert, H.J. Kals, F.J. van Houten, Tool paths and cutting technology in computer-aided process planning. Int. J. Adv. Manuf. Technol. **11**, 186–197 (1996)
20. S.K. Ong, A.Y.C. Nee, Fuzzy-set-based approach for concurrent constraint setup planning. J. Intell. Manuf. **7**(2), 107–120 (1996)
21. L. Wang, H.-Y. Feng, N. Cai, Architecture design for distributed process planning. J. Manuf. Syst. **22**(2), 99–115 (2003)
22. L. Wang, M. Holm, G. Adamson, Embedding a process plan in function blocks for adaptive machining. CIRP Ann. Manuf. Technol. **59**(1), 433–436 (2010)
23. A. Azab, H.A. ElMaraghy, Mathematical modeling for reconfigurable process planning. Ann. CIRP. **56**(1), 467–472 (2007)
24. K. Ueda, N. Fujii, R. Inoue, An emergent synthesis approach to simultaneous process planning and scheduling. Ann. CIRP. **56**(1), 463–466 (2007)

25. S.T. Newman, A. Nassehi, Machine tool capability profile for intelligent process planning. Ann. CIRP. **58**(1), 421–424 (2009)
26. S.T. Newman, A. Nassehi, R. Imani-Asrai, V. Dhokia, Energy efficient process planning for CNC machining. CIRP J. Manuf. Sci. Technol. **5**(2), 127–136 (2012)
27. Z. Xu, M. Cai, L. Lin, J. Dong, Study of networked manufacturing oriented cooperative CAPP system, in *Proceedings of the 9th International Conference on Computer Supported Cooperative Work in Design* (Coventry, UK, 2005)
28. C.F. You, C.H. Lin, Java-based computer-aided process planning. Int. J. Adv. Manuf. Technol. **26**(9–10), 1063–1070 (2005)
29. Z.M. Qiu, Y.P. Chen, Z.D. Zhou, S.K. Ong, A.Y.C. Nee, Multi-user NC machining simulation over the WWW. Int. J. Adv. Manuf. Technol. **18**, 1–6 (2001)
30. W. Liu, G. Peng, An internet-enabled setup planning system, in *Proceedings of 3rd International Conference on Information Technology and Applications* (Sydney, 2005)
31. R. Agrawal, S.K. Shukla, S. Kumar, M.K. Tiwari, Multi-agent system for distributed computer-aided process planning problem in e-manufacturing environment. Int. J. Adv. Manuf. Technol. **44**(5–6), 579–594 (2009)
32. L. Wang, N. Cai, H.-Y. Feng, Z. Liu, Enriched machining feature based reasoning for generic machining process sequencing. Int. J. Prod. Res. **44**(8), 1479–1501 (2006)
33. L. Wang, Y. Song, Q. Gao, Designing function blocks for distributed process planning and adaptive control. Eng. Appl. Artif. Intell. **22**(7), 1127–1138 (2009)
34. L. Min, B. Li, S. Zhang, Modeling integrated CAPP/PPS systems. Comput. Ind. Eng. **46**(2), 275–283 (2004)
35. J. Ciurana, M.L. Garcia-Romeu, I. Ferrer, M. Casadesus, A model for integrating process planning and production planning and control in machining processes. Rob. Compu. Integr. Manuf. **24**(4), 532–544 (2008)
36. D.N. Sormaz, B. Khoshnevis, Generation of alternative process plans in integrated manufacturing systems. J. Intell. Manuf. **14**(6), 509–526 (2003)
37. S. Kernbaum, S. Heyer, S. Chiotellis, G. Seliger, Process planning for IT-equipment remanufacturing. CIRP J. Manuf. Sci. Technol. **2**(1), 13–20 (2009)
38. X. Xu, L. Wang, S.T. Newman, Computer-aided process planning—a critical review of recent developments and future trends. Int. J. Comput. Integr. Manuf. **24**(1), 1–31 (2011)
39. L. Wang, J. Ma, F.-Y. Feng, Web-DPP: towards job-shop machining process planning and monitoring. Int. J. Manuf. Res. **6**(4), 337–353 (2011)

# Chapter 5
# Cloud-Enabled Distributed Process Planning

## 5.1 Introduction

Manufacturing companies nowadays face an intense global competition. To remain competitive, companies are urged to apply a responsive and adaptable approach. Manufacturing responsiveness is referred to the ability of a production system to respond to disturbances that impact upon production goals, and consequently, its ability to adapt to changing market conditions [1]. The changing market conditions can highly affect small-to-medium-sized enterprises (SMEs) that are mainly engaged in job-shop operations coupled with operation complexity. Customised products with large varieties and small batch sizes as well as the shortened product lifecycles contribute to complexities. Specifically, SMEs active in machining and metal-cutting sector who normally deal with complex and intensive process planning problems are experiencing more shop-floor uncertainties today than ever before, e.g. frequent product changeover, urgent job insertion, job delay, broken or missing tools, and unavailability of machines, fixture and labour shortage [2]. Therefore, it is critical that SMEs can cope with such dynamic environments.

Despite the existence of computer-aided technologies (CAX: CAD/CAPP/CAM, etc.) for more than three decades and their accomplishments in many different aspects of design, process planning, and manufacturing, they are still not adaptive to dynamic manufacturing shop floor environments [3]. Particularly, in case of a change at the shop floor level, the manufacturing procedure has to be re-assessed by expert process planners according to the newly introduced limitations (for example a machine breakdown, fixture shortage or even a prioritised order). However, such evaluations are usually time consuming and can reduce the efficiency of a system, as the machines have to remain idle until the new alternative plans are generated. On the other hand, the increasing trend of outsourcing, joint ventures, and cross-border collaborations have led to a job environment geographically distributed across organisational and even national boundaries [2]. As the manufacturing industry becomes more and more globalised and mass customisation



L. Wang and X. V. Wang, *Cloud-Based Cyber-Physical Systems in Manufacturing*,
https://doi.org/10.1007/978-3-319-67693-7_5

becomes a norm for many industrial sectors, process planning systems need to become more adaptive, distributed, agile and integrated [4]. Also, assigning machining tasks to available manufacturing resources over time (or scheduling) in an integrated way is another aspect of complexity [5]. Therefore, a two-layer hierarchical structure for adaptive distributed process planning was proposed consisting of supervisory planning and operation planning [3].

Distributed process planning (DPP) is introduced as a link between product design and NC control and should be capable of transferring and processing the design data so as to meet the requirements of subsequent NC machining. The supervisory planning layer as the top shop-floor layer of this architecture design is responsible for generating a generic and machine-neutral process plan from product design data and machining features. The steps for developing such process plan are product data analysis, machining feature recognition or feature parsing, setup generation and feature sequencing. The operation planning layer as the bottom controller layer on the other hand is responsible for transforming the generic plan into a machine-specific one where all the cutting parameters, machine specifications, G-codes, etc. are specified. Function blocks (FBs) have also been introduced as one of the main enabling technologies for such system [6, 7], mainly due to their portability, reusability and adaptability in real-time control applications. Function blocks can encapsulate a generic process plan generated during supervisory planning and transfer it to the operation planning. Furthermore, they are provided with know-hows and logics to complete and generate some parts of the process plan at the operation planning level. FB design and some necessary FBs including machining feature FB, event switching FB, communication FB and management FB are reported in [8] with their relevant input and output data. Moreover, a framework has been proposed in [9] that by using the function block concept as well as the proposed architecture of DPP in [3] allows for collaborative manufacturing, including distributed process planning, dynamic scheduling, real-time monitoring, and remote control through a shared web-based environment. However, by making some modifications, adjustments and extensions, and by using the cloud concept, the DPP framework can be presented in a shared cloud environment where different partners have access to different modules of the system.

Because of the dynamic changes in the market and different shop floor uncertainties, SMEs active in the metal cutting industry face many cost and time minimisation challenges on a daily basis. One of the challenges in machining is to machine as many part features as possible in a single setup on a single machine in order to reduce setup time and cost. Furthermore, lack of appropriate machining quality against the customer's requirements is another factor that can increase the cost due to material wastes. Recently, multi-tasking machines have been introduced and are widely used nowadays due to their various advantages, particularly the cycle time reduction resulted from the integrated machining capabilities in one single machine. Moreover, the multi-tasking machines may possess special functionalities that can automatically carry out many manual tasks such as part transition (to transfer the part from one fixturing status to another) or part switching (to remove or move the finished parts to another position and load new parts for machining). This further reduces the non-value-adding time. Transferring a manual task into an automated task can also decrease the cycle time and as a result contribute to the cost and time minimisation.

This chapter presents an extended FB-enabled adaptive process planning approach for mill-turn parts (parts with both milling and turning features) in multi-tasking machines with special functionalities, which can be implemented in a cloud-based distributed process planning framework. The extension presented in this chapter is based on a proven method, including process planning functionalities to support turning features and combination of them with milling features in different aspects. The concept of *machine mode* is introduced to identify different machining and configuration states in a multi-tasking machine. Switching between each pair of different machine modes, if possible, needs to be addressed specifically and special actions should be performed for the transition. Knowing the special functionalities of the multi-tasking machine, all valid combinations of machine mode transitions can be categorised as either manual or automatic. This information is further used for setup merging through a cost estimation and optimisation process. Special FBs for handling the above activities are explained through a case study.

The rest of this chapter is organised as follows: Sect. 5.2 gives an overview of multi-tasking machines and related research studies on the process planning of mill-turn parts. Section 5.3 presents the methodology and its extensions to the previous adaptive distributed process planning systems. The methodology is employed in a case study in Sect. 5.4, followed by discussions in Sect. 5.5. Finally, Sect. 5.6 concludes the chapter.

## 5.2 Multi-tasking Machines and Mill-Turn Parts

Diversified, low-volume production has always been a great challenge for SMEs. Setup planning is one of the key elements in such a diversified production system. Especially, reducing the setup time as well as reducing the number of setups can be challenging. In order to increase the profit and decrease both the time and cost of production, SMEs have to apply different methods to reduce their setup times and the number of setups. The concept of SMED (single-minute exchange of die) was introduced by [10] in 1985 where different methods for reducing the setup time was suggested for companies with different production strategies (different product variants and batch sizes). As a solution, parallel machine tools were introduced which had the capability of performing multiple operations simultaneously by using multiple cutting tools or turrets (at least two) [11]. Their introduction had a revolutionary impact on cycle time reduction and material removal rate increment of machining tasks. Process planning and sequencing on parallel machines has been a major challenge and many researchers have worked on this subject such as [11, 12]. Despite the advantages of parallel machine tools, dynamic interaction and vibration control between the tools in such an environment can be challenging and might affect the quality of the machined part [13].

In a similar approach to decrease the cycle time and increase productivity, multi-function or multi-tasking machines are introduced [14]. Multi-tasking is defined as "an ability to execute simple and complex turning, milling, drilling, boring, reaming and tapping operations on a specially designed machine with less human intervention such as workpiece setup change, tool change, etc." [14]. As a consequence of less intervention, the number of setup changes can be minimised and a higher quality of the machined part is guaranteed as there is no loss of datum due to re-location and re-clamping. Djassemi [15] has studied multi-tasking machines (MTM), their applications and advantages, and defined multi-tasking machines as "machines with five or more axes of motion (capable of utilising any combination of x, y, z, a, b, and/or c-axes), and equipped with two or more tool systems and spindles and can operate in synchronous or asynchronous machine modes". Based on this definition, it can be interpreted that MTM can perform parallel machining operations. Djassemi also presented the taxonomy of multi-tasking machines as shown in Fig. 5.1.

Although multi-tasking machining refers to multiple machining functionalities (milling, turning, drilling, etc.), it does not necessarily require simultaneous machining capabilities as in parallel machining. Therefore, it is necessary to make a distinction between parallel machining and multi-tasking machining. In other words, the method presented in this chapter uses multi-tasking machining with the definition of multiple functionalities that are not being performed simultaneously but can each be individually active. A comprehensive survey on multi-tasking machines (noted as multi-functional machine tools) used for metal cutting, their kinematic configurations, control and programming technologies can be found in

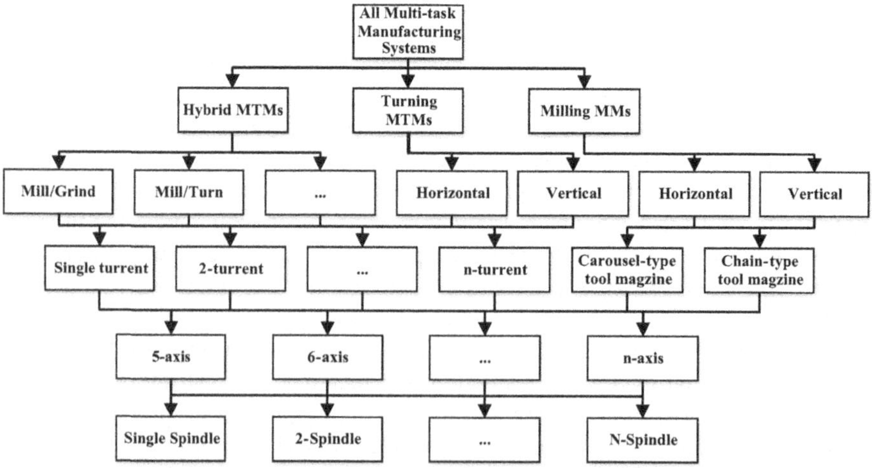


**Fig. 5.1** Taxonomy of multi-tasking machining systems, adapted from [15]

[16]. Selvaraj et al. [17] defined multi-tasking machines, as "a class of new CNC machines that enable to combine different family of operations on to a single machine in single workpiece setup without manual intervention". Park [18] has also discussed the importance of multi-tasking in today's industries. Multi-machines (i.e. moving from one machine to another for finishing a part) and multi-tasking machines have been compared in terms of time, cost, number of setups involved, etc. for the manufacture of aircraft components [14]. It was concluded that, multi-tasking machines can reduce the number of required machines for machining a part, shop floor space, number of setups, setup time, tool change time, and the number of operators. A general methodology for the calculation and comparison of time and cost of manufacturing for a class of mechanical system components using both multi-machines and multi-tasking machines is presented in [17].

Some of the major contributions of multi-tasking machines can be presented as follows, which can altogether contribute to the increase of profitability, productivity and resource efficiency in production.

1. *Better quality and consistency*: In contrast to the conventional production systems where multiple machines were used for the manufacturing of one part and the part had to be transferred between machines, multi-tasking machines allow the machining to be performed on one single machine. As a result and due to less re-location and re-fixturing of the part, the risk of stacked tolerances can be eliminated. Consequently, in an ideal environment with similar conditions, a better quality of final product can be achieved.
2. *Production flexibility*: Having multiple axes and spindles allows a better adoption to changes in market demand. In addition, less fixturing equipment, and labour involvement would be very beneficial for diversified production (high varieties of products with small batch sizes) where tool changeover time can be significant [15].
3. *Setup and setup time reduction*: The number of setups can be reduced for those parts that require multiple setups. This is mainly due to the capability of using multiple axes and spindles. The capability to change from one machining function to another while the part is fixed in a specific setup can reduce the setup time. In addition, some special functionalities of multi-tasking machines, e.g. automatic part transition between different fixturing states, can further contribute to setup time reduction.
4. *Process simplification*: Multi-tasking machines allow the machining process of complex parts in one single machine and as a result facilitating the machining process [15]. The research reported in [19] presented some examples of complex-shaped workpieces machined on an integrated multi-tasking machine.

Multi-tasking machines are now significantly popular in metal-cutting industry and have become increasingly sophisticated in design and configuration. Despite their numerous advantages, multi-tasking machines possess programing challenges because of their complex configuration and multiple machining functions [20]. This problem has led the software industry to develop a number of simulation, CAM, and verification systems to aid this complex process. The advances of virtual machine tool technology [21] in recent years have simplified the verification of NC programs and collision avoidance through real-time modelling and simulation. A smart decision-making system that supports multi-tasking machining is necessary for adaptive process planning in an integrated machining environment. According to [22], the introduction of STEP-NC highlights a need for new programming methods and software tools for multi-tasking machining of components, which requires further research to support industrial developments.

Mill-turn machine tools are a subset or a class of multi-tasking machines that can perform both milling and turning operations [23]. Consequently, the term "mill-turn part" was assigned to those parts with both rotational turning and prismatic milling features [24]. Mill-turn parts are also referred as prisronal parts in [25]. In order to produce a mill-turn part, both milling and turning operations are required. Mill-turn machine tools allow mill-turn parts to be machined and finished in one single machine instead of being transferred between two (or even more) milling and turning machines. Miska [26] refers to "mill-turn machine tools" as "turning centres" and recognises them as suitable alternative to machining centres for workpieces that require cylindrical as well as prismatic machining [27]. Different research studies have been performed on mill-turn machining centres and mill-turn parts. In [27], computational techniques for determining the Maximum Turnable State (MTS) of a mill-turn part are presented with the applied rule of thumb that "the total percentage of milled features should be between 30 and 40% for the part to be cost effectively machined on a mill-turn". In other words, the authors mentioned the fact that it is more efficient to remove material by turning than by milling in a mill-turn machine. Machining non-coaxial parts on mill-turn machines has been studied in [28]. A methodology for automating the process planning and NC code generation for a widely encountered class of freeform features that can be machined on a 3-axis mill-turn centre has been proposed in [29]. Feature recognition in mill-turn parts is usually a complex task as many feature interactions between turning and milling features occur [23]. To solve this problem, the authors of [30] presented a machining volume generation method for recognition of interacting prismatic and rotational features.

Due to their multiple functionalities, multi-tasking machines have also affected process planning research [31]. A feature-based geometric reasoning system for part modelling and process planning as applied to mill-turn machined parts is proposed in [32], which applied feature recognition system based on convex decomposition and the mapping method to relate the negative feature volumes to machining process classes. A concurrent analysis model for analysing machining sequences and fixturing setups of mill-turn parts has been proposed in [24] where the best machining sequences can be found with the minimum number of machine changes and fixturing setup changes. An intelligent expert process planning system has been proposed in [25] for five-axis mill-turn parts where a feature-based process planning approach is applied using both variant and generative approaches, and a new machining features classification is reported. In addition, a new group code has been defined based on a survey from different industries in order to classify mill-turn parts. Furthermore, the authors mentioned that a machining feature representation scheme consists of open and hidden layers that store different types of data from information of the initial stock and geometrical data to the machining processes and sequencing rules. A three-module knowledge-based approach is then applied for process planning. Authors in [33] have developed a CAM system that can recognise the machining features of a mill-turn part and automatically generate its tool paths. Another automatic process planning system for mill-turn parts has been proposed and developed in [34, 35], which by applying machining feature recognition, can generate alternative machining plans and identify the one with the shortest processing time. A graph-based process planning system for multi-tasking machines was reported in [36] where the manufacturing features are recognised based on graph isomorphism and geometrical rules, and feasible machining sequences are generated. Finally, optimal machining plans are identified according to the user-defined cost evaluation.

Common to all process planning systems, one difficult problem is how to solve the complex iterations between interim feature geometry and process parameters in each individual machining operation and transformation among different machining operations, especially for mill-turn machining. Li et al. [37, 38] converted this to the elimination of non-machining configuration spaces or C-spaces, making the complex iteration problem solvable.

## 5.3 Methodology

This chapter aims to present an extension to Cloud-DPP (cloud-based function-block enabled adaptive distributed process planning) methodology [6] that supports (1) mill-turn parts, (2) process planning for multi-tasking machining centres specifically mill-turns class, and (3) special functionalities in some machine tools such as part transfer to further reduce the total number of setups. Furthermore, applying this methodology can facilitate the complex programing and planning step required for multi-tasking machining. The extension has been made to the adaptive distributed process planning approach reported in [3] where a two-layer process planning is suggested. The supervisory planning generates a generic process plan featuring a generic setup plan with sequenced machining features, which will be later merged and adapted to available machines. The process plan represented by a function-block network is then deployed to the machine controllers for execution. The process plan is finally detailed by feature-based operation planning algorithms at execution time. In this chapter, the proposed extensions are:

- generic setup planning for turning features in addition to milling,
- definition of machine modes and machine mode transitions to extend machine tool model so as to include multi-tasking machines,
- indicating the possible fixturing states of the workpiece using the candidate setup frames,

- defining new function blocks for handling mode transitions, sub-setups, and turning features, and
- developing an adaptive setup planning and task assignment algorithm that performs a semi-optimal decision making on merging generic setups together and assigning them to one or more of available machines.

Generic setups for turning features are similar to 3-axis setups of milling features applying the following rule that the turning features with similar turning vector will be grouped together [39]. Also, generic turning setups precede generic milling setups. Sequencing of the generic turning setups and assigned turning features are to be performed in a similar way as was introduced by the authors in [40] based on [41]. Details about the rest of the extensions are described in this section.

### 5.3.1 Machine Modes

In multi-tasking machines research, the term *machine mode* was used for addressing the state or functionality of a machine during a machining operation [11]. In this section, a *machine mode* represents a defined state of a machine with a specified machining function and active kinematic mechanisms. The definition of a machine mode data model consists of the following:

- Functionality of the machine: Determines what machining function is active at a specific state of machining.
- Kinematics of the machine: A machine consists of different kinematic configurations that can be either active or inactive in different states of the machine. In each machine mode the active kinematic chain is defined which consists of the data about involved moving links, joints and their freedom of movement.
- Part fixturing reference frame in the machine: The machine mode also provides information about the fixturing reference frame based on which the part will be positioned on the machine. By knowing the fixturing reference frame in relation to the machine coordinate system, the accessibility to the part in that specific configuration can be assessed. Moreover, the active fixturing reference frame of the machine mode is required for geometric calculations of the tool paths. The orientation of this frame is important in particular and is to be defined by two X and Z unique vectors of the frame represented in the machine coordinate system.

The combination of the above information forms a model representing a multi-tasking machine in different states to be used in planning and control algorithms of the system. In other words, a machine can be active in different machine modes that are each represented by the set of the mentioned attributes. For example, Table 5.1 presents a machine mode data model with information of the machine kinematics (degrees of freedom of rotational axes and the orientations of the axes).

AL, AU, BL, BU, CL and CU refer to upper and lower boundaries of the rotary axes A, B and C, respectively. Moreover, [Ix, Iy, Iz] and [Kx, Ky, Kz] are unit vectors of the fixturing reference frame according to the machine coordinate system.

## 5.3.2 *Machine Mode Transitions*

Once the different machine modes have been defined, the transition between the modes during the machining process should also be defined. First of all, the possibility of switching from one mode to another has to be determined. Secondly, the possibility of an automated transition among the valid modes (by analysing the machine's special functionalities) can be specified. Special functionalities are those extra functionalities that are embedded in the machine in addition to its main expected functionalities and can increase the efficiency of machining in different ways. For example, a two-spindle mill-turn machine is expected to be capable of both turning and milling functions. However, functionality of an automated part transition between the main and sub spindles can be added to the machine by equipping the machine with hydraulic chucks and additional controllable axis. As a result, the sub spindle can approach towards the main spindle, take the part and retract away. This additional functionality can decrease the setup time and increase the quality as the setup changeover is performed automatically. Another special functionality that can be added is for example an automated switch between parts by means of a rotary table. The machine mode transition can be represented by a $n \times n$ square matrix where $n$ is the number of existing machine modes The values of transitions can be assigned with either A (automated), M (manual) or X (not applicable, on the main diagonal of the matrix).

## 5.3.3 *Setup Frames*

To involve fixturing of a workpiece, the term *setup frame* is introduced. Setup frames are those candidate frames defined on the workpiece that can provide guidance on orientation of potentially possible workpiece fixturing on the machine.

**Table 5.1** Representation of a machine mode data model

| Machine mode ID | Type | Kinematics | AL | AU | BL | BU | CL | CU |
|---|---|---|---|---|---|---|---|---|
| 2 | Milling | XYZ-CB | 0 | 0 | 0 | 90 | 0 | 360 |
| Machine mode ID | | | Ix | Iy | Iz | Kx | Ky | Kz |
| 2 | | | 1 | 0 | 0 | 0 | 0 | 1 |

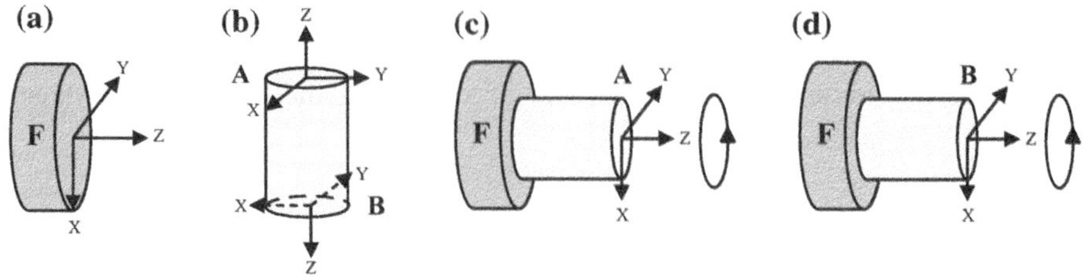


**Fig. 5.2** **a** Fixture's coordinate system; **b** candidate setup frames (A and B); **c** and **d** part fixed according to setup frames A and B, respectively

All potentially possible fixturing alternatives of the workpiece for each machine mode are all possible combinations where orientation of one candidate setup frame of the workpiece matches that of the fixturing reference frame of the machine mode. As an example, Fig. 5.2b shows a cylindrical workpiece and both **A** and **B** are its given candidate setup frames. **F** is the fixturing reference frame of a machine mode for milling (Fig. 5.2a). Two possible ways of fixturing the workpiece on the machine are known by orienting the part in the way that candidate setup frames have similar orientation as the fixturing reference frame (Fig. 5.2c, d).

## *5.3.4 Setup Planning and Setup Merging*

Setup planning is one of the sub-tasks of the supervisory planning in adaptive DPP. An initial generic and machine-neutral setup plan is generated at this level. To do so, the same approach suggested in [40] has been applied. In case the machine tool has more than three machining axes, the generated 3-axis setups will be merged due to higher machine reachability. Since multi-tasking machines normally possess more than 3 degrees of freedom, a setup merging is necessary for generating an optimal process plan. To perform setup merging, the following information is required:

- Generic setups: The 3-axis setups including milling features with the same TADs (tool approach directions) and generic turning setups including turning features with the same turning vector.
- Precedence graph: A graph indicating the precedence relations among available generic setups and machining features belonging to the same generic setups can be provided based on a number of requirements such as the machinability rule or datum references among machining features [40].
- Available machines and machine modes.
- Candidate setup frames of the workpiece.

A cost matrix can be generated based on the transition matrix. The cost matrix estimates the cost of a transition between two different modes. The main factor for estimating the cost would be the time of transition (which is also affected by the possibility to automate the transition). Other factors such as energy consumption, power, stability, and accuracy of the employed kinematics, etc. can affect the cost of transition. The time can therefore be translated in terms of cost for the valid mode transitions. However, this section does not provide guidelines for cost matrix generation but uses the following approach. Overall cost of an adaptive setup plan can be calculated as summation of the cost of each assigned setup conditions according to the previous assigned setup conditions in the sequence. For $S_i$ and $S_{i+1}$, two consecutive generic setups in an adaptive setup plan, the assigned cost increases in the following order (when $S_i$ and $S_{i+1}$ are):

1. Merged into the same setup group in one machine mode.
2. Merged into different setup groups in different machine modes of the same machine with automatic transition.
3. Merged into different setup groups in different machines with automatic transition.
4. Merged into different setup groups in different machine modes of the same machine without automatic transition.
5. Merged into different setup groups in different machines without automatic transition.

Using the inputs mentioned earlier as well as the cost estimations, the greedy algorithm is applied in setup merging. The greedy algorithm follows the problem solving heuristic of making the locally optimal choice at each stage with the hope of finding a global optimum. However, this method does not necessarily provide a global optimum but usually yields local optimal solutions that approximate a global optimal solution in a reasonable time.

### *5.3.5 New FBs and FB Network Generation*

Function blocks (FBs) are necessary elements in Cloud-DPP and are responsible for adaptive execution control of the machining process. AutoTrans FB is developed for handling machine mode initialisation and transition between machine modes. The developed Sub Setup FB handles 3-axis sub-setups in the context of a merged setup assigned to a specific machine. Also for every involved milling/turning machining feature type, one specific FB is developed. These FBs encapsulate machining know-hows of those features and are able to control the machining process. The system then generates an FB network that is deployed to the chosen machine for execution. The generated function blocks are illustrated in Fig. 5.3.

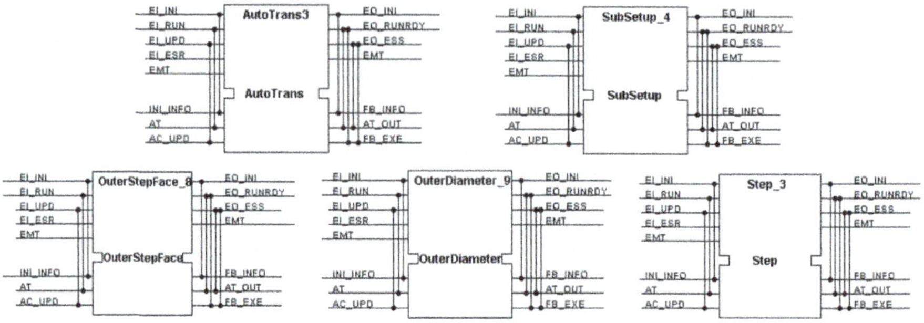


**Fig. 5.3** Function blocks for handling multi-tasking machining process

## 5.4 Case Study

The selected workpiece (Fig. 5.4) consists of both turning (2 outer step faces and 1 outer diameter) and milling (4 steps, 1 chamfered blind hole, and 2 rectangular pockets) features. The part is machined from a cylindrical stock. The selected multi-tasking machine is a Mazak Integrex e410h with two turning spindles, one milling spindle, seven controllable axes, and the special functionality of workpiece transfer between its main and sub spindles (linear movement of axis W). The configuration of the machine and its movable axes is shown in Fig. 5.5. According to the methodology, the following steps are required for setup planning.

- *Three-axis setup generation and sequencing*

The ten existing features are grouped into generic setups according to their TADs or turning vectors. They are then sequenced based on available mandatory referencing or accessibility rules. As a result of this step, six setups are generated. The setups and their sequences are depicted in Fig. 5.6.

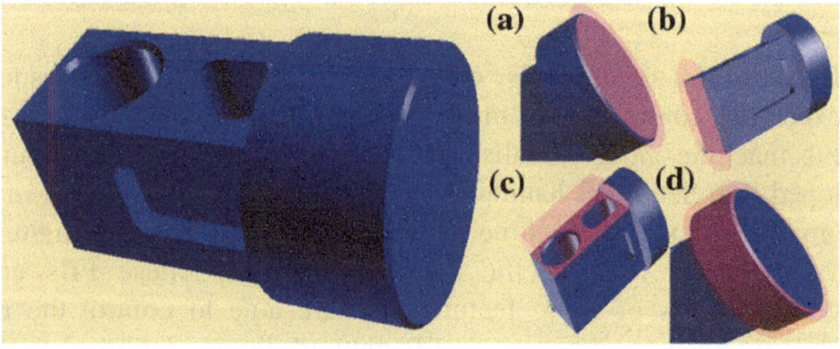


**Fig. 5.4** A mill-turn part with some highlighted milling and turning features: outer step face turning features (**a** and **b**); step milling feature (**c**); and outer diameter turning feature (**d**)

**Fig. 5.5** A Mazak multi-tasking machine (top), and its axes configuration (bottom)

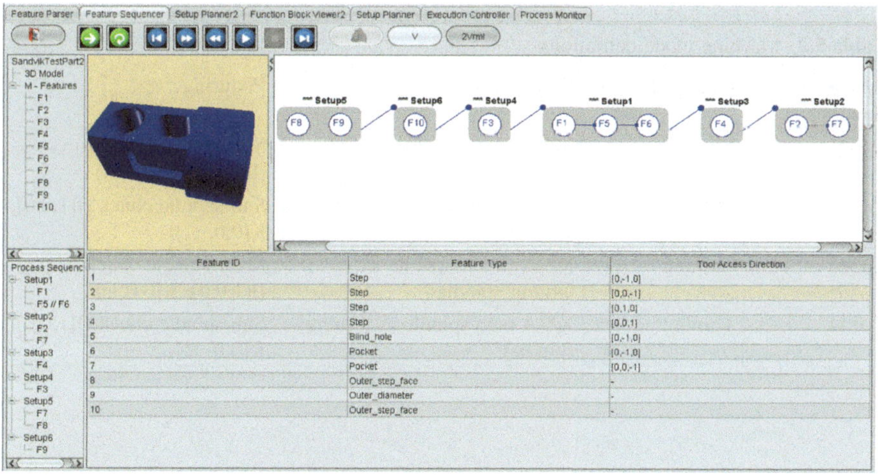


**Fig. 5.6** Generic setup planning and sequencing

- *Candidate setup frames indication*

Since the stock has a cylindrical shape, it can be fixed in the turning chuck on either one of its two ends. Therefore, two different candidate setup frames can be defined for the workpiece.

- *Machine mode and transition matrix definition*

Four different machine modes are defined according to the capabilities of the available multi-tasking machine, as listed in Table 5.2. Each machine mode represents a specific state of the machine that is different from the rest of the states in at least one of the attributes of functionality, kinematics, or fixturing. The transition matrix is presented in Table 5.3 where A and X represent automated and not applicable transitions, respectively. Note that in case that W axis does not exist, the transitions between main and sub spindles (such as mode 1–mode 2) would have been manual.

- *Adaptive setup planning*

Figure 5.7 shows the result of the adaptive setup planning process where the greedy algorithm and the heuristic rules presented in the previous sections have been applied. The six existing generic setups have been merged into three setups. Two of the new setups are marked automatic due to either pure function switching with no need for re-fixturing or special functionality such as part transfer. Therefore, only one setup needs to be performed manually.

- *Function block generation and deployment*

According to the results of generic process planning, adaptive setup planning and job assignment, a function block network (Fig. 5.8) is generated in which the instances of the previously defined function blocks [8] and newly introduced ones in Sect. 5.3.5 are used. Once deployed to a machine, FBs can automatically and

**Table 5.2** Machine mode definitions

| Machine modes | Functionality | Kinematics | Fixturing |
|---|---|---|---|
| MM1 | Milling | XYZ-CA (main spindle as C axis) | Main spindle chuck (I[1,0,0], K[0,0,1]) |
| MM2 | Milling | XYZ-CA (sub spindle C2 as C axis) | Sub spindle chuck (I[1,0,0], K[0,0,−1]) |
| MM3 | Turning | XZ-A (main spindle as turning spindle) | Main spindle chuck (I[1,0,0], K[0,0,1]) |
| MM4 | Turning | XZ-A (sub spindle as turning spindle) | Sub spindle chuck (I[1,0,0], K[0,0,−1]) |

**Table 5.3** Transition matrix

| Transition matrix | MM1 | MM2 | MM3 | MM4 |
|---|---|---|---|---|
| MM1 | X | A | A | A |
| MM2 | A | X | A | A |
| MM3 | A | A | X | A |
| MM4 | A | A | A | X |

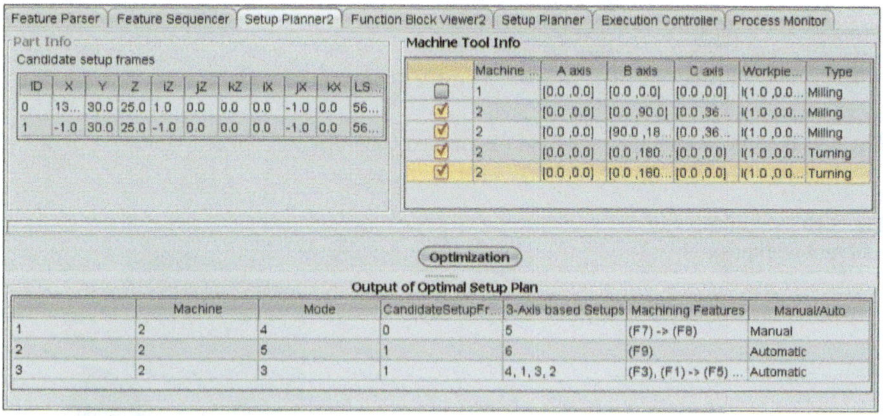


**Fig. 5.7** Setup merging

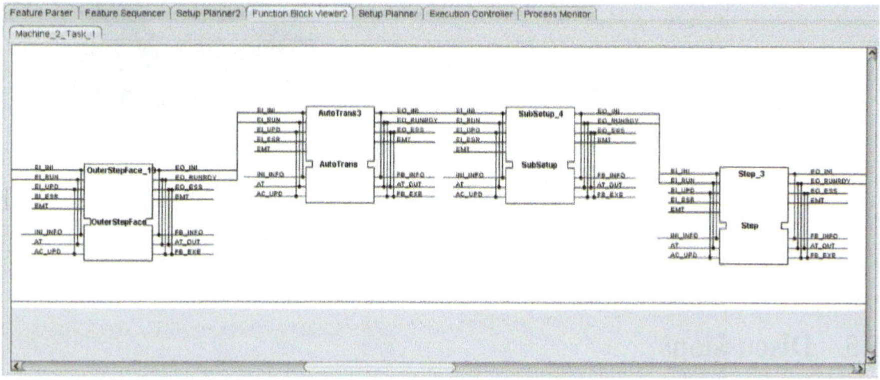


**Fig. 5.8** Generated FB network

adaptively control and execute the machining process of the mill-turn part according to the available tools on the machine and other conditions (Fig. 5.9). Finally, in Fig. 5.10, the machined part on the multi-tasking machine is revealed.

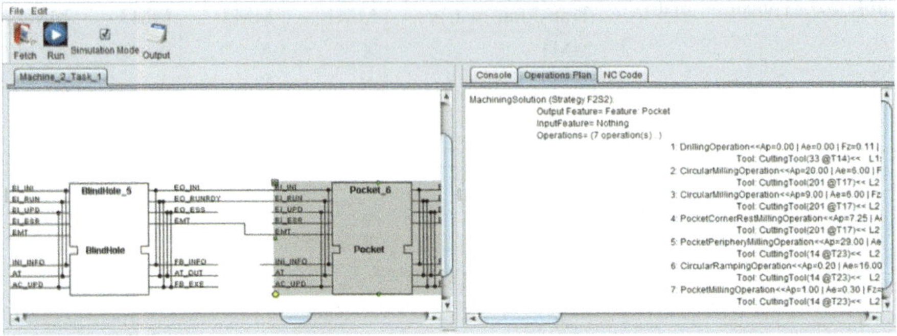


**Fig. 5.9** Process execution on a hybrid FB-enabled controller

**Fig. 5.10** Machined mill-turn workpiece

## 5.5 Discussions

To be able to compare the solution with other existing solutions, the comparison criteria need to be determined. The main objective of Cloud-DPP is adaptability and thus considers process planning as an embedded part of run-time production system to achieve the objective. Therefore, adaptiveness should be considered as the main comparison criterion. Considering the class of parts represented by the test part in the case study, the present system possesses the equivalent functionalities of other systems presented in [25, 33–36], except machining feature recognition that is out of the scope of process planning. However, none of those systems are designed to achieve the adaptability that the Cloud-DPP offers. This is due to the fact that Cloud-DPP is designed by considering the adaptability as the essential principle and therefore employs a different multi-layer process planning and execution method with feed-back loops from shop floor monitoring data. In comparison, other aforementioned systems are mainly based on traditional process planning and CAM practice which follows unidirectional data/information flow with no adaptability. Moreover, Cloud-DPP supports process planning and execution functionalities for machines with special capabilities such as automatic part transfer among their redundant subsystems. This can significantly increase the productivity by reducing the number of setups and relax planning and programming complexities. This is not possible for other aforementioned systems. Another unique aspect in contrast is that Cloud-DPP as an embedded part of a production system is intended to seamlessly provide and consume process planning and process execution services on the cloud to enable collaborative work in cloud manufacturing.

## 5.6 Conclusions

A cloud-based function block enabled distributed adaptive process planning and execution system for machining of mill-turn parts is presented in this chapter. The contribution has been in form of extending the current Cloud-DPP methodology for supporting mill-turn parts machining on multi-tasking machines with special functionalities, and validating the methodology through a case study. Turning features are introduced to the system and generic process planning algorithms are outlined accordingly. The concept of machine mode is used for representing the machining function, its kinematic configuration, and fixturing reference. Transition matrix indicates the possible transitions among different machine modes as well as the characteristics of the transition. Such characteristics can be specified according to the special functionalities of the multi-tasking machines. Greedy algorithm is then used for adaptive setup planning and job assignment according to a cost function representing the optimisation objectives. Tool orientation accessibility analysis is performed to check validity of assigning setups to different machine modes and the possibility of merging generic setups together. Finally, five function blocks are explained that can handle machine mode initiations and transitions, adopting generic setups in a specific machine mode, and detailed operation planning and execution for newly introduced turning features. A case study is carried out with a mill-turn part machined on a mill-turn multi-tasking machine for system validation.

As a future work, the genetic algorithm should be applied to find near-optimal solutions in adaptive setup planning. Research on more sophisticated fixturing models and dynamics should also be considered.

## References

1. M. Jeremy B., D.C. McFarlane, Assessing the responsiveness of existing production operations. Int. J. Oper. Prod. Manage. **19**(8), 765–784 (1999)
2. L. Wang, M. Givehchi, Web-DPP: an adaptive approach to planning and monitoring of job-shop machining operations, in *Proceedings of the 7th CIRP International Conference on Digital Enterprise Technology* (2011)
3. L. Wang, H.Y. Feng, N. Cai, Architecture design for distributed process planning. J. Manuf. Syst. **22**(2), 99–115 (2003)
4. X. Xu, L. Wang, S.T. Newman, Computer-aided process planning—a critical review of recent developments and future trends. Int. J. Comput. Integr. Manuf. **24**(1), 1–31 (2011)
5. W. Shen, L. Wang, Q. Hao, Agent-based distributed manufacturing process planning and scheduling: a state-of-the-art survey. IEEE Trans. Syst. Man Cybern. Part C Appl. Rev. **36**(4), 563–577 (2006)
6. L. Wang, An overview of function block enabled adaptive process planning for machining. J. Manuf. Syst. **35**, 10–25 (2015)
7. L. Wang, M. Holm, G. Adamson, Embedding a process plan in function blocks for adaptive machining. CIRP Ann. Manuf. Technol. **59**(1), 433–436 (2010)
8. L. Wang, H.Y. Feng, C. Song, W. Jin, Function block design for adaptive execution control of job shop machining operations. Int. J. Prod. Res. **47**(12), 3413–3434 (2009)
9. L. Wang, Wise-shopfloor: an integrated approach for web-based collaborative manufacturing. IEEE Trans. Syst. Man Cybern. Part C Appl. Rev. **38**(4), 562–573 (2008)
10. S. Shingo, *A Revolution in Manufacturing: The SMED System* (Productivity Press, 1985)
11. N.C. Chiu, S.C. Fang, Y.S. Lee, Sequencing parallel machining operations by genetic algorithm. Comput. Ind. Eng. **36**(2), 259–280 (1999)
12. D.E.R.E.K. Yip-Hoi, D. Dutta, A genetic algorithm application for sequencing operations in process planning for parallel machining. IIE Trans. **28**(1), 55–68 (1996)
13. E. Budak, A. Comak, E. Ozturk, Stability and high performance machining conditions in simultaneous milling. CIRP Ann. Manuf. Technol. **62**(1), 403–406 (2013)
14. P. Selvaraj, E. Thirumal, P. Radhakrishnan, Multi-tasking machines: a new approach to increase the productivity of aircraft components manufacture. Int. J. Comput. Appl. Technol. **27**(1), 24–30 (2006)
15. M. Djassemi, Emergence of multitasking machining systems: applications and best selection practices. J. Manuf Technol. Manage. **20**(1), 130–142 (2008)
16. T. Moriwaki, Multi-functional machine tool. CIRP Ann. Manuf. Technol. **57**(2), 736–749 (2008)
17. P. Selvaraj, P. Radhakrishnan, M. Adithan, Study of economics of multitasking with reference to a class of mechanical system components. Int. J. Manuf. Technol. Manage. **22**(4), 301–323 (2011)
18. K.H. Park, The growing importance of multitasking. Manuf. Eng. **148**(5), 152–152 (2012)
19. M. Mori, M. Fujishima, O. Yohei, 5 axis mill turn and hybrid machining for advanced application. Procedia CIRP **1**, 22–27 (2012)
20. Siemens, Master the Challenges of Programming Multi-Function Machines (2014), http://m.plm.automation.siemens.com/en_us/Images/4746_tcm1224-54502.pdf. Accessed 20 Dec 2014
21. Y. Altintas, C. Brecher, M. Weck, S. Witt, Virtual machine tool. CIRP Ann. Manuf. Technol. **542**, 115–138 (2005)
22. R.S.U. Rosso, S.T. Newman, S. Rahimifard, The adoption of STEP-NC for the manufacture of asymmetric rotational components. Proc. Inst. Mech. Eng. Part B J. Eng. Manuf. **218**(11), 1639–1644 (2004)
23. S. Li, J.J. Shah, Recognition of user-defined turning features for mill/turn parts. J. Comput. Inf. Sci. Eng. **7**(3), 225–235 (2007)

24. Y.J. Tseng, C.C. Liu, Concurrent analysis of machining sequences and fixturing set-ups for minimizing set-up changes for machining mill-turn parts. Int. J. Prod. Res. **39**(18), 4197–4214 (2001)
25. K. Waiyagan, E.L. Bohez, Intelligent feature based process planning for five-axis mill-turn parts. Comput. Ind. **60**(5), 296–316 (2009)
26. K.H. Miska, Driven tools turn on turning centers. Manuf. Eng. **104**(5), 63–66 (1990)
27. D. Yip-Hoi, D. Dutta, Finding the maximum turnable state for mill/turn parts. Comput. Aided Des. **29**(12), 879–894 (1997)
28. Y.S. Lee, C.J. Chiou, Unfolded projection approach to machining non-coaxial parts on mill-turn machines. Comput. Ind. **39**(2), 147–173 (1999)
29. N. Date, P. Krishnaswami, V.S.K. Motipalli, Automated process planning method to machine A B-spline free-form feature on a mill–turn center. Comput. Ind. Eng. **56**(1), 198–207 (2009)
30. Y.J. Tseng, S.B. Joshi, Recognition of interacting rotational and prismatic machining features from 3-D mill-turn parts. Int. J. Prod. Res. **36**(11), 3147–3165 (1998)
31. B. Hedrick, R.J. Urbanic, H.A. ElMaraghy, Multi-tasking machine tools and their impact on process planning, in *CIRP 2004 International Design Conference*, Cairo, Egypt, 2004
32. Y.S. Kim, Y. Kim, F. Pariente, E. Wang, Geometric reasoning for mill-turn machining process planning. Comput. Ind. Eng. **33**(3), 501–504 (1997)
33. K. Kubota, T. Kotani, K. Nakamoto, T. Ishida, Y. Takeuchi, Development of CAM system for multi-tasking machine tools. J. Adv. Mech. Des. Syst. Manuf. **4**(5), 816–826 (2010)
34. K. Dwijayanti, H. Aoyama, Basic study on process planning for turning-milling center based on machining feature recognition. J. Adv. Mech. Des. Syst. Manuf. **8**(4) (2014)
35. K. Dwijayanti, H. Aoyama, Development of automatic process planning system for turning-milling center with 2 turrets system. Precis. Eng. J. **81**(5), 471–480 (2015)
36. J. Zhu, M. Kato, T. Tanaka, H. Yoshioka, Y. Saito, Graph based automatic process planning system for multi-tasking machine. J. Adv. Mech. Des. Syst. Manuf. **9**(3) (2015)
37. Y. Li, X. Liu, J.X. Gao, P.G. Maropo, A dynamic feature information model for integrated manufacturing planning and optimization. CIRP Ann. Manuf. Technol. **61**(1), 167–170 (2012)
38. Y. Li, W. Wang, X. Liu, Y. Ma, Definition and recognition of rib features in aircraft structural part. Int. J. Comput. Integr. Manuf. **27**(1), 1–19 (2014)
39. L. Wang, N. Cai, H.Y. Feng, J. Ma, ASP: an adaptive setup planning approach for dynamic machine assignments. Autom. Sci. Eng. IEEE Trans. **7**(1), 2–14 (2010)
40. M. Givehchi, A. Haghighi, L. Wang, Generic machining process sequencing through a revised enriched machining feature concept. J. Manuf. Syst. **37**(2), 564–575 (2015)
41. Z. Liu, L. Wang, Sequencing of interacting prismatic machining features for process planning. Comput. Ind. **58**(4), 295–303 (2007)

# Chapter 6
# Adaptive Machining Using Function Blocks

## 6.1 Introduction

As stated in [1], in general, the main requirements of a Cyber-Physical System (CPS) are: high predictive skills and capabilities, real-time intelligent distributed computation and control approaches, and reliable and secure network connection. CPS has recently gained a lot of attention and has shown promising applications in different fields of healthcare, traffic control, etc. [2]. Different attempts have been made for modelling CPS [3] and the 5C architecture has recently been proposed for CPS in Industry 4.0 manufacturing systems [4]. In the field of manufacturing, different applications of CPS have also been studied [5]. A CPS consists of a "cyber" part and a "physical" part (refer to Chap. 2 for details on CPS). In a CPS, sensors or other communicating tools embedded in physical entities are responsible for real-time data acquisitions. The data is then sent to the computational part of the system through a network/cloud environment, allowing decentralised control. The computational part is responsible for monitoring and controlling the actions of the physical entity through the embedded communicating equipment such as actuators. However, humans as the third part of the system can also interact with the system when necessary/desired through embedded user interfaces. This whole process is performed dynamically through feedback loops from the physical world to the cyber world and vice versa. Duc to the specific characteristics of CPS, modelling a CPS requires techniques that can address both the software and the hardware [6]. Interestingly, holon shares a similar feature as CPS and has two sides representing both the digital and the physical attributes of an object (e.g. a device, a machine, or a fixture). This distributed architecture of holon, makes holonic paradigm a suitable approach to constructing and modelling a CPS system in form of a holarchy. While a CPS connected to smart sensor networks can learn over time, a holarchy is also capable of learning through evolutionary self-organisation. Both CPS and Holons and holonic manufacturing systems (HMS) allow the integration of the physical world with the computational world where decisions are adaptively made according



L. Wang and X. V. Wang, *Cloud-Based Cyber-Physical Systems in Manufacturing*,
https://doi.org/10.1007/978-3-319-67693-7_6

to the physical inputs, and are transferred to the physical entities in order to optimise the performance of the system. However, the holonic concept and representation must be realised in real-world applications. To implement this concept, agents and function blocks are the key enablers. To address their specific characteristics, these two technologies are adopted at different levels of control. The cyber part of a CPS (or a holarchy) in most cases is non-real-time and can be implemented using the multi-agent approach. Once it comes to the physical part of the CPS, real-time behaviours must be respected and controlled. Function blocks are most suitable for the low-level device monitoring and control. This chapter describes the application of function blocks at different control levels of a holarchy representing a CPS for adaptive machining. See more details on holons, agents and function blocks in [1].

At the shop-floor level, not all the events are entirely predictable. Therefore, CPS has to be able to operate in a random environment where unexpected conditions may occur at any time. Failing to respond and adapt to disturbances and subsystem failures can affect the system's performance and cause critical issues. In other words, at the shop-floor level, reaction to changes should occur according to specific timeframes and within strict deadlines e.g. change or halt of a robot movement due to human interferences should occur within milliseconds to avoid injuries. Therefore, a holon representing a physical equipment would require a hard real-time control system that can control the behaviours of the physical entity within a strict deadline. The IEC 61499 standard using function blocks for decentralised control has shown promising applications in this area and has been recognised by the HMS community. Aside from the process encapsulation, function blocks as reusable modules are used for real-time, distributed, intelligent and event-driven control, and execution of processes on physical equipment. The embedded algorithms in a function block can be controlled through its execution control chart, and the internal variables and algorithms can be tuned to match the environment conditions. For example, when a tool breakage occurs during machining, function blocks can automatically modify the current process and adjust to the new conditions by triggering the right algorithms (i.e. finding an alternative option to finish the machining process with the remaining tools). Moreover, with the help of function blocks, the process can be immediately paused at any moment and resumed when necessary (e.g. robot and human interference). This quick self-adjustment process not only saves a lot of time (in contrast to traditional re-planning scenarios) but also prevents system suspension or physical damages/human injuries, which can impose unnecessary costs. This ability makes function block (FB) a good candidate for low-level control and execution. Low-level control is the closest control layer to the physical equipment on the shop floor and is responsible for both executing the transferred plans and handling disturbances in a real-time manner. FB-enabled controllers are necessary for implementing this idea but few FB-enabled commercial controllers are currently available in the market. One solution would be to integrate an FB runtime environment hosted in a frontend computer with a commercial controller and consider the whole combination as an FB-enabled controller [7].

## 6.2 Function Block Concept

### *6.2.1 Function Blocks*

An FB is a block that encapsulates functionality. According to the IEC 61499, a function block instance has the following characteristics:

- Type name and instance name,
- Event inputs, interface of an FB receives events from an event connection, and can affect the execution of one or more algorithms,
- Event outputs, interface of an FB issues events to an event connection,
- Data inputs, interface of an FB receives data from a data connection and corresponds to the input variables,
- Data outputs, interface of an FB supplies data to a data connection and may correspond to output variables,
- Internal data mapped also as internal variables,
- Functional characteristics, which are divided into the Execution Control and the Internal Algorithms.

Figure 6.1 shows how all these characteristic features can be mapped to an FB. Taking IEC 61499 into account, the function block type specifications should include: (1) type name; (2) the number, names, type names and order of events inputs and events outputs; and (3) the number, names, data type and order of data input, data output and internal variables.

### *6.2.2 Function Block Types*

In the IEC 61499 standard, the basic unit for encapsulating and reusing Intellectual Property (IP = "know-how") is the function block type. In object-oriented terms, this is a class defining the behaviour of possibly multiple instances. It includes event inputs and outputs as well as the more traditional data inputs and outputs, to provide for synchronisation between data transfer and programme execution in distributed systems.

#### 6.2.2.1 Basic Function Blocks

As its name implies, the basic function block type is the "atom" out of which higher-level "molecules" are constructed. With IEC 61499 compliant software tools, software developers can encapsulate IP in form of algorithms. Execution of these algorithms is triggered by Execution Control Charts (ECCs), which are event-driven state machines. Figure 6.2a illustrates a basic function block with event input (EI), event output (EO), data input, data output, and internal variables.

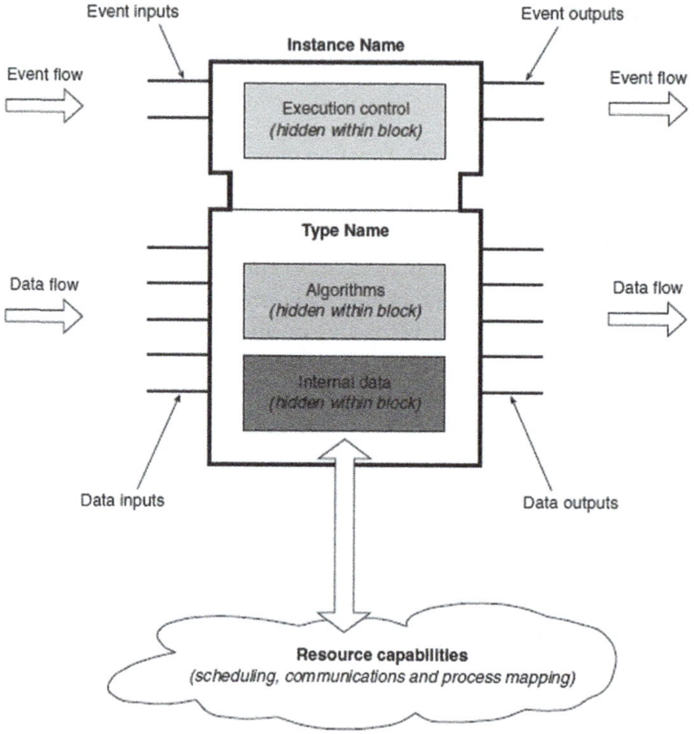


**Fig. 6.1** Characteristic features of a function block

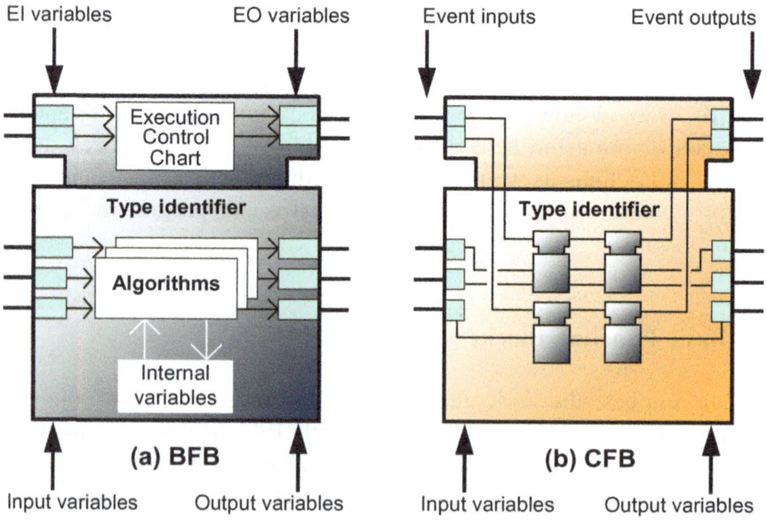


**Fig. 6.2** Function block types: **a** basic FB, **b** composite FB

#### 6.2.2.2 Composite Function Block

Function blocks can be connected via event and data connections to form FB networks. The event connections and behaviour of every single function block completely determine the joint behaviours of the networks. Such networks can be encapsulated into composite function blocks for future reuse as shown as in Fig. 6.2b. Similar to basic function blocks, composite function blocks have interface with input and output event and data variables. Composite function blocks do not have internal variables. Elements of an FB network can be other composite function blocks. Thus, FB applications can be structured hierarchically and the levels of hierarchy are unlimited.

Moreover, composite function blocks do not have ECCs. However, an additional component block with a composite function block can play this role as illustrated in Fig. 6.3.

The Behaviour of a composite function block is purely determined by a network of function block instances. A composite function block is simply a container for a network of other function blocks. The container as such performs no specific actions except for setting input and output variables and for the activities of its components. The network can include basic, service interface and composite function block types, and function block applications can be hierarchical as mentioned earlier.

A basic FB is able to represent a small task, having a similar behaviour as an electronic device or circuit, and can solve simple problems, but joining different FBs (including composite FB), a more complex problem can be solved.

#### 6.2.2.3 Service Interface Function Block and Sub-application

Service interface FB, as the name suggests, is an interface function block, which allows interfering between the FB domain and external services [8], such as

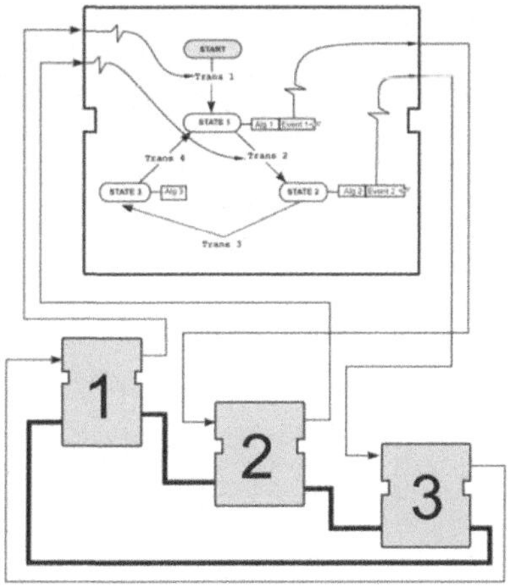


**Fig. 6.3** ECC in a composite function block

hardware target or remote device (PLC, microcontroller, etc.). Explained in an easy way, they can be described as readers and writers.

Sub-application is another category of function block similar to composite FB, but they can be distributed to run on more than one resource. It is constructed from networks of basic FBs and composites FBs that also could contain sub-application of lower level inside. This type of block can be distributable [8].

### 6.2.3 *Execution of Function Block*

The execution model of a function block describes eight phases that defines the behaviour of a basic FB when it is running. In this case, a scheduling function is used in order to ensure that each phase is executed in a correct order and at the correct priority [8]. The phases are the illustrated in Fig. 6.4.

- Data inputs of the function blocks have a stable value;
- An event which is associated with a data input is arrived to the event input of the FB;
- The execution controller indicates to the scheduling function that it has a signal and it is ready to execute an algorithm;
- After a time, the scheduling function executes the algorithm;
- Algorithm produces an output value after processing the input values and if there are internal variables, which also can be changed;

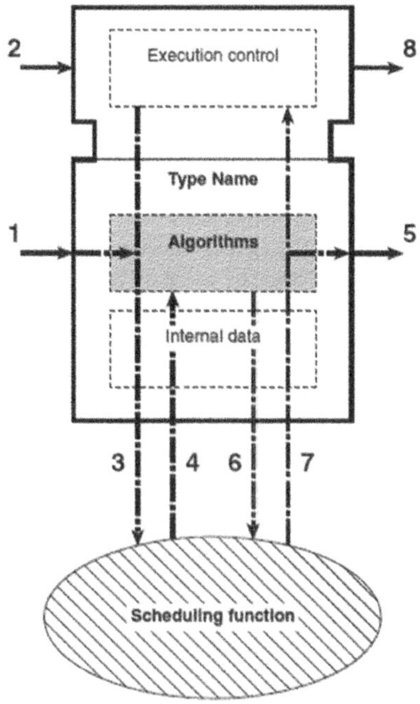


**Fig. 6.4** Execution of function block

- The internal algorithm sends a signal to the scheduling function notifying that the execution is finished;
- The scheduling function invokes the FB's execution control, informing that the algorithm has finished its execution;
- The execution control creates an output event in the FB's output event interface according to the execution of the internal algorithm.

### *6.2.4 Internal Behaviour of Function Block*

The internal behaviour of a basic FB takes into account the algorithm bodies and the algorithm execution control. A basic FB usually contains one or more algorithm; however, there are cases in which an FB only uses an ECC without any algorithm. Each algorithm is invoked by the scheduling function that depends on an event input. As it is explained before, one of the most important parts of the behaviour of an FB consists of the relation between the events and the algorithms, and this is joined by a concept named Execution Control Chart (ECC), the configuration of which was developed in this [8].

An Execution Control Chart (ECC) is the graphical or textual representation among the relations of Event Inputs, Event Outputs and algorithms [4], including execution control states, transitions and actions. An example of a basic function block and its ECC is depicted in Fig. 6.5.

IEC 61499 [8] defines the following characteristics for the ECC:

- it resides in the upper portion of an FB,
- it has one initial Execution Control (EC) state, represented graphically with a double outlined shape like START in Fig. 6.5,
- there will be one or more EC states, which are represented graphically with a single outlined shape, and the states could have one or more associated EC actions, and

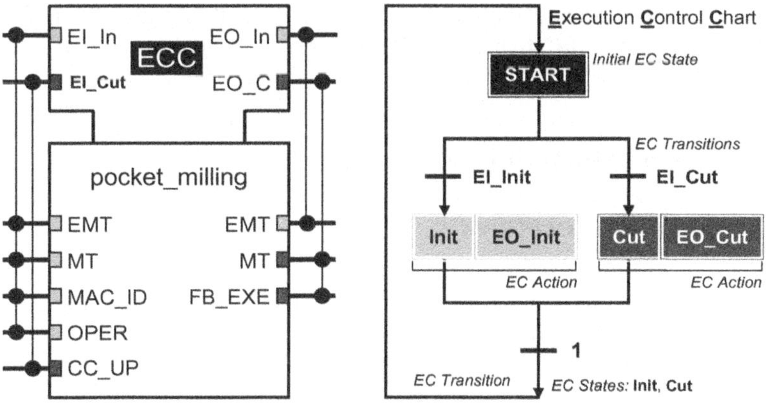


**Fig. 6.5** A sample basic function block and its ECC

- the ECC can use but not modify variables declared in the function block type specification.

Taking all these characteristics into account, it can be said that an ECC is divided into EC states, EC transitions and EC actions.

- EC states: An EC state is a part of the ECC which is defined by the IEC 61499 as the "situation in which the behaviour of a basic FB with respect to its variables is determined by the algorithm associated with a specified set of execution control actions". There are two types of EC states:

  - Initial EC state, and
  - Common EC state.

  The initial EC state is the state in which the ECC starts when it is executed in the beginning, and after an EC transition, the state will be changed to a common EC state, but also can be returned to the initial EC state. An EC state can have one or more EC actions.
- EC transitions: An EC transition is a Boolean expression, part of the ECC that allows "jumping" between an EC state to another. This Boolean expression can be used with Event Input variable, input variable, output variable, or internal variable.
- EC actions: An EC action is an "element associated to EC state that identifies algorithm(s) to be executed and event(s) to be issued upon completion of execution of the algorithm". The differences between EC states and EC actions can be seen in Fig. 6.5.

An ECC executes algorithms as "*finite set of well-defined rules for the solution of a problem in a finite number of operations*". The algorithms are invoked following some rules. When an FB is not executing any algorithm and an Event Input occurs, EC transitions in the ECC are evaluated following the active EC state. If there is not a true condition, no action will be performed, and the FB will wait until another Event Input arrives. If there is a true condition, EC action will be performed, and an algorithm could be invoked after a request to the scheduling function that will schedule the execution of the algorithm's operation. Once the actions are completed, EC transitions will be evaluated again.

## 6.3 Enriched Machining Features

### *6.3.1 Machining Features*

Machining features are mapped to FBs in distributed process planning (DPP, see Chap. 4 for details). Machining processes are carried out by executing the relevant FBs with MFs embedded.

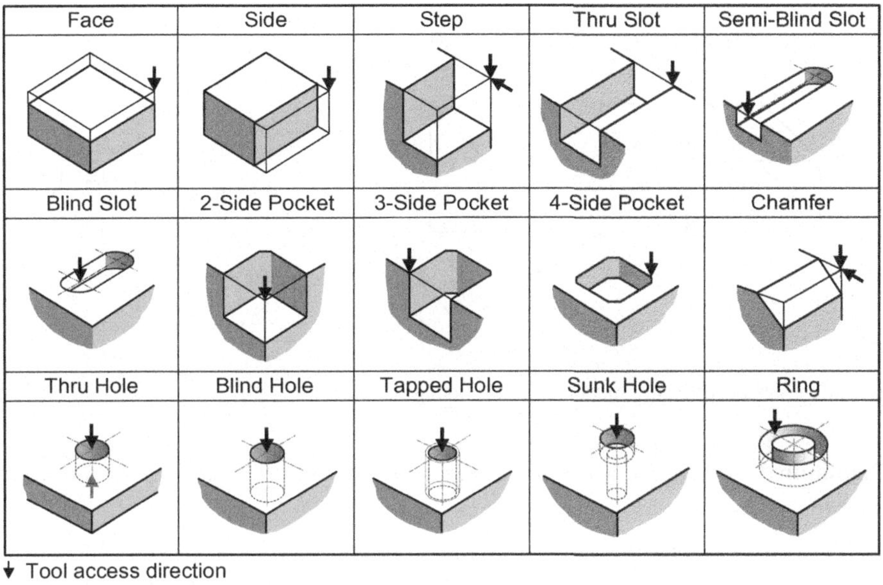


**Fig. 6.6** Typical machining features

Machining features are those shapes, such as step, slot, pocket and hole, which can be easily achieved by the available machining resources and defined machining technologies. Different from design features, as standard shapes that can be machined, each machining feature holds a set of loosely-coupled information on how to fabricate it, such as cutting tool type, machine-independent machining sequence, tool path generation logic, cutting strategy, and suggested cutting conditions, which provide an indication as to what kind of operations and tools will be required to manufacture the feature. Some typical machining features are shown in Fig. 6.6. Since milling and drilling operations are dominant in machining, only milling and drilling features on prismatic workpieces are covered in this chapter.

Within the context, each machining feature can be represented by its geometric feature, surface feature, volume feature, and loosely-coupled cutting information. A geometric feature is a topological unit that holds the main information of the machining feature itself, such as geometry, dimension, and tolerance; a surface feature captures the attributes and the relationship of faces defining the surface of the machining feature; a volume feature is the solid volume enclosed in the machining feature. Figure 6.7 shows the combined feature models of a machining feature *step*.

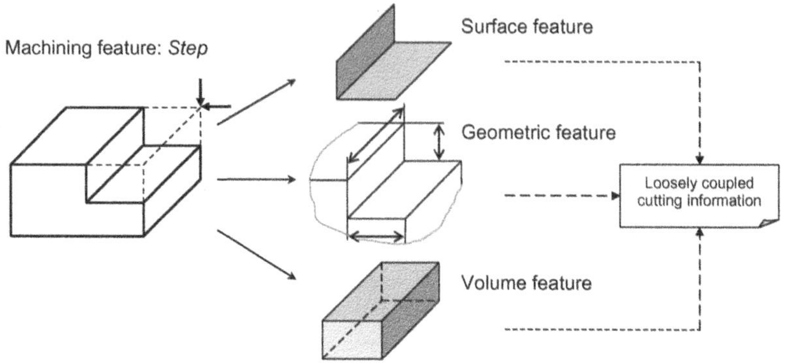


**Fig. 6.7** Embedded feature models of a machining feature

**Definition 1** *A machining feature is a machinable unit that satisfies the following conditions*:

- Its removal creates a portion of the part surface without destroying the part.
- It can be removed from the workpiece by one or more operations in one setup with a three-axis milling machine or machining centre.
- It not only contains design information, such as geometric dimensioning and tolerancing (GD&T), surface finishing, but also possesses a set of loosely-coupled information on how to fabricate it.
- It can be modelled by surface feature, geometric feature, volume feature, and supported by a loosely-coupled machining knowledge base.

Here, it is focused on machining process sequencing, and is assumed that the machining feature list of a part is given. Such a feature list can be obtained either by adopting third-party feature recognition solutions or incorporating machining feature based design methodology—designing a part in the same way of 'machining' by subtracting machining features from its blank.

## 6.3.2 *Enriched Machining Features*

### 6.3.2.1 Maximum Machining Volumes

Traditionally, machining (e.g. milling, turning) is the process that removes materials from its blank or raw material. Therefore, raw materials must be considered in process sequencing. The materials to be removed from a blank are in the shape of machining features. Their volumes are usually bigger than the desired ones as far as the blank is concerned. This chapter treats the task of machining process sequencing as the task of machining feature sequencing, if a machining feature list is given. In other words, it is the task of deciding how the materials (in the shape of

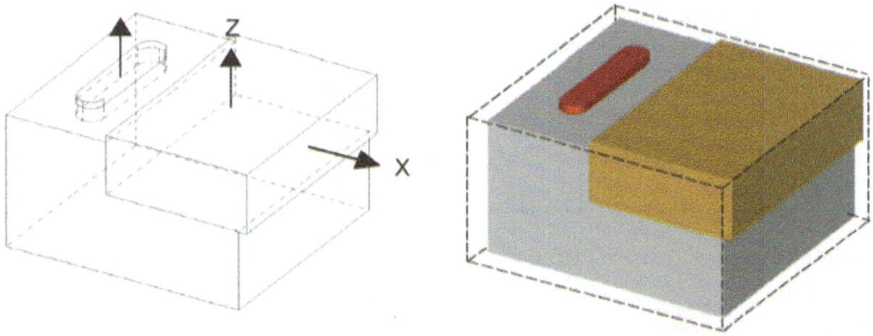


**Fig. 6.8** Maximum machining volumes

machining features) should be removed sequentially to achieve a final part. A concept of maximal feature for feature recognition is used here. This concept is extended as the maximum machining volume (MMV) for machining feature sequencing by combining the tool access directions of machining features in Fig. 6.6.

**Definition 2** *The MMV of a machining feature is the volume to be removed to create the machining feature directly from the raw material along its defined tool access direction(s) without destroying the part.*

Figure 6.8 shows the MMVs of two machining features (a step and a blind slot) for a given raw material (shown in dashed lines), which are generated by extending their volumes to the surfaces of the raw material along X and Z directions, accordingly. MMVs are used extensively in the enriched MF-based reasoning to identify a sequential order amongst machining features.

#### 6.3.2.2 Intermediate Machining Volumes

The machining features achieved from design side often cannot be used in process planning directly unless the part is simple in shape. The reason behind this is that the information of machining features is static, which only represents the final requirements of a part; whereas the machining process is rather dynamic. During the machining process, the shape of a workpiece keeps changing because of the material removal. To capture the dynamic change of a workpiece for feature recognition in 2.5D components, it considers a workpiece change as an intermediate workpiece for the next step feature recognition. The updated workpiece also helps in recognition of machining features for roughing, semi-finishing or finishing operations. Rather than updating the entire workpiece, it targets the dynamic change of each individual machining feature for machining sequence determination. A concept of intermediate machining volume (IMV) is therefore introduced to reflect the dynamic change of a machining feature during machining operations.

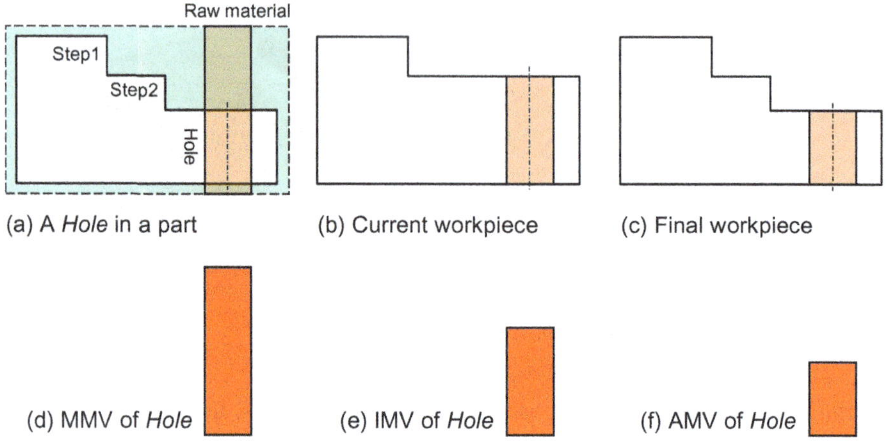


**Fig. 6.9** Intermediate machining volumes

**Definition 3** *An IMV of a machining feature is the intersection of its MMV and the current workpiece.*

Figure 6.9 shows the concept of IMV through a hole. The IMV of the machining feature *hole* varies between its MMV and its actual machining volume (AMV) along the machining process of the part. The upper limit of the IMV is the volume (or MMV) to be removed from the raw material; the lower limit of the IMV is the volume feature (or AMV) of this machining feature. Collectively, the change of IMVs of machining features demonstrates the change of a workpiece while the workpiece is gradually taking its shape during the machining operation. Together with MMVs, intermediate machining features play an important role in feature sequencing.

#### 6.3.2.3 Machining Feature Interference

Machining Limit Value (MLV) is a dimensional value that is related to both the minimum size of an MF and the interference size of the MF. MLV is used to decide the size of cutting tools and is obtained by comparing two parameters, the minimum size and the interference size of an MF [9]. The two parameters are defined as follows.

- *MF minimum size*: It includes several geometrical dimensions of material removal volume of MF, as shown in Fig. 6.10a. Also, the dimensional values are closely related to MF types, and for instance MF width and MF height are the MF minimum size of a semi-blind slot. They provide a reference for the decision of cutting tool diameter and cutting edge length.
- *MF interference size*: It refers to the relationship between the neighbouring MFs, and includes minimum horizontal distance (MHD) and minimum vertical

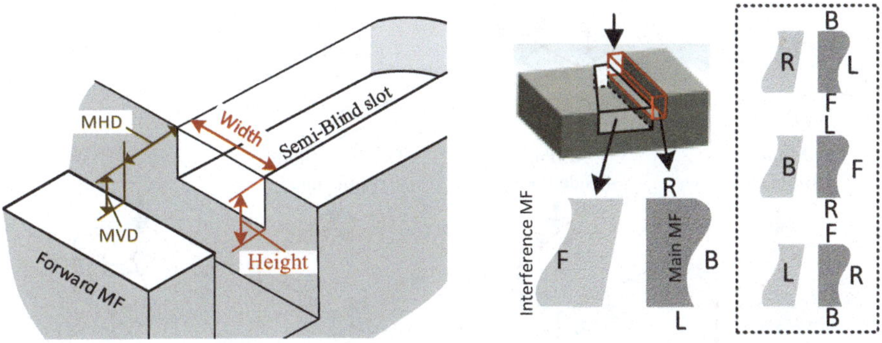


(a) Interference parameters between MFs (b) Position definition of interference

**Fig. 6.10** Machining feature interference

distance (MVD), as shown in Fig. 6.10a, the semi-blind slot MF and the forward MF. MHD is defined as the minimum distance at the horizontal direction between two neighbouring MFs, and MVD is defined as the minimum distance at the vertical direction between the two MFs. There are 4 sides of interference at most, i.e. forward (F), backward (B), rightward (R) and leftward (L), as shown in Fig. 6.10b.

MF minimum size and MF interference size are related to MFs. Based on the geometrical properties of all the MFs, the current MFs are divided into three classes: closed, semi-closed and open MFs, as shown in Fig. 6.11, respectively. MF minimum and interference sizes are thus determined. Consequently, based on the comparison of the two sizes, MLV can be calculated.

- *Closed MF*: Fig. 6.11a shows a series of closed MFs. Here, the MF is created by removing the material that is located in a closed space, i.e. interference MF does not exist. Therefore, MLV is only related to MF minimum size.
- *Semi-closed MF*: Fig. 6.11b shows four types of semi-closed MFs. In this class, an MF is created by removing the material that has a semi-closed surrounding space, i.e. there may be some interference MFs in the side of open space. Therefore, MLV depends on both the MF minimum size and the interference size.
- *Open MF*: Four types of open MFs, as shown in Fig. 6.11c, are generated by removing the material that is located in an open surrounding space. Therefore, MLVs are only decided by MF interference size.

MFs are widely used in design and manufacturing for the ease of information retrieval and processing. However, only the geometry information of MFs is insufficient for DPP due to the dynamic nature of underlying machining processes. The shape of a workpiece evolves at different stages along its machining process. In order to embody the dynamic changes of MFs in DPP, extra needed information is analysed and an EMF (enriched machining feature) concept is introduced.

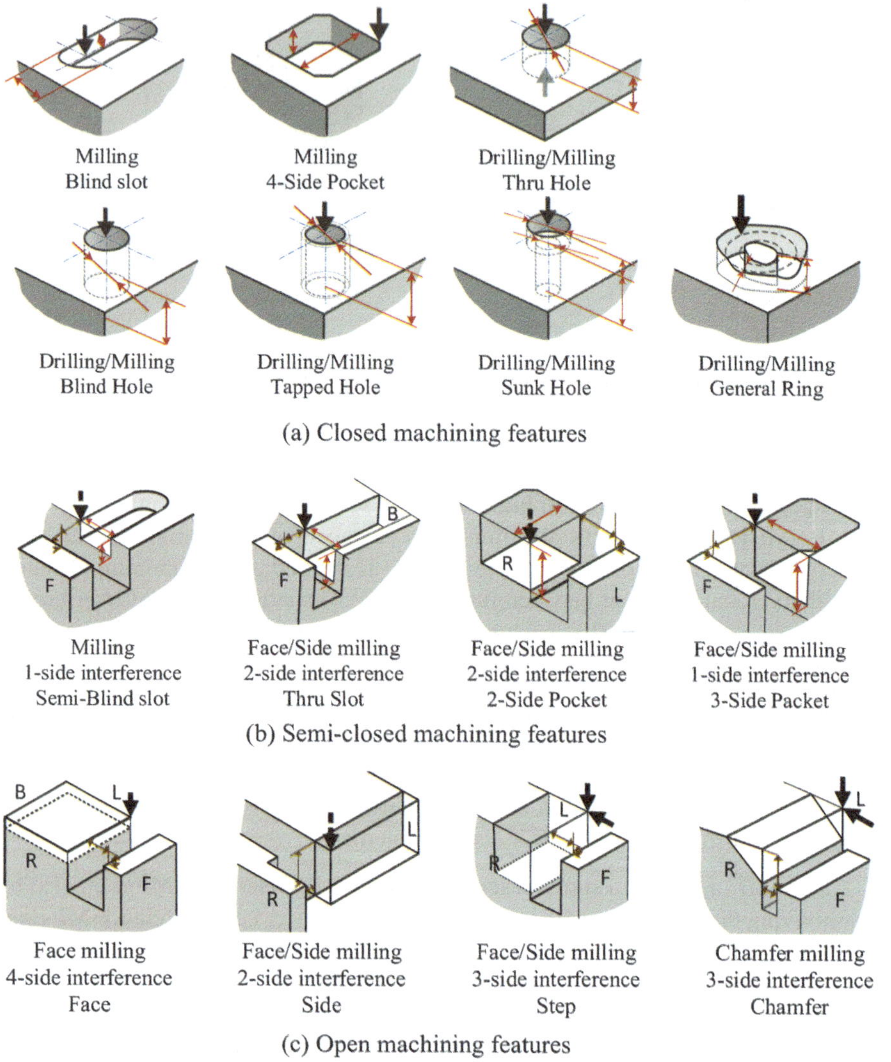


**Fig. 6.11** Machining feature classes

#### 6.3.2.4 Enriched Machining Features

For the purpose of effective process planning, particularly for machining process sequencing, the IMV is combined with each machining feature to reflect the dynamic change of its machining volume. A combined machining feature is named EMF in this research.

**Definition 4** *An EMF is an entity that possesses information of the machining feature itself (e.g. GD&T, suggested tool type, tool access directions, tool path generation logic and suggested cutting conditions,* etc.*) and its current IMV* [10].

### 6.3.2.5 Representation Scheme of an EMF

The detailed representation of an EMF, especially its surface feature, volume feature, and IMV, is formulated using the basic geometric entity—surface. Here, a surface refers to a basic individual face shape, such as planar surface, cylindrical surface. Jointly, they define the geometry of the EMF. A surface is termed *real* when the inside of its boundary is solid or *imaginary* when the boundary is enveloping an empty area.

Taking the four-side pocket F10 (Fig. 6.12) as an example, its detailed representation can be described as follows, where S27–35 are real surfaces and S101–111 are imaginary surfaces.

Representation Scheme of 4-side pocket {

Feature ID: F10;
Feature type: 4-side pocket;
Reference feature/Reference face: none;
Main surface: S27;

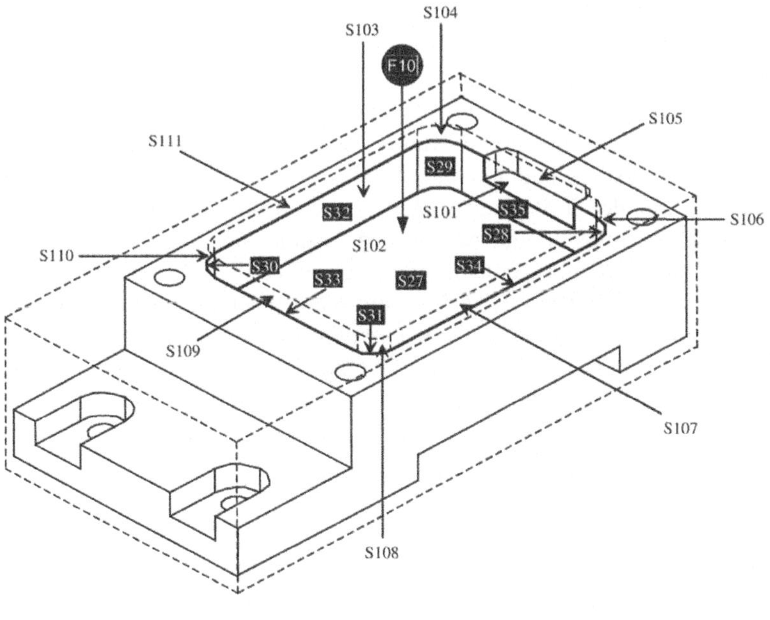


**Fig. 6.12** Test part with a four-sided pocket

Associated surfaces: S28, S29, S30, S31, S32, S33, S34, S35;
Length: 110, Tol.: 0.01;
Width: 80, Tol.: 0.01;
Depth: 20, Tol.: 0.01;
Corner radius: 10;
Volume feature: (S27, S28, S29, S30, S31, S32, S33, S34, S35, S101, S102);
Primary tool access direction: [0, 0, −1];
Secondary tool access direction: none;
Tool type: 2-Flute-End-Mill with diameter no more than 20;
Tool path generation logic: loop (wall) and zigzag (island);
Shell of IMV: (S27, S28, S29, S30, S31, S32, S33, S34, S35, S101, S103, S104, S105, S106, S107, S108, S109, S110, S111);
IMV: (S27, S28, S29, S30, S31, S32, S33, S34, S35, S101, S102, S102, S103, S104, S105, S106, S107, S108, S109, S110, S111);

} End of Representation Scheme.

## *6.3.3 Generic Machining Process Sequencing*

The task of machining process sequencing is treated as the task of EMF sequencing in this book, provided that a list of all machining features of a part is given. Within the context, the task can be accomplished by grouping the EMFs first into setups and then sequencing the EMFs in each setup using EMF-based reasoning approach.

### 6.3.3.1 EMF Grouping

In a mechanical design, functional requirements of a part are normally expressed by geometrical dimensions and tolerances. To eliminate as much machining error stack-up as possible, it is suggested that the machining features with certain functional relationships should be grouped together and machined in one single setup, based on an appropriate datum reference frame. A datum reference frame is a reference coordinate system used to secure other machining features in the same part, and is determined by the functional relationships (e.g.//, ⊥, ⊕, etc.) among the machining features. The EMF grouping includes three steps: choosing datum references, finding a primary locating direction, and grouping EMFs into appropriate setups.

- Step 1: *Choosing datum references.* One of the relationships among EMFs is the datum dependency precedence given in the representation scheme of an EMF as *reference feature* and/or *reference face*, which expresses the position, orientation or profile tolerance requirements of the EMF. By tracing the reference feature/face of each EMF, a primary datum reference frame and its dependency precedence of multiple datum references (if any) can be identified. The first item of the sorted results of datum dependency is the primary datum reference. The EMF grouping must be arranged according to the sequence of the datum reference frame and their dependency.

- Step 2: *Finding a primary locating direction*. Here, a primary locating direction is the surface normal $\vec{V}$ of the primary locating surface (*PLS*), which usually serves as the primary datum reference for determining the spatial position and orientation of a workpiece and constrains at least three degrees of freedom. It should align with or be orthogonal to the Z-axis of a machine tool, depending on the configuration of the machine tool.
- Step 3: *Grouping EMFs into appropriate setups*. Based on the primary locating direction $\vec{V}$ (setup orientation) determined in Step 2, the EMF grouping can be accomplished by searching for those EMFs whose tool access directions $T_{EMF}$ are opposite to $\vec{V}$, and grouping them into setup $ST_{\vec{V}}$.

To be generic, the setups at this stage are planned for 3-axis machines, as their configurations form the basis of other machines with more axes. In other words, the 3-axis-based setup planning makes a process plan generic and applicable to other machines with different configurations. However, a setup merging is required for 4- or 5-axis machines, after a specific CNC machine is selected. The machining environment where the workpiece is to be machined is considered as a constraint during setup merging. This setup merging is straightforward if a machine's configuration is given, and can be carried out by the Execution Controller before downloading them to specific machines. Fixturing information is also integrated with DPP. Specific fixturing constraints are input to DPP, against which each machining feature is checked during EMF grouping into a specific setup. Details on setup planning for fixturing are to be described separately.

#### 6.3.3.2 EMF Sequencing

The EMF sequencing normally consists of two parts: multiple setup sequencing and EMF sequencing within each setup. The issue of multi-setup sequencing is addressed implicitly when selecting locating directions (primary, secondary, etc.) for the EMF grouping, in terms of the generalised accuracy grade and critical datum reference. The true challenge of EMF sequencing is now shifted to how to sequence EMFs within each setup, when their machining sequence cannot be determined simply by the datum relationships and manufacturing constraints among the EMFs. An EMF-based geometric reasoning approach is adopted by tracking and comparing the IMV against AMV (or volume feature) of each EMF. By applying the following five reasoning rules sequentially, a machine-neutral sequence plan with multiple setups can be created. For example, in the case shown in Fig. 6.9, the IMV of the *hole* varies between its MMV and its AMV along the machining process. As a rule of thumb, if the IMV of an EMF equals the AMV of the EMF, it is the time to machine the EMF.

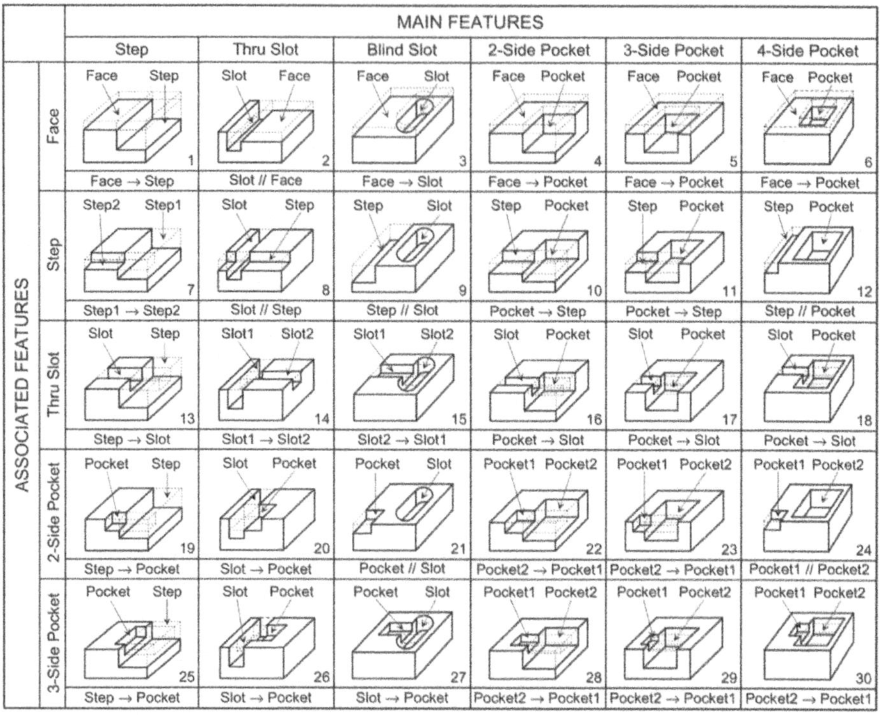


**Fig. 6.13** Enriched machining feature sequencing results after applying Rule 1

**Rule 1** *During sequencing, when the IMV of a machining feature equals the AMV of the machining feature or IMV = AMV, this machining feature is ready for machining.*

Applying Rule 1 to the case shown in Fig. 6.9, it is easy to conclude a sequence of Step 1 → Step 2 → Hole for machining. Figure 6.13 shows 30 typical cases after applying Rule 1.

This rule works effectively for EMF sequencing with feature interactions. However, after applying Rule 1, there still exist some cases that cannot be handled by this rule, in which the sequence of two machining features is remained in parallel (as shown in Fig. 6.13), such as Case 8: Thru slot + Step. In this case, if the thru slot is cut first, the Step will be divided into two smaller ones, which is against the definition of a machining feature being *a basic single machinable shape*.

**Rule 2** *If the IMV of machining feature A is to be divided into more than one piece as a result of the machining operation of machining feature B, the machining feature A should be cut first.*

In addition to the feature-splitting case encountered in Rule 2, there are cases that incorrect sequences may result in different types of machining features, such as Case 2: Slot + Face. In this case, if the Face is milled first, the Face feature in

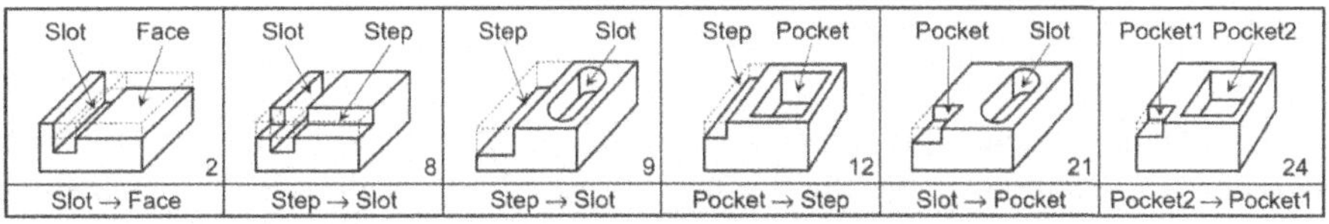


**Fig. 6.14** Sequenced results of the six parallel cases

machining is actually changed to a Step. This is not allowed, as different EMF types require different machining data (tool type, tool access direction, and tool path pattern, etc.). Rule 3 is therefore established to prevent such ill cases.

**Rule 3** *If a machining feature is to be changed to another feature type as a result of its own machining operation, this machining feature is not ready and should be cut later.*

The remaining parallel cases after applying Rule 1 to Rule 3 do not have feature interactions and their machining sequences are non-critical. They are further handled by adopting the knowledge of best practices or know-how of operators. One rule commonly used by machinists is that the bigger volume is to be removed first, because removing a bigger volume generally produces greater cutting force and cutting heat that result in more deformation and poorer surface quality, especially for large workpieces.

**Rule 4** *A bigger machining volume is to be cut first.*

Figure 6.14 shows the EMF sequencing results after applying the reasoning Rules 2–4 to those parallel cases remained in Fig. 6.13.

Although Rules 1–4 are applied sequentially during EMF sequencing, Face and Side features are handled differently, except Case 2. These two types of EMFs usually cover large surface areas and are frequently used as datum references. They are normally removed first in each setup. In addition, the tool type information embedded in each EMF is used to group the sequenced EMFs in each setup to minimise the tool-change time.

**Rule 5** *In a setup, the machining features sharing the same tool types are grouped into clusters.*

By applying the five rules, a machine-neutral sequence plan can be generated. These rules cover all critical EMF sequences of a prismatic part. The remained parallel sequences, if any, are non-critical and will be up to the controller-level operation planning to determine.

## 6.4 Adaptive Machining Feature Sequencing

For adaptive machining feature sequencing, a reachability-based approach [11] is introduced in this section.

### *6.4.1 Reachability-Based Approach*

In graph theory, reachability refers to the ability to get from one vertex to another in a graph. Using similar logic, the reachability-based approach uses the reachability matrix for representing an MF sequence in a graph structure. Reachability matrix is developed from the adjacency matrix that is formed from the MFs path graph [12]. The method has been used in operation sequencing, considering different aspects of operations, e.g. dimensional tolerance, geometrical tolerance, surface finish, accuracy, and cost [13, 14]. In order to use the reachability matrix for MF sequencing, it is necessary to introduce two basic application conditions:

- MFs obtained from the workpiece can be regarded as the nodes in a path graph, and
- The path can represent the relationship between MFs.

Based on the two conditions, MF path graph can be mapped. It is assumed that MFs have already been obtained during MF recognition. Thereafter, the path can be mapped according to the relationships of MFs. Figure 6.15 shows the reachability based MF sequencing approach, where MF path graph, adjacency and reachability matrices are the three key elements. First, MF path graph is mapped according to the four mapping principles (MPs), i.e. Basic MP, Cutting Tool MP, Cross-Setup MP, and Particular Requirement MP. These MPs are derived from the MF information of the workpiece, selected cutting tools, setup plans, and machining requirements. Adjacency matrix can then be obtained from the MF path graph. Reachability matrix can thus be calculated based on the generated adjacency matrix. In addition, if there are dynamic links due to the similarity of cutting tools in MF path graph, the remaining MFs will be rescheduled after the relevant MF is cut. Moreover, if "No Pointing End-MFs (NPE-MFs)" exist in the MF path graph, they need to be sequenced through a two-step NPE-MF sequencing process. Finally, combining all the results, the final sequence strategy can be determined.

Given that the reachability-based method aims to provide the MF sequence after setup planning, the test part and setup plans are depicted beforehand to make the process clearer. Figure 6.16 shows (a) a test part with 29 MFs, and (b) 3-axis and 4-axis based setup plans and the MFs under the same cutting tool. In SP of the DPP

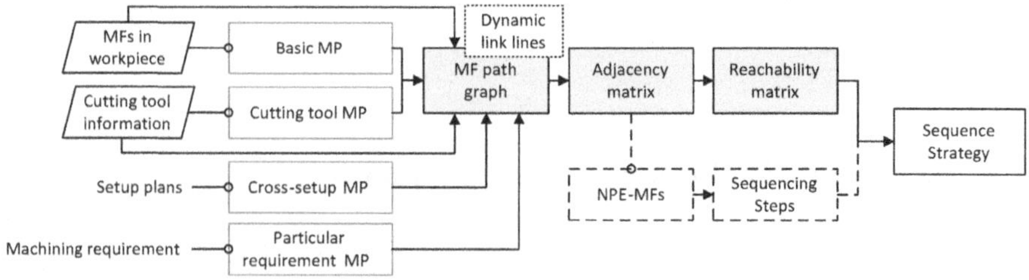


**Fig. 6.15** Reachability based MF sequencing method

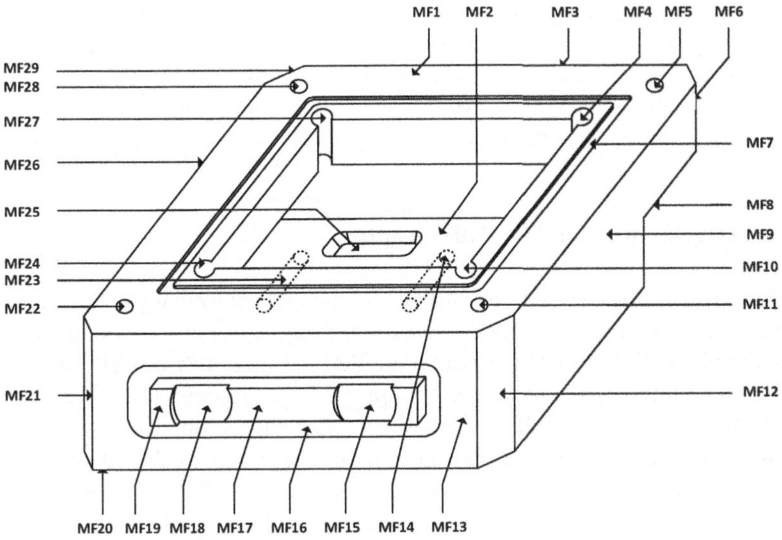


(a) A test part consisting of 29 machining features

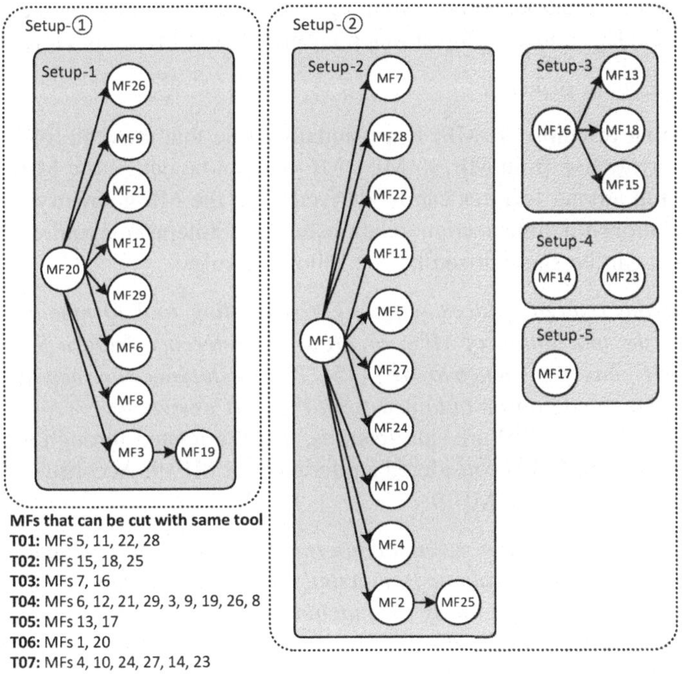


(b) Setup plans and MFs with similar cutting tool

**Fig. 6.16** Information of the test part and its setup plans

system, the setup plan is generated for a 3-axis machine tool, as Setups-1, 2, 3, 4, 5 shown in Fig. 6.16b. In the execution control process, a 4-axis machine tool is chosen. Therefore, after adaptive setup merging, the test part can be cut in 2 setups; Setup-① and Setup-②. Here, Setup-① only consists of Setup-1, while Setups 2, 3, 4, and 5 are grouped into Setup-②.

#### 6.4.1.1 Development of MF Path Graph

MF path graph is generated according to the MF information, e.g. geometry information, tolerance requirement, and machining operation. Different rules have been defined for MF sequencing which are derived directly from MF information. These rules are the key references for mapping MF path graph. Once these rules are applied, the precedence of MFs among each other will be obtained where the MFs with higher precedence for machining will be connected to the MFs with lower precedence using an outgoing arrow. As a result, the MF path graph can be obtained. Note that after applying these rules, some MFs can be grouped together. In this case, the sequencing approach discussed here will be performed on the grouped MFs. However, within the group, the sequence of MFs is determined by their precedence relation before grouping and/or the volume rule (explained in Sect. 6.3.3.2). The four developed mapping principles (MPs) are as follows.

(1) *Basic mapping principle*

Basic mapping principle (B-MP) is a mandatory rule that must be followed, i.e. if there is an arrow line from $MF_1$ to $MF_2$, $MF_2$ cannot be cut before $MF_1$. This type of relationship among features can be derived from the MF geometry information, position relationship, intersection relationship and tolerance requirement. These requirements can be summarised by the following rules:

**Rule B1** *If the reference faces required by a cutting tool to machine $MF_1$ are removed by the machining of $MF_2$, or $MF_1$ is a reference feature for machining $MF_2$, then $MF_1$ has precedence to $MF_2$* [15]. *These references are usually defined at the design stage and can be obtained from the part design.*

In Fig. 6.17, MFs 3, 19 are side features, and are related through a parallelism requirement where MF3 is considered a reference for MF19. According to Rule B1, MF3 should be cut before MF19.

**Rule B2** *During sequencing, when the intermediate machining volume (IMV) of an MF equals the machining volume (the actual machining volume: AMV) of the MF, or IMV = AMV, this MF is ready for machining* [16].

In Fig. 6.18, MF1 (face) and MF2 (4-side pocket) can be sequenced using Rule B2. Before MF1 is cut, IMV of MF2 is the sum of the volumes shown inside the green and blue lines. Therefore, IMV > AMV (only the volume shown in blue lines) and as a result, MF2 cannot be cut. Whereas after MF1 is cut, IMV of MF2 would be the blue part (equal to AMV) and MF2 can be cut. Hence, MF1 should be cut before MF2.

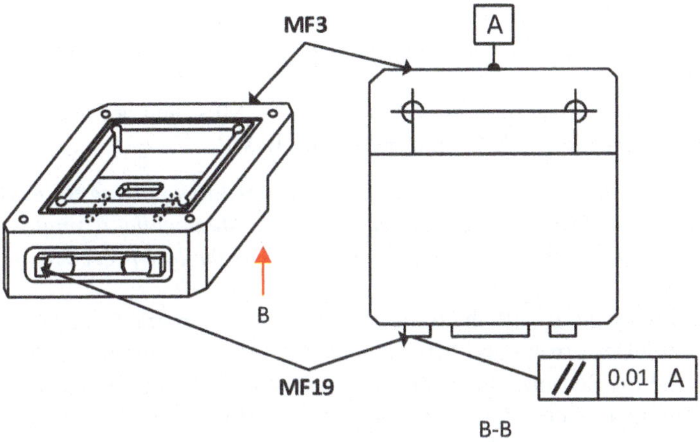


**Fig. 6.17** Sequencing machining features with reference relationship

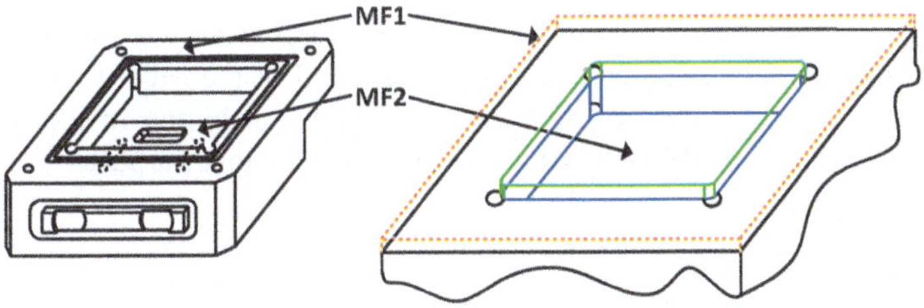


**Fig. 6.18** Machining feature sequencing related to intermediate machining volume

**Rule B3** *If the IMV of $MF_1$ is to be divided into more than one piece as a result of the machining operation of $MF_2$, then $MF_1$ should be cut first* [16].

In Fig. 6.19, MF18 (brown area) is a hole, and MF16 (red area) is a slot. If MF18 is cut first, MF16 would be divided into two parts. Hence, after applying Rule B3, MF16 should be cut before MF18.

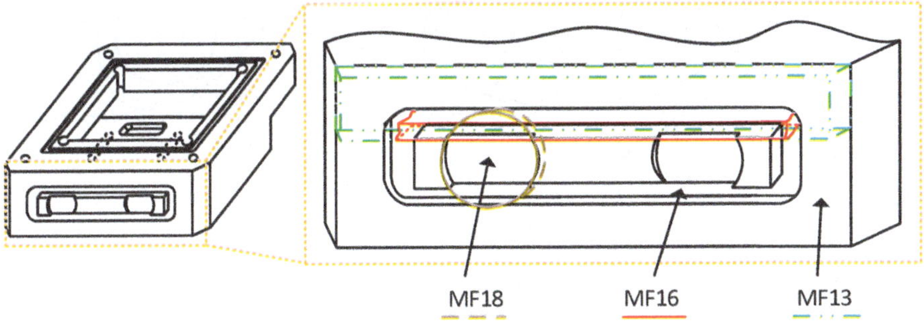


**Fig. 6.19** Machining feature sequencing related to intersection

**Rule B4** *If an MF is to be changed to another feature type as a result of its own machining operation, this machining feature is not ready and should be cut later* [16].

Rules B1-B4 specify the basic relationships of MFs, and are used to determine the sequence between the MFs and obtain the B-MP.

**B-MP** *If the sequence of two MFs can be determined by applying Rules B1, B2, B3 and B4, these two MFs can be mapped by B-MP. Meanwhile, a single-arrow line should be mapped from the high-precedence MF to the low-precedence one in the MF path graph.*

After applying the B-MP, the basic schema of MF path graph can be obtained, as shown by the black arrows and MF elements in Fig. 6.20. Moreover, all of MFs can be linked and divided into different layers. The first MF(s) to be machined will be placed in the first layer, its neighbouring MFs are located in the second layer, and other layers can be obtained in a similar way, as shown in Fig. 6.20.

(2) *Particular requirement mapping principle*

Each workpiece consists of different particular structures which require specific machining strategies different from the traditional approaches e.g. machining of thin walls or material splice structures. In such a case, all of the MFs involved in those structures need to be cut together with special machining methods. Hence, Particular Requirement MP (PR-MP) is introduced for this purpose.

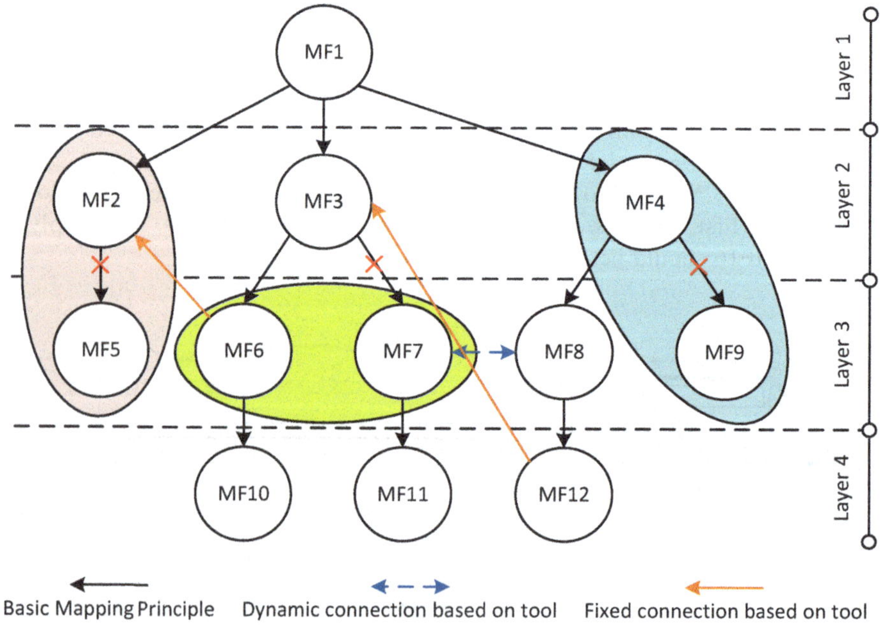


**Fig. 6.20** Machining feature path graph

**PR-MP** *All MFs involved in a particular structure should be merged together and treated as one unique MF in the sequencing process. Therefore, there is no need to map arrows among these MFs.*

The MF2 → MF5 area in Fig. 6.20 illustrates one scenario where PR-MP is applied. After applying PR-MP, related MFs are represented by the related MF located in the highest layer (MF2 in this case), i.e. all of the low-layer MFs are removed in the sequencing process. Also, all of the arrows between these MFs are removed. In addition, all of the arrows pointing to these MFs need to be mapped to the related MF located in the highest layer. As shown in Fig. 6.20, both MF5 and the arrow connecting MF2 to MF5 are removed from the sequencing process.

(3) *Cutting tool mapping principle*

In the current manufacturing environment, more and more multi-functional cutting tools can be found in the market. As a result, a single type of cutting tool is capable of machining different types of MFs. This issue should be considered in the MF sequencing process in order to reduce the number of tool changes. Also, sometimes different unexpected conditions might occur during the machining process, especially when a cutting tool is broken and no similar cutting tool is available to replace it. In this case, the MF sequence needs to be changed in case another type of tool is selected for machining. Hence, the cutting tool MPs should be considered. However, depending on the positions of the involved MFs (similar or different layers), different types of mappings have to be taken into account.

- Same layer mapping principle

When the MFs in the same layer could be cut with the same cutting tool, Same Layer Mapping Principles (SL-MPs) should be used.

**SL-MP 1** *If MFs in the same layer could be cut with the same cutting tool, and have the same up-layer MF, they should be grouped as one MF for sequencing, and represented by the smallest ID.*

The yellow area in Fig. 6.20 shows a scenario where SL-MP 1 can be applied. After applying this MP, MF6 represents the merged MF6 and MF7. In addition, the arrows pointing to those MFs that should be removed from the sequencing process are removed (e.g. the arrow from MF3 to MF7).

As shown in Fig. 6.16b, MFs 6, 12, 21, 29, 3, 9, 26, 8 are in Setup-1, positioned in the same layer, and can be machined by the same cutting tool T04. Therefore, according to SL-MP 1, they can be treated as one unique MF in the sequencing process. Hence, the MF path graph for Setup-1 can be revised as shown in Fig. 6.21, where MF3 represents MFs 6, 12, 21, 29, 3, 9, 26, 8. Thereafter, it is easy to achieve the MF sequence, MF20 → MF3 (MF3, 6, 12, 21, 29, 9, 26, 8) → MF19.

**SL-MP 2** *If MFs in the same layer could be cut with the same cutting tool, and have different up-layer MFs, they should be linked with a dynamic two-arrow line. The sequencing strategy is that as long as one of the linked MFs can be cut (i.e. one of the up-layer MFs has been cut), the two-arrow line among linked MFs should be changed to a single-arrow line pointing from the second MF to the current*

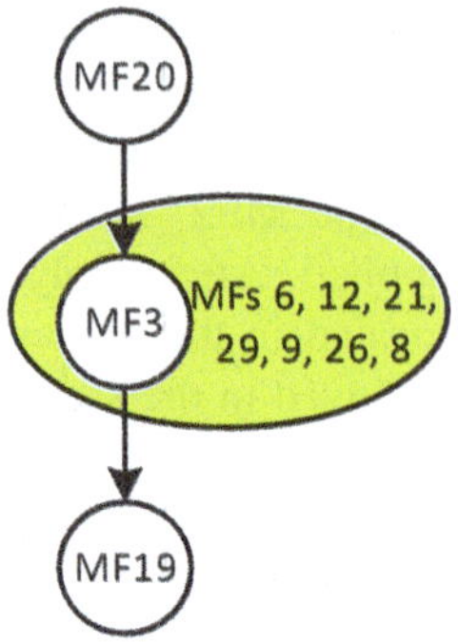


**Fig. 6.21** Path graph of MFs in Setup-1

*machinable MF. In this case, MF sequence should be rescheduled where a first-layer MF has to be selected among the remaining MFs. Generally, first-layer MF should be the MF with the least reference lines (arrows pointing to it from other MFs). However, if there were more than one MF with the same least reference lines, the Volume Rule should be used.*

**Volume Rule** *A bigger machining volume is to be cut first between MFs.*

In Fig. 6.20, MF7 and MF8 can be cut with the same tool and have different up-layer MFs. A dynamic arrow (dotted blue line) represents their relationship. In case MF3 (or MF4) is cut, the rescheduling process should be targeted (SL-MP 2) and the first-layer MF has to be selected. If MF3 or MF4 is cut, both MF4/MF3 and MF2 can be candidates for first-layer MF since they have similar least number of reference lines, i.e. 1. As can be seen, since both MFs have similar reference lines, the Volume Rule should be used for selecting the first-layer MF. Therefore, the MF with the largest machining volume will become the first-layer MF. Finally, the sequencing process will be carried out based on the new MF path graph.

- Different layer mapping principle

When the MFs in different layers could be cut with the same cutting tool, different layer mapping principles (DL-MPs) should be used.

**DL-MP 1** *If two MFs are in neighbouring layers, and there is a single-arrow line between them, they should be grouped together for sequencing.*

For example, as shown in Fig. 6.20, MF4 and MF9 can be cut with the same cutting tool; they are merged together (the blue area) and represented by MF4 for the sequencing purpose.

**DL-MP 2** *If two MFs are in neighbouring layers, and there is no line between them, a single-arrow line should be mapped from the lower-layer MF to the upper-layer MF.*

As shown in Fig. 6.20, MF2 and MF6 could be cut using the same tool, and therefore a single-arrow line (orange line) is mapped from MF6 to MF2.

**DL-MP 3** *If two MFs are not in neighbouring layers, and there are no path ways from the upper-layer to the lower-layer, a single-arrow line should be drawn from the lower-layer MF to the upper-layer MF.*

In Fig. 6.20, MF3 and MF12 can be cut with the same cutting tool, and a single-arrow line (orange line) is therefore mapped from MF12 to MF3.

(4) *Cross-setup mapping principle*

In the DPP system, the setup plan in SP is for 3-axis based setups; therefore, when a 4-axis or 5-axis machine tool is selected, some of the 3-axis setups need to be merged in a new setup [17]. In this case, MF sequence needs to be adjusted, and all of the first-layer MFs of the merged 3-axis setups can be the candidates of the first-layer MF in the new setup. Hence, Cross-Setup Mapping Principle (CS-MP) is introduced to decide the first-layer MF among the first-layer MFs in all of the original setups.

**CS-MP** *If MF sequencing is carried out for cross-setups, the first-layer MF in a new setup should be the one with the most links among the first MFs of the original setups. In other words, the MF that has precedence to more MFs should be selected in order to ensure that more MFs can be exposed for sequencing. In case there are more than one MF with similar number of the most lines, Volume Rule should be applied and the biggest machining volume among the MFs should be selected as the first-layer MF.*

In Fig. 6.16, in the new Setup-②, MF1, MF16, MF14 (23), MF17 are the first-layer MFs in the original Setups-2, 3, 4, 5. MF1 has the most linked lines, and after applying CS-MP, it becomes the first-layer MF in the new Setup-②.

#### 6.4.1.2 Basic Algorithm of Reachability Matrix

(1) *Adjacency matrix*

Adjacency matrix $\boldsymbol{M}$ is derived from the generated MF path graph, and can be used for calculating the reachability matrix $\boldsymbol{M_r}$. Matrix $\boldsymbol{M}$ is shown as follows:

$$\begin{array}{c} \\ 1 \\ \vdots \\ n \end{array} \begin{array}{c} \begin{array}{ccc} 1 & \cdots & n \end{array} \\ \begin{bmatrix} x_{11} & \cdots & x_{1n} \\ \vdots & x_{ij} & \vdots \\ x_{n1} & \cdots & x_{nn} \end{bmatrix} \end{array} \tag{6.1}$$

where $x_{ij}$ is either 0 or 1, depending on the relationship between MFs $i$ and $j$. When there is an arrow line from MF $i$ to $j$, $x_{ij} = 1$; otherwise, $x_{ij} = 0$.

(2) *Reachability matrix calculation*

Based on the adjacency matrix $\boldsymbol{M}$, the reachability matrix $\boldsymbol{M_r}$ can be obtained through the step-by-step procedure explained below. To compute the matrix $\boldsymbol{M_r}$

from the adjacency matrix, the 0 and 1 elements will be processed using the Boolean arithmetic concept which differs from ordinary arithmetic only in that:

$$1 + \text{anything} = 1 \tag{6.2}$$

In general, to go from $\boldsymbol{M}_{k-1}$ to $\boldsymbol{M}_k$ for every row in $\boldsymbol{M}_{k-1}$ that has a 1 in column $k$, add row $k$ to that row. Continue until $\boldsymbol{M}_n$ is obtained, and the reachability matrix $\boldsymbol{M}_r = \boldsymbol{M}_n$.

(3) *Sequence calculation*

According to the reachability matrix, the sum $\boldsymbol{S_R}$ of every row can be calculated, and the MF sequence can be scheduled by following the value of $\boldsymbol{S_R}$ from the biggest to the smallest.

$$\begin{array}{cccc} & 1 \quad \cdots \quad n & & S_R \\ \begin{matrix} 1 \\ \vdots \\ n \end{matrix} & \begin{bmatrix} y_{11} & \cdots & y_{1n} \\ \vdots & y_{ij} & \vdots \\ y_{n1} & \cdots & y_{nn} \end{bmatrix} & & \sum_{j=1}^{n} y_{ij} \end{array} \tag{6.3}$$

where $y_{ij} = 0$ or 1, based on the calculation result of reachability matrix.

#### 6.4.1.3 Sequencing for no Pointing End Machining Feature

No Pointing End Machining Features (NPE-MF) are located at the end of MF path graph. Therefore, they do not have any outgoing arrows (precedence) to other MFs and have only one incoming arrow from another MF, e.g. MF10 and MF11 shown in Fig. 6.20. The sequence of these MFs cannot be obtained by reachability matrix calculation.

In addition, NPE-MFs can also be obtained from the final adjacency matrix (without dynamic two-arrow lines in the path graph) and according to the every-row sum $\boldsymbol{S_{AR}}$ and every-column sum $\boldsymbol{S_{AC}}$. Any MF with the following condition: $\boldsymbol{S_{AR}} = 0$ and $\boldsymbol{S_{AC}} = 1$, is considered a NPE-MF. Therefore, checking $\boldsymbol{S_{AR}}$ and $\boldsymbol{S_{AC}}$ of an MF is one way of recognising NPE-MFs, which can easily be performed by computer.

$$\begin{array}{cccc} & 1 \quad \cdots \quad \text{n} & & S_{AR} \\ \begin{matrix} 1 \\ \vdots \\ n \end{matrix} & \begin{bmatrix} x_{11} & \cdots & x_{1n} \\ \vdots & x_{ij} & \vdots \\ x_{n1} & \cdots & x_{nn} \end{bmatrix} & & \sum_{j=1}^{n} x_{ij} \\ S_{AC} & \sum_{i=1}^{n} x_{ij} & & \end{array} \tag{6.4}$$

The sequence of NPE-MFs cannot be obtained from the calculation of the reachability matrix. Therefore, the sequence needs to be determined according to the characteristics of those MFs. First, it is important to consider whether they are in the same 3-axis based setup in order to reduce the number of setup changes. Second, if they are not in the same setup, it has to be determined whether they can be cut with the same tool in order to reduce the number of tool changes. The following steps have been adopted for sequencing NPE-MFs while considering the above concerns.

- **Step 1**: If there are NPE-MFs in a setup including the last MF (the last MF that can be sequenced before this step), the machining should start from the MFs without the same tool relationships with other NPE-MFs. The sequence for them is organised from the biggest machining volume to the smallest one. Then the MFs with the same tool should be cut immediately.
- **Step 2**: For the remaining MFs mentioned in Step 1, if the last MF is in a different setup from Step 1, apply Step 1 again. If the last MF is in the setup of Step 1, the MFs with the biggest machining volume should be cut first.
- **Step 3**: If there are MFs that can be machined by the same tool of the last machined MF, all of those MFs should be cut accordingly, before re-applying Step 1. Otherwise, go back to Step 1.

After applying Steps 1–3, the sequence of all of the NPE-MFs can be obtained.

## 6.4.2 *Case Study*

In order to demonstrate the effectiveness of the proposed method in this chapter, it is necessary to compare it with the DPP approach [18]. The test part depicted in Fig. 6.16a is used for the case study. The MF sequence of the test part and the number of tool changes are calculated and presented for both methods.

### 6.4.2.1 Current Method

MF sequence in DPP can be generated based on the rules within 3-axis setup in SP, which does not change during execution control. Figure 6.16b shows the results of the setup plan, where Setups-1, 2, 3, 4 and 5 are based on 3-axis machining.

- In Setup-1, the MF sequence is MF20 → MF3 (including MFs 6, 12, 21, 29, 9, 26, 8) → MF19, as shown in Fig. 6.21. Two types of cutting tools are needed. Therefore, the number of tool changes $C_1$ is 2.
- In Setup-2, the MF sequence is MF1 → MF2 → (MFs 4, 10, 24, 27) → MF25 → MF7 → (MFs 5, 11, 22, 28), and six types of cutting tools are used. Therefore, the number of tool changes $C_2$ is 6.

- In Setup-3, the MF sequence is MF16 → MF13 → (MFs 15, 18), which can be machined by three types of tools. Therefore, the number of tool changes $C_3$ is 3.
- Similarly, in Setup-4, MFs 14 and 23 can be cut by the same tool. Therefore, the number of tool change $C_4$ is 1.
- In Setup-5, only one MF (MF17) exists, which uses a different tool from the one used in Setup-4. Therefore, the number of tool change $C_5$ is also 1.

After setup merging, Setup-① and Setup-② are obtained, as shown in Fig. 6.16b. According to the MF sequences in the original setups, the new MF sequence can be determined as follows:

- Setup-①: MF20 → MF3 (MFs 6, 12, 21, 29, 9, 26, 8) → MF19
- Setup-②: MF1 → MF2 → MF4 (MFs 10, 24, 27) → MF25 → MF7 → MF5 (MFs 11, 22, 28) → MF16 → MF13 → MF15 (MF18) → MF14 (MF23) → MF17.

#### 6.4.2.2 Reachability Matrix

Setup-① only consists of Setup-1, and MF sequence is shown in Fig. 6.16b. Therefore, the number of tool changes is 2.

Setups-2, 3, 4 and 5 are merged into Setup-②. In this case, MFs sequence is to be adjusted. According to the present method, the MF path graph should be mapped first. Figure 6.22 shows the MF path graph obtained by applying B-MP, SL-MP 1-2, and DL-MP 2 in CT-MPs.

Adjacency matrix can be obtained, as shown in Eq. (6.5), according to the relationship of MFs shown in Fig. 6.22.

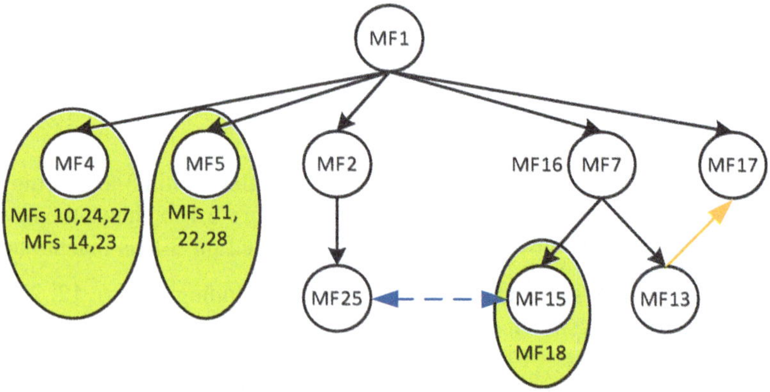


**Fig. 6.22** MF path graph in Setup-②

$$\begin{array}{c} \\ MF1 \\ MF2 \\ MF4 \\ MF5 \\ MF13 \\ MF15 \\ MF7 \\ MF17 \\ MF25 \end{array}
\begin{array}{c} \begin{array}{ccccccccc} MF1 & MF2 & MF4 & MF5 & MF13 & MF15 & MF7 & MF17 & MF25 \end{array} \\
\left[\begin{array}{ccccccccc}
0 & 1 & 1 & 1 & 0 & 0 & 1 & 1 & 0 \\
0 & 0 & 0 & 0 & 0 & 0 & 0 & 0 & 1 \\
0 & 0 & 0 & 0 & 0 & 0 & 0 & 0 & 0 \\
0 & 0 & 0 & 0 & 0 & 0 & 0 & 1 & 0 \\
0 & 0 & 0 & 0 & 0 & 0 & 0 & 0 & 0 \\
0 & 0 & 0 & 0 & 0 & 0 & 0 & 0 & 0 \\
0 & 0 & 0 & 0 & 1 & 1 & 0 & 0 & 0 \\
0 & 0 & 0 & 0 & 0 & 0 & 0 & 0 & 0 \\
0 & 0 & 0 & 0 & 0 & 0 & 0 & 0 & 0
\end{array}\right] \end{array} \tag{6.5}$$

Based on the adjacency matrix and the reachability matrix calculation steps, the reachability matrix can be determined as follows:

$$\begin{array}{c} \\ MF1 \\ MF2 \\ MF4 \\ MF5 \\ MF13 \\ MF15 \\ MF7 \\ MF17 \\ MF25 \end{array}
\begin{array}{c} \begin{array}{ccccccccc} MF1 & MF2 & MF4 & MF5 & MF13 & MF15 & MF7 & MF17 & MF25 \end{array} \\
\left[\begin{array}{ccccccccc}
0 & 1 & 1 & 1 & 1 & 1 & 1 & 1 & 1 \\
0 & 0 & 0 & 0 & 0 & 0 & 0 & 0 & 1 \\
0 & 0 & 0 & 0 & 0 & 0 & 0 & 0 & 0 \\
0 & 0 & 0 & 0 & 0 & 0 & 0 & 0 & 0 \\
0 & 0 & 0 & 0 & 0 & 0 & 0 & 1 & 0 \\
0 & 0 & 0 & 0 & 0 & 0 & 0 & 0 & 0 \\
0 & 0 & 0 & 0 & 1 & 1 & 0 & 1 & 0 \\
0 & 0 & 0 & 0 & 0 & 0 & 0 & 0 & 0 \\
0 & 0 & 0 & 0 & 0 & 0 & 0 & 0 & 0
\end{array}\right] \end{array}
\begin{array}{c} SUM \\ 8 \\ 1 \\ 0 \\ 0 \\ 1 \\ 0 \\ 3 \\ 0 \\ 0 \end{array} \tag{6.6}$$

According to the reachability matrix, the sequence is derived as: MF1 → MF7 → MF2 or MF13 → MF4 (MFs 5, 15, 17, 25), on the basis of which, after machining MF1 and MF7 (MFs 7 and 16), MF15 can be cut. In this case, SL-MP 2 is applied so that the remaining MFs are adjusted and the dynamic line is changed to a single-arrow line from MF25 to MF15. Also, MF13 and MF17 are grouped according to SL-MP 1, and MF13 (MFs 13, 17) becomes the first-layer MF based on SL-MP 2. Therefore, the new MF path graph can be mapped, as shown in Fig. 6.23.

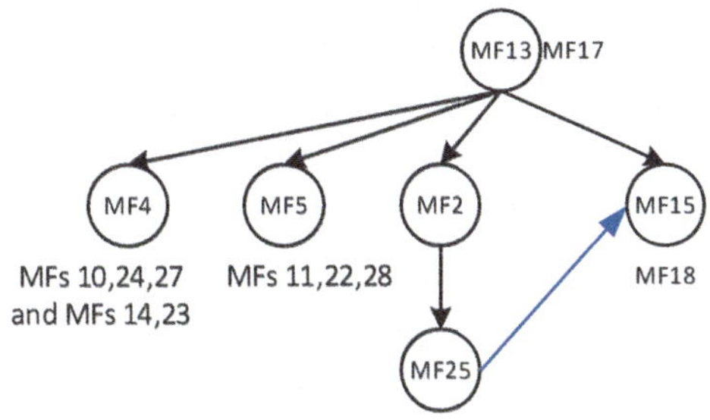


**Fig. 6.23** Rescheduled MF path graph in Setup-②

Based on the above revised MF path graph, adjacency matrix is updated. Since the new graph includes no dynamic lines, NPE-MFs can be determined. According to the definition of NPE-MFs, MFs 4 and 5 belong to this category.

$$\begin{array}{l} \\ \text{MF2} \\ \text{MF4} \\ \text{MF5} \\ \text{MF13} \\ \text{MF15} \\ \text{MF25} \\ S_{AC} \end{array}
\begin{array}{c} \begin{array}{cccccc} \text{MF2} & \text{MF4} & \text{MF5} & \text{MF13} & \text{MF15} & \text{MF25} \end{array} \\ \begin{bmatrix} 0 & 0 & 0 & 0 & 0 & 1 \\ 0 & 0 & 0 & 0 & 0 & 0 \\ 0 & 0 & 0 & 0 & 0 & 0 \\ 1 & 1 & 1 & 0 & 1 & 0 \\ 0 & 0 & 0 & 0 & 0 & 0 \\ 0 & 0 & 0 & 0 & 1 & 0 \end{bmatrix} \\ \begin{array}{cccccc} 1 & 1 & 1 & 0 & 2 & 1 \end{array} \end{array}
\begin{array}{c} S_{AR} \\ 1 \\ 0 \\ 0 \\ 4 \\ 0 \\ 1 \\ \\ \end{array} \tag{6.7}$$

Reachability matrix can thereafter be calculated:

$$\begin{array}{l} \\ \text{MF2} \\ \text{MF4} \\ \text{MF5} \\ \text{MF13} \\ \text{MF15} \\ \text{MF25} \end{array}
\begin{array}{c} \begin{array}{cccccc} \text{MF2} & \text{MF4} & \text{MF5} & \text{MF13} & \text{MF15} & \text{MF25} \end{array} \\ \begin{bmatrix} 0 & 0 & 0 & 0 & 1 & 1 \\ 0 & 0 & 0 & 0 & 0 & 0 \\ 0 & 0 & 0 & 0 & 0 & 0 \\ 1 & 1 & 1 & 0 & 1 & 1 \\ 0 & 0 & 0 & 0 & 0 & 0 \\ 0 & 0 & 0 & 0 & 1 & 0 \end{bmatrix} \end{array}
\begin{array}{c} \text{SUM} \\ 2 \\ 0 \\ 0 \\ 5 \\ 0 \\ 1 \end{array} \tag{6.8}$$

According to the result of the reachability matrix, the sequence can be refined as: MF13 → MF2 → MF25 → MF4, MF5 or MF15. Since MF15 (MFs 15 and 18) and MF25 can be machined with the same tool, after machining MF25, MF15 should be cut immediately. Therefore, the sequence can be finalised as: MF13 → MF2 → MF25 → MF15 → MF4 or MF5.

For sequencing the NPE-MFs (MFs 4, 10, 24, 27), MFs 14, 23, and MFs 5, 11, 22, 28, after applying the method, the following sequence is obtained: MFs 5, 11, 22, 28 are cut first, followed by MFs 4, 10, 24, 27, and finally MFs 14, 23 are cut. According to the cutting tool information shown in Fig. 6.16b, the number of tool changes is 7.

Therefore, from the calculations above, the final MF sequences are,

- Setup-①: MF20 → MF3 (MFs 6, 12, 21, 29, 9, 26, 8) and MF19
- Setup-②: MF1 → MF7 (MF16) → MF13 (MF17) → MF2 → MF25 → MF15 (MF18) → MF5 (MFs 11, 22, 28) → MF4 (MFs 10, 24, 27) and MF14 (MF23).

The number of tool changes for machining the whole part are 9 (Setup-①: 2, Setup-②: 7). From the results of the case study, the number of tool changes is reduced from 13 to 9, leading to a considerable performance improvement.

## 6.5 Adaptive Setup Merging and Dispatching

The *supervisory planning* [19] within DPP can generate a generic sequence plan. At the same time, a setup plan is created for 3-axis machines, as their configurations form the basis of other machines having more axes. However, setup merging is required for a 4- or 5-axis machine to best utilise the capability of the higher-end machine tool, after the non 3-axis machine is selected. This is explained through an example.

According to the geometric reasoning rules [16], a 3-axis based generic setup plan of a test part (shown in Fig. 6.24a) with 26 machining features can be generated. It consists of 5 setups, each of which contains a set of partially-sequenced machining features, as shown in Fig. 6.24b. The light grey areas are setups and the dark grey areas indicate the feature groups sharing the same cutting tools. Each 3-axis based setup can be represented by a unique unit vector $\boldsymbol{u}$ indicating its tool-access direction (TAD). When a 5-axis machine {*X*, *Y*, *Z*, *A* (around *X*), *B* (around *Y*)} is selected, more than one of the 3-axis based setups of the test part may have a chance to be machined in one *final* setup through setup merging.

The setup merging examines whether other setups can be included in a *final* setup by checking the unit vector $\boldsymbol{u}$ of each setup against the tool orientation space (TOS) of the selected machine. The procedure is straightforward by following two steps and their iterations, i.e. (1) aligning the locating direction of a *final* setup to the spindle axis *Z*, and (2) searching for an orientation that includes a maximum number of 3-axis based setups by rotating the part around the spindle axis *Z*. This merging process is repeated for all setups until a minimum number of 5-axis based final setups can be reached. Since the first step can be done easily by using matrix transformation, we only provide details on the second step due to page limitation.

Figure 6.25a shows a typical scenario, where a setup has been aligned with -*Z* axis and another 3-axis based setup with a tool access direction $\boldsymbol{u_i}$ $(x_i, y_i, z_i)$ is under consideration. The goal is to rotate the vector $\boldsymbol{u_i}$ (or the test part) around *Z* and at the same time determine a mergable range (or ranges) within $2\pi$, where $\boldsymbol{u_i}$ can fit in the TOS of the machine. The TOS is represented as a spherical surface patch denoted by **EFGH** in Fig. 6.25a.

As shown in Fig. 6.25a, the spherical coordinates of $\boldsymbol{u_i}$ are $(1, \gamma_i, \theta_i)$. By rotating $\boldsymbol{u_i}$ around *Z*, a circle $\boldsymbol{C_i}$ is obtained.

$$\begin{cases} x_i = \sin\theta_i \cos\gamma_i \\ y_i = \sin\theta_i \sin\gamma_i \\ z_i = -\cos\theta_i \end{cases} \tag{6.9}$$

where, $\theta_i$ is a constant and $\gamma_i \in [0, 2\pi]$. The $\boldsymbol{C_i}$ may intersect with the spherical surface patch **EFGH** defined by

**Fig. 6.24** A test part with 5 setups after applying geometric reasoning rules

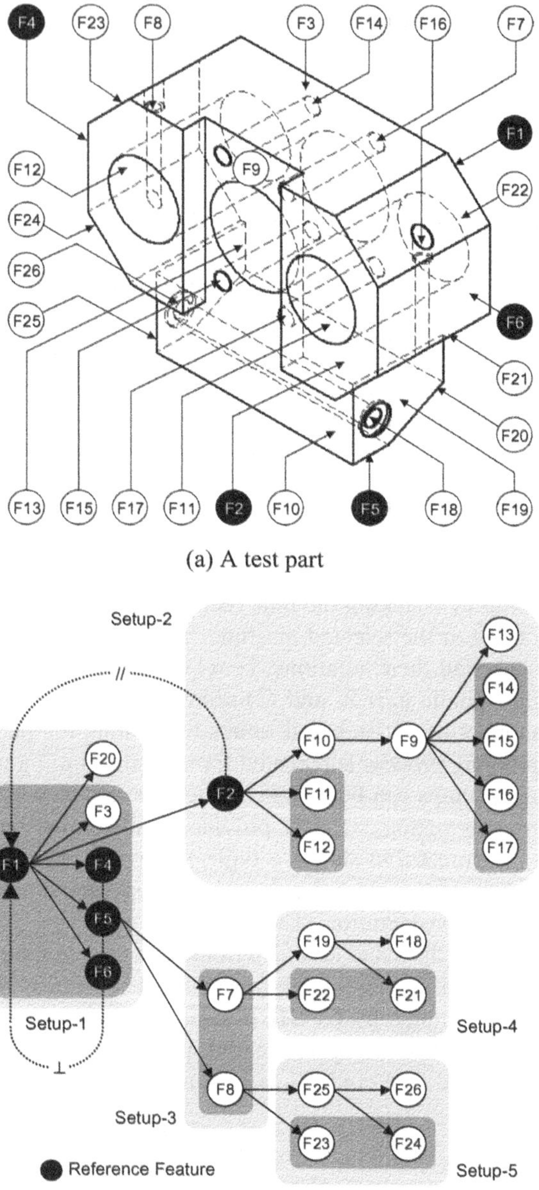


(a) A test part

(b) 3-axis based setups

$$\mathbf{EF}: \quad \phi_A = \Phi_A^+, \quad \phi_B \in \left[\Phi_B^-, \Phi_B^+\right] \tag{6.10}$$

$$\mathbf{FG}: \quad \phi_B = \Phi_B^+, \quad \phi_A \in \left[\Phi_A^-, \Phi_A^+\right] \tag{6.11}$$

$$\mathbf{GH}: \quad \phi_A = \Phi_A^-, \quad \phi_B \in \left[\Phi_B^-, \Phi_B^+\right] \tag{6.12}$$

$$\mathbf{HE}: \quad \phi_B = \Phi_B^-, \quad \phi_A \in \left[\Phi_A^-, \Phi_A^+\right] \tag{6.13}$$

where, $\left[\Phi_A^-, \Phi_A^+\right]$ and $\left[\Phi_B^-, \Phi_B^+\right]$ are the motion ranges of axes $A$ and $B$, respectively. For $\phi_A = \Phi_A^+$ and $\phi_B \in \left[\Phi_B^-, \Phi_B^+\right]$,

$$|z| = \sqrt{\frac{(\cos(\Phi_A^+))^2}{1 + (\cos(\Phi_A^+) * \tan(\varphi_B))^2}}, \quad \phi_B \in \left[\Phi_B^-, \Phi_B^+\right] \tag{6.14}$$

If $|z_i| < |z|_{\min}$, the segment $\mathbf{EF}: \left\{\phi_A = \Phi_A^+, \phi_B \in \left[\Phi_B^-, \Phi_B^+\right]\right\}$, and the circle $\boldsymbol{C_i}$ has no intersection. If $z_i < 0$ and $|z_i| > |z|_{\max}$, the segment **EF** and circle $\boldsymbol{C_i}$ intersect over the entire range of $[0, 2\pi]$. Otherwise, if $z_i < 0$ and $|z|_{\min} < |z_i| < |z|_{\max}$, **EF** and

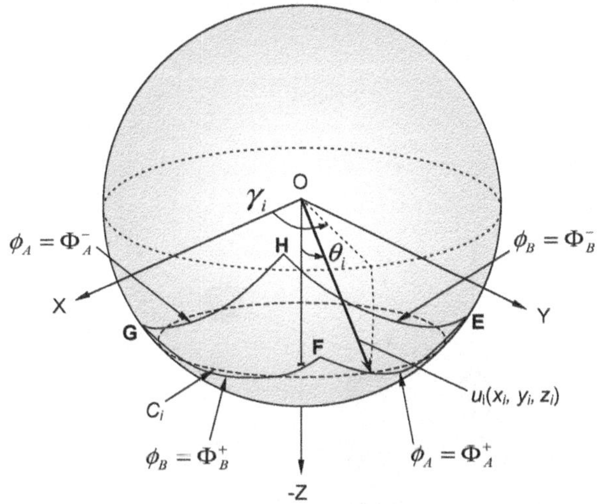


(a) Searching for setup mergability in TOS

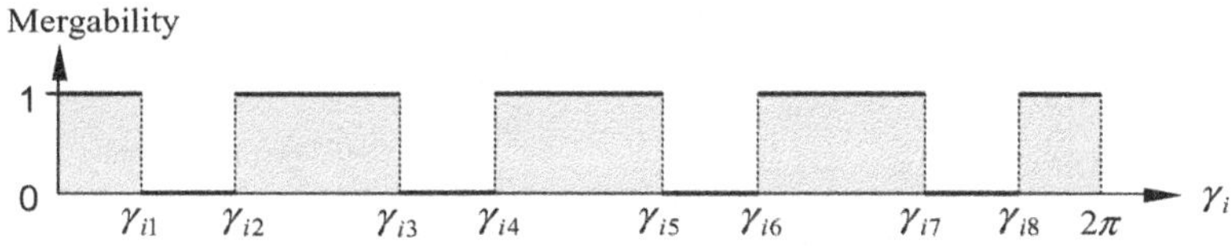


(b) Mergable range of a setup with TAD $\boldsymbol{u_i}$

**Fig. 6.25** Setup merging for a 5-axis machine

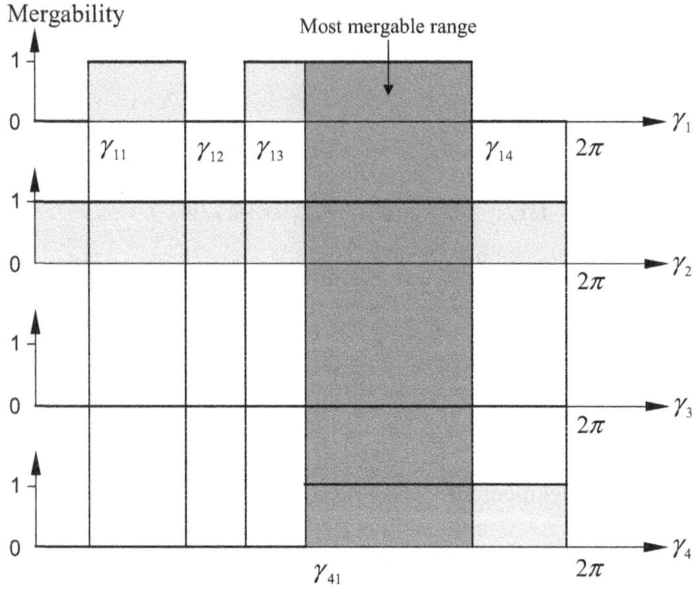


**Fig. 6.26** Determination of a most mergable range

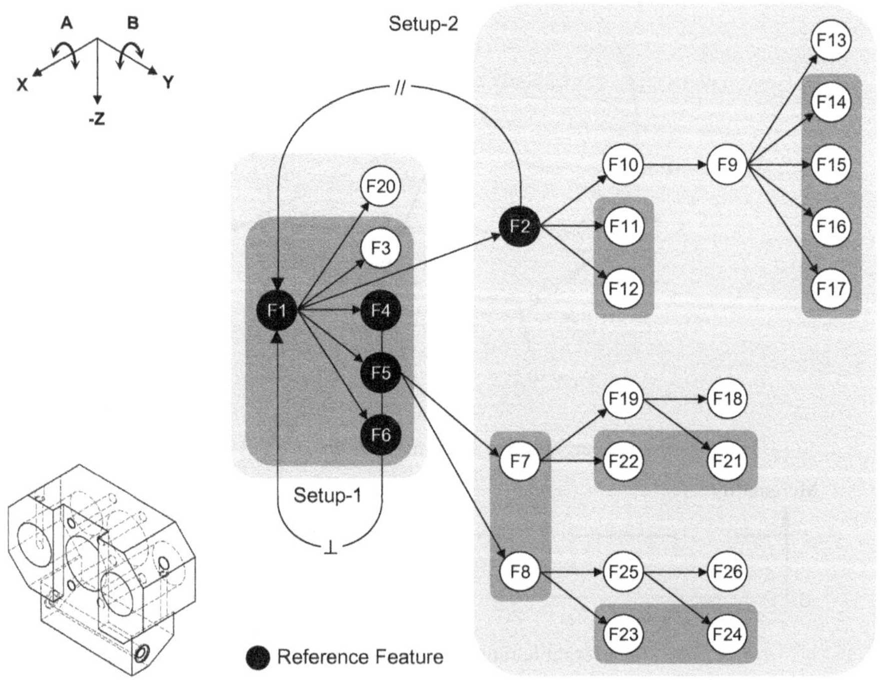


**Fig. 6.27** Results of setup merging for a 5-axis machine

$C_i$ intersect with each other along the edge of the TOS. Figure 6.25b gives the mergable range of the case shown in Fig. 6.25a, which can be calculated for every 3-axis based setup.

As shown in Fig. 6.26, a pose (position and orientation) of the test part that provides the most overlapping mergable range determines a 5-axis based setup.

Figure 6.27 depicts the result of the test part after the *five* generic setups in Fig. 6.24b have been merged to *two* setups (light grey areas) for the 5-axis machine. This final setup plan in the form of two composite function blocks can then be dispatched to a chosen machine for machining.

## 6.6 Conclusions

This chapter introduces a function block enabled distributed process planning (DPP) system including supervisory planning and operation planning. The former is done in a central computer in the cyber world and the latter is carried out in a machine controller in the physical world. Particular focus is given to machining process sequencing, which is treated as sequencing of machining features in varying setups. Tool change reduction is also considered in the process planning in improve the overall machining performance. For interested readers, further reading is referred to the following references.

## References

1. L. Wang, A. Haghighi, Combined strength of holons, agents and function blocks in cyber-physical systems. J. Manuf. Syst. **40**(Part 2), 25–34 (2016)
2. J. Shi, W. Arndt, F. Hu, J. Tang, Isolating—a new resampling method for gene order data, in *2011 IEEE Symposium on Computational Intelligence in Bioinformatics and Computational Biology (CIBCB)* (2011), pp. 135–140
3. P. Derler, E.A. Lee, A.S. Vincentelli, Modeling cyber–physical systems. Proc. IEEE **100**(1), 13–28 (2012)
4. J. Lee, B. Bagheri, H.A. Kao, A cyber-physical systems architecture for industry 4.0-based manufacturing systems. Manuf. Lett. **3**, 18–23 (2015)
5. L. Wang, M. Törngren, M. Onori, Current status and advancement of cyber-physical systems in manufacturing. J. Manuf. Syst. **37**(Part 2), 517–527 (2015)
6. E.A. Lee, Cyber-physical systems—are computing foundations adequate? October, 1(Jan 2006), pp. 1–9
7. S. Horbach, *Advances in Sustainable and Competitive Manufacturing Systems* (2013)
8. V. Vyatkin, IEC 61499 as enabler of distributed and intelligent automation: state of the art review. IEEE Trans. Industr. Inf. **7**(4), 768–781 (2011)
9. W. Ji, L. Wang, A. Haghighi, M. Givehchi, X. Liu, W. Ji, An enriched machining feature based approach to cutting tool selection. Int. J. Comput. Integr. Manuf. 1–10 (2017)
10. M. Givehchi, A. Haghighi, L.H. Wang, Generic machining process sequencing through a revised enriched machining feature concept. J. Manuf. Syst. **37**, 564–575 (2015)

11. W. Ji, L. Wang, A. Haghighi, M. Givehchi, X. Liu, A reachability based approach for machining feature sequencing. J. Manuf. Syst. **40**, 96–104 (2016)
12. N. Deo, *Graph Theory with Applications to Engineering and Computer Science* (Prentice-Hall, New Jersey, 1974)
13. R. Meenakshi Sundaram, Process planning and machining sequence. Comput. Ind. Eng. **11** (1–4), 184–188 (1986)
14. P. Prabhu, S. Elhence, H. Wang, R. Wysk, An operations network generator for computer aided process planning. J. Manuf. Syst. **9**(4), 283–291 (1990)
15. Z. Liu, L. Wang, Sequencing of interacting prismatic machining features for process planning. Comput. Ind. **58**(4), 295–303 (2007)
16. L. Wang, N. Cai, H.-Y. Feng, Z. Liu, Enriched machining feature-based reasoning for generic machining process sequencing. Int. J. Prod. Res. **44**(8), 1479–1501 (2006)
17. L. Wang, J. Ma, H.-Y. Feng, An adaptive and optimal setup planning system, in *4th IEEE Conference on Automation Science and Engineering* (2008), pp. 67–72
18. L. Wang, H.-Y. Feng, N. Cai, Architecture design for distributed process planning. J. Manuf. Syst. **22**(2), 99–115 (2003)
19. L. Wang, N. Cai, H.-Y. Feng, Function blocks enabled dynamic set-up dispatching and execution monitoring. Int. J. Comput. Integr. Manuf. **22**(1), 3–12 (2009)

# Chapter 7
# Condition Monitoring for Predictive Maintenance

## 7.1 Introduction

Predictive maintenance is based on condition monitoring and prognosis. Condition monitoring observes the current status of a situation, whereas prognosis refers to forecasting the likely outcome of the situation which typically involves two inherently related steps. First, analytical models are established to summarise the historical evolution of the situation (e.g., variation in stock price, deterioration of machine conditions, or spread of infectious disease) in a quantitative manner. These models are then modified by updated information to predict the future development of the situation [1]. The predicted value is associated with a confidence level, which results from the uncertainty involved in the prediction process.

Prognosis has been investigated for a wide range of applications, including disease [2] and epidemiology prediction [3], weather forecasting [4], and maintenance scheduling [5] (see Fig. 7.1). In the context of manufacturing, prognosis has been used to identify short-term and long-term actions or decisions to estimate the remaining useful life (RUL) of a tool, machine or system [6–8] based on the conditions monitored and diagnosis obtained [9, 10]. It provides a scientific and technological basis for maintenance scheduling, asset management, and more reliable system design [11, 12].

The operational reliability of industrial machines and assets significantly influences the sustainability of manufacturing [13] and competitiveness of the industry. Because the operational reliability of a machine system decreases as the duration of its operation progresses, ensuring reliability during the designed lifecycle of the machine becomes a critical task for maintenance. In traditional time-based maintenance, machine inspections are performed periodically independent of a machine's current condition. Although such an approach is effective in reducing equipment failures, it generally does not provide information on the RUL of a machine. Furthermore, time-based maintenance can be expensive with the increasing complexity of machines and equipment in modern manufacturing.



L. Wang and X. V. Wang, *Cloud-Based Cyber-Physical Systems in Manufacturing*,
https://doi.org/10.1007/978-3-319-67693-7_7

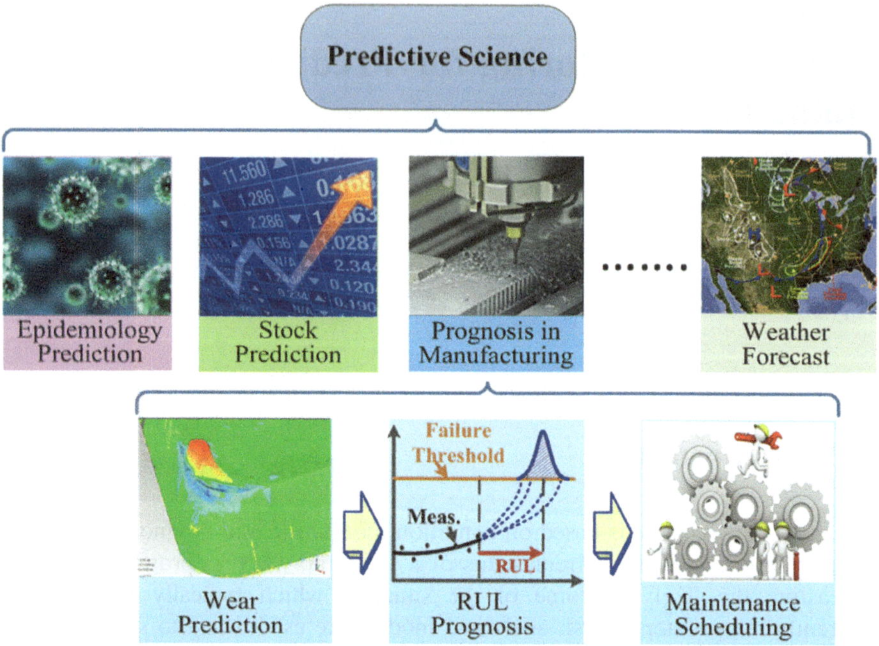


**Fig. 7.1** Predictive science and its application in manufacturing [1]

Addressing this challenge, condition-based maintenance (CBM) has been developed as a maintenance strategy that schedules maintenance actions based on the machine conditions without interrupting normal machine operations [14]. Fault diagnosis is a critical part of this process that links the identified abnormal behaviours in a machine to possible root causes [15]. Maintenance actions may then be performed based on the identified failure type and underlying mechanism. With the advancement of predictive science, prognosis has been increasingly recognised as a valuable complement to CBM in manufacturing. This has led to a more efficient maintenance approach termed intelligent preventive maintenance (IPM), which minimises the machine down time, maintenance cost, and reliance on human experience for maintenance scheduling.

The remainder of this chapter reviews the historical development of prognosis theories and techniques and projects their future growth enabled by the emerging cloud infrastructure. Techniques for cloud computing are highlighted, as well as their influence on cloud-enabled prognosis. Finally, this chapter discusses the envisioned architecture and challenges of cloud-enabled prognosis in manufacturing.

## 7.2 Fundamentals of Prognosis

Failure in a machine progresses through several stages from failure initiation to eventual functional failure. Predictive techniques can help determine how quickly a machine's functional degradation is expected to progress from its current state to its final failure [16]. Figure 7.2 illustrates the relationship between maintenance cost and reliability of machines [11].

Prognosis and preventive maintenance can specifically [17]:

- Increase system safety, improve operational reliability, and extend service life of machines;
- Increase maintenance effectiveness and optimisation of logistic supply chains; and
- Reduce maintenance costs created by repair-induced failures or unnecessary replacement of components.

Research on prognostic technologies has grown and provides the basis for prognosis-centred maintenance. Jardine et al. [8] summarised technologies for diagnosis and prognosis that implement CBM. Peng et al. [11] reviewed typical prognostic techniques and presented a strengths-and-weaknesses analysis of the candidate techniques. Si [18] discussed statistical approaches. Sikorska et al. [19] compared different modelling options for RUL estimation, from the perspective of industry and business applications. Baraldi [20] investigated the capabilities of prognostic approaches to deal with various sources of uncertainty in RUL prediction, focusing on particle filtering (PF) and bootstrap-centred techniques. Heng et al. [21] and Sun et al. [17] discussed the potential benefits, challenges, and opportunities associated with rotating machinery prognosis.

Depending on the types of data and information needed to characterise the systems of interest and predict its future behaviour, prognosis techniques can be classified into three categories: physics-based, data-driven, and model-based (see Fig. 7.3).

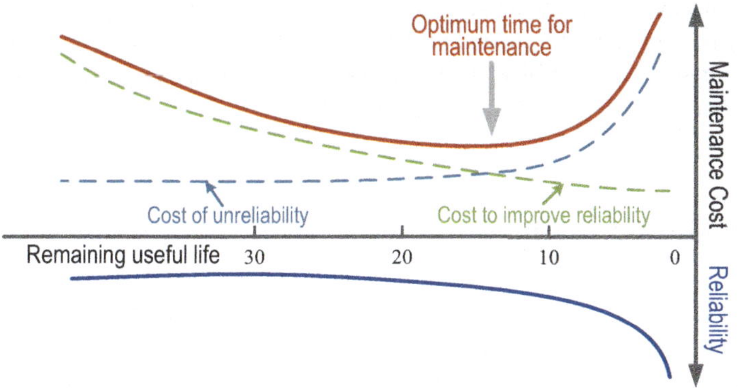


**Fig. 7.2** Relationship between RUL, reliability, and maintenance cost, adapted from [11]

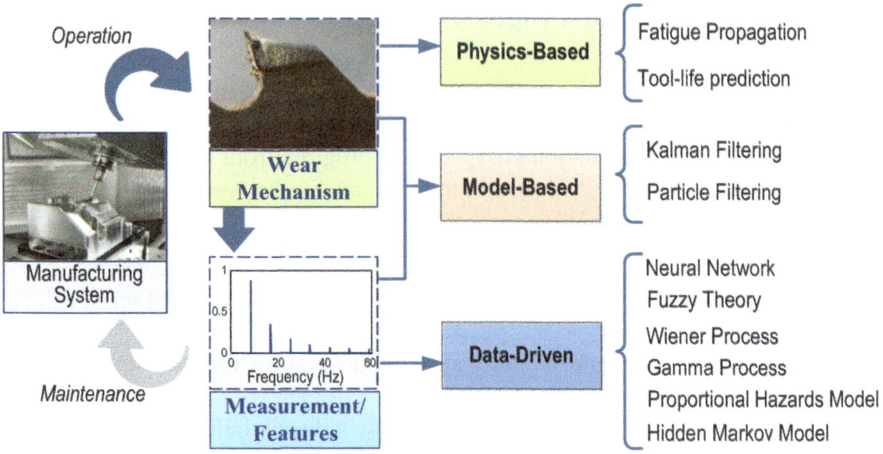


**Fig. 7.3** Classification of prognosis methods [1]

Physics-based techniques describe the system behaviour with empirical formulae, for which the related parameters are determined experimentally. In comparison, data-driven methods rely exclusively on historical data, and numerically establish the relationship between a machine's current damage state and future health state. Data-driven methods can be further divided into artificial intelligence-based (AI) and statistical methods. In contrast to AI-based methods, the relationship between current and future states in statistical methods is presumed to have specific probability distributions, for which the parameters are obtained through regression or maximum probability distribution algorithms. Model-based prognostic techniques combine the above-mentioned two methods to improve the prediction accuracy and robustness. In addition to this classification scheme, prognosis techniques can also be specified by how uncertainty is handled in the prediction process, in terms of deterministic or probabilistic properties. Compared to deterministic methods, probabilistic methods regard machine health states and observations as probability distributions instead of a defined value. Accordingly, damage degradation can be modelled as evolution of the distributions. Furthermore, the results of prognosis, such as future state or RUL, are also presented as probability distributions, with which confidence intervals for evaluation of prognosis results can be obtained.

## 7.3 Prognostic Methods

Prognosis determines the expected progression of degradation in a machine or its components from its current state to functional failure, and the confidence associated with the prediction. The confidence level quantifies the uncertainty that affects the RUL prediction [22].

Machine-specific data plays an essential role in prognosis. Data used in various prognostic models can be categorised into condition monitoring data and event data. Condition monitoring data refer to the data measured by sensors (e.g. force, vibration, acoustic emission, or temperature) that are reflective of the current health condition or state of the machines [10]. Characteristic features can be extracted from the raw data and used as input to establishing analytical models for RUL estimation. Event data includes information on what happened (e.g. installation, breakdown, and overhaul) and what was done (e.g. component change and preventive maintenance) to the machine or component.

### 7.3.1 Physics-Based Models

Physics-based approaches provide a reliable and accurate estimate of all modelling options by estimating the RUL using a mathematical representation of the physical behaviour of the degradation processes. The difficulty is that this process requires detailed and complete knowledge of the system behaviour, which is not readily available for many manufacturing systems. Moreover, the majority of coefficients or parameters involved in the physical models need to be determined experimentally, which makes physical models application specific.

A common approach to assessing machine performance and degradation is to evaluate tool wear or tool life, which directly correlates with the parameters for machining (e.g. cutting speed, temperature, and feed rate). Among physical models describing tool life, an important branch is based on Taylor's tool life equation. As described in Mills and Redford [23], Taylor's basic equation relates tool life to cutting speed in a reverse exponential relationship, $VT^n = C$, where the exponential coefficient $n$ is experimentally determined.

On the other hand, tool wear rate models provide information about wear growth rate (volume loss per unit contact area/unit time) due to some wear mechanisms (e.g. abrasive wear, adhesive wear, diffusion wear). It has been indicated experimentally that the cutting velocity and the index of diffusion coefficient have the most significant effect on tool wear rate [24]. The tool wear rate model can be seen as a particular type of crack growth model or fatigue spall progression model. Generally, a crack growth model is characterised by the stress intensity factor at the tip of a crack $K = f(a, \sigma)$, where $a$ is the half crack length and $\sigma$ is the nominal stress. Theoretically, the crack is assumed to not propagate when $K$ is smaller than a threshold value. After exceeding the value, the crack growth rate will be governed by a power law, such as Paris' law $\mathrm{d}a/\mathrm{d}N = C\Delta K^m$, where $C$ and $m$ are material parameters [25]. However, Paris' law does not account for the mean stress effects and is only valid under conditions with uniaxial loading and "long cracks".

More recently, Fan et al. [26] proposed a mathematical model for the wear analysis of the slide guideway under cutting conditions by revealing the inherent interactions between cutting force, wear, and deformation of the slide guideway, geometric errors, and final accuracy degradation of machine tools. Like other physics-based models, these techniques require experimental estimation of various model parameters.

### 7.3.2 AI-Based Data-Driven Models

Data-driven methods utilise information extracted from historical data to numerically establish a relationship between the current damage state and future state, including AI-based and statistical methods.

Among AI-based methods, artificial neural network (ANN) and fuzzy logic are the common approaches in RUL estimation to determine the next measurements or extract feature indices based on the values measured at several preceding time units [27]. A neural network applies historical data to train a model, which is in turn used for prediction. Fuzzy logic compares the transformed input to a series of fuzzy rules to obtain the prediction.

ANN provides an estimated output for the remaining useful life of a machine or component based on measured condition-monitoring data or event data rather than a physical understanding of the wear or failure mechanism. Because ANN is a purely data-based method, it is insensitive to linear or nonlinear characteristics of a studied system and does not require an analytical expression of the system behaviour. Its drawbacks include that: (a) it requires a comprehensive data set to train the model; (b) its performance relies largely on the selected model (network architecture, activation function, etc.); and (c) it provides no uncertainty quantification on the estimated output. It should also be noted that one developed neural network generally cannot be extended to other neural network architectures, other kinds of machining operations, or other materials or tools. ANN is not able to process linguistic and inaccurate input data. To overcome this problem, past research has focused on integrating ANN with other methods, e.g. expert systems [28], Bayesian inference [29], and fuzzy logic.

Fuzzy logic is a technique for arriving at a definite conclusion using linguistic rules rather than empirically derived *if-then* rules. Compared to traditional expert systems and other estimation techniques, fuzzy systems enable: (1) modelling system behaviour in continuous mathematics of fuzzy sets rather than discrete statements (true or false) and offering a reasonable compromise between rigorous analytical modelling and purely qualitative simulation; and (2) qualitative and imprecise reasoning statements to be incorporated with rule bases, which enables these systems to process vague, imprecise, and noisy inputs.

For system behaviour and state forecasting, a fuzzy system estimates future system states based on the information collected from previous states. To differentiate the impacts of inputs at different times on the next step value prediction, information weights are added to previous states. However, fuzzy logic systems have a major drawback in that fuzzy rules are developed by experts; therefore, fuzzy logic cannot be applied when there is no sufficient knowledge and experience for a problem.

### 7.3.3 Statistical Data-Driven Models

With presumed known knowledge of fault propagation characteristics, statistical techniques assume that system performance degradation or fault deterioration follow a certain distribution, such as Gaussian, Wiener, or Gamma distribution. Unknown variables in the distributions that determine the moments are estimated through regression, given available observations. Once the probability distribution is determined, future state and RUL can be predicted from the current state, through evaluation of the distribution. The benefit of this modelling concept is that prediction results are probability distributions instead of deterministic values, hence a confidence interval can be provided.

The Wiener process is a stochastic regression model with random noise that can be used for modelling degradation processes and RUL prediction. It was first proposed to model the movement of small particles in fluids and air with small fluctuations, and can now be used to model the path of degradation processes where successive and accumulative fluctuations in degradation can be observed. Regarding practical applications, it has the following assumptions or limitations:

- The estimation of degradation uses only the current measurement data. This assumption however can introduce problems.
- It was designed to model the non-monotonic motion of small particles. Thus, it is inappropriate to process the monotonic machine degradations.
- The mean representation of modelled degradation, $\lambda t$, is linear, and thus the application limitation exists when handling nonlinear situations.

In contrast to the Wiener process, which is a non-monotonic process, the Gamma process monotonically models gradual degradation accumulating over time, such as wear, crack growth, and corrosion [30]. The major advantage of degradation modelling using a Gamma process is the straightforward mathematical calculation. However, the strict assumptions of the Gamma process limit its applications: (1) the Gamma process is only appropriate to characterise a monotonic degradation process; (2) due to its independent increment property, the estimation of a future state is independent of the historical behaviour, which is similar to the assumption of a Wiener process; and (3) the noise involved in the Gamma process that is used to quantify the estimation uncertainty must follow the Gamma distribution.

Hidden Markov model (HMM) is defined as a combination of two stochastic processes. The underlying stochastic process is a finite-state homogeneous Markov chain that is not observable (i.e. hidden), which affects another stochastic process that produces a sequence of observations [31]. A HMM is characterised by five elements: number of model states; number of distinct observation symbols; an initial state distribution; a state transition probability distribution; and an observation symbol probability distribution [32]. HMM deals with three basic problems:

- Computing the probability of an observation sequence given the specific model;
- Identifying the most likely state sequence that might produce the observation sequence; and
- Adjusting the parameters of the model to maximise the likelihood of the given observation sequence.

When conducting RUL estimation, the implementation of HMMs includes two stages: training and predicting. Typically, each HMM can only represent two states: normal and failed. Thus, if the entire life of a piece of equipment is segmented into *M* distinct sequential ranges, *M* different HMMs should be trained to characterise each range. The presentation of temporally ordered observation sequences from such a process would yield the sorts of log-likelihood trajectories.

If one HMM results in the largest log-likelihood for a given observation sequence acquired within one duration, this HMM can be declared as the best estimate describing the process during this duration, as shown in Fig. 7.4 [33]. Once parameters in HMMs are determined, RUL prediction is fulfilled by forecasting the progression of health states from the current state (the largest likelihood HMM) to the failure state using transition probability between states and sojourn time in each state (the duration of staying in one state) [34].

Regular HMMs tend to be limited in their ability to represent complex systems. More importantly, in the absence of labelled state and measurement data, the unsupervised training process is computationally tedious. In addition, regular HMMs do not have intrinsic transition probabilities between underlying states since each HMM represents a distinct health state. Hence, they require additional methods to calculate health-state transition probabilities to be utilised in RUL estimation.

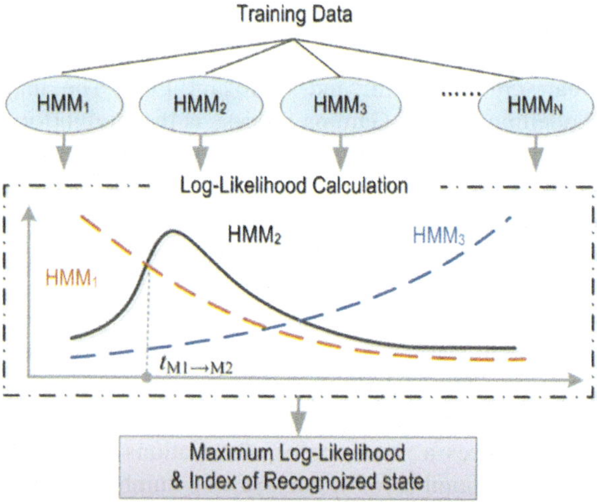


**Fig. 7.4** Log-likelihood for different HMMs, adapted from [1, 33]

### 7.3.4 Model-Based Approach

Model-based technique uses probability distribution for its formulation. Unlike the statistical data-driven method that regards a single variable (such as extracted signal feature or one specific failure) as probability distribution, it considers all the related variables (such as system states and measurements) as distributions. Based on the relevant physical mechanisms, state evolution models and measurement models that relate sensor output to the underlying machine states are established. Subsequently, the machine state can be inferred given new measurements, by means of estimating the posterior probability density function (PDF) [35]. For RUL prediction, once a posterior PDF is determined, the RUL is defined as the conditional expected time to failure, given the current state. In addition, model-based approach is capable of evaluating uncertainty due to process and measurement noise when quantifying accuracy, precision, and confidence (see Fig. 7.5). Accuracy is a measure for how close a point estimate of the failure time is to the actual one, whereas precision is a measure for the narrowness of an interval in which the remaining life falls. In comparison, confidence is the probability of the actual RUL falling between the bounds as defined by the precision [36].

A general way to estimate and update the posterior PDF is Bayesian inference. Based on the assumptions of selected models and noise, RUL prognosis based on Bayesian inference can be implemented by Kalman and particle filtering methods. Figure 7.6 summarises various methods for calculating the posteriori distribution under the framework of Bayesian inference.

The Kalman filter (KF) is a computationally efficient recursive data processing technique used to optimally estimate the underlying state of a dynamic system given a set of noisy measurements in the way that minimises the mean squared error (MSE) of predictions [37]. The general process of Kalman filter includes state and covariance prediction and update as shown in Fig. 7.7.

It should be noted that the state estimate is just the conditional expectation and the covariance of the estimation error is actually the same as the covariance of the state. KF is based on the Gaussian-Markov process assumption that both process and measurement noise are zero-mean white stochastic processes. Meanwhile, the

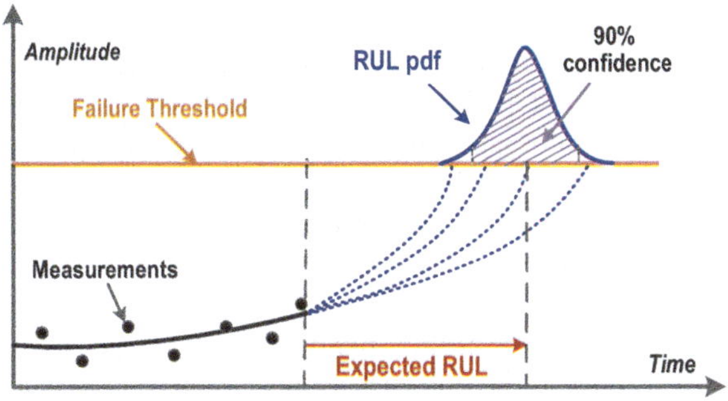


**Fig. 7.5** RUL estimation and associated uncertainty quantification [1]

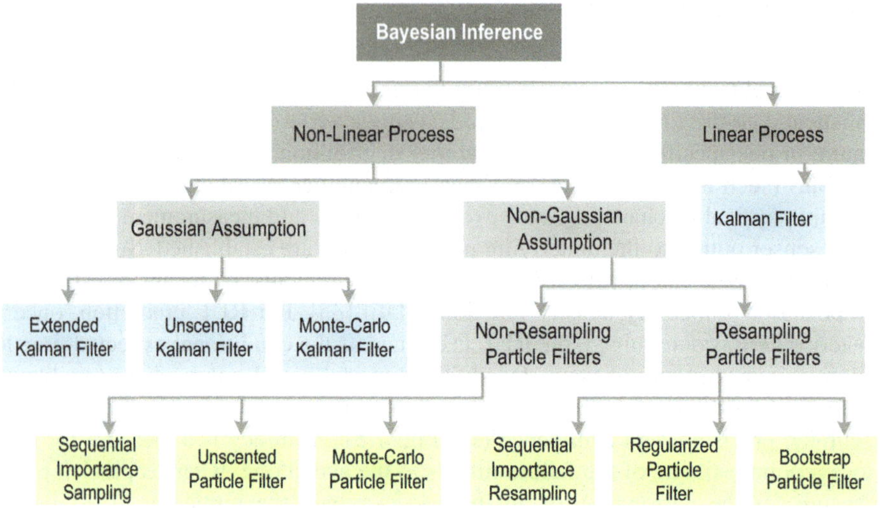


**Fig. 7.6** Overview of methods for calculating posteriori PDF [19]

**Fig. 7.7** General flow of Kalman filter process [38]

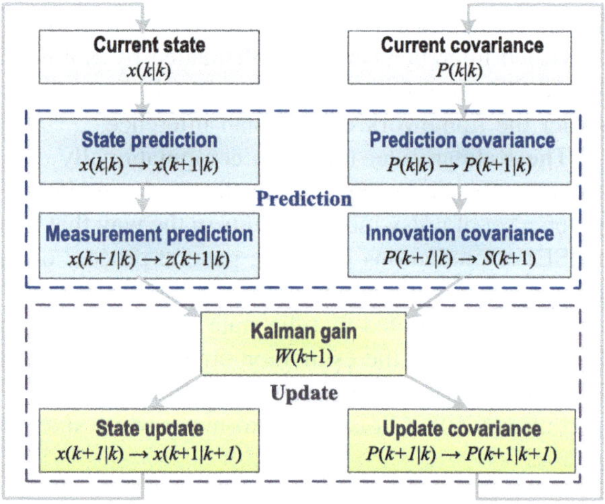


initial state, process, and measurement noise are assumed to be mutually independent. Under this assumption, the KF is the optimal minimum MSE state estimator [38]. For an observable time-invariant system, the state estimation covariance will be finite and the filter will finally converge to a steady state. However, this introduces another limitation besides Gaussian-Markov assumption that the estimation model for time-variant system degradation can be unstable and its estimations divergent.

An alternative to KF under Bayesian inference models and without requiring strict modelling hypotheses such as linearity and Gaussian assumptions, are Particle filters (PF). The PF process provides a different approach to estimating the posterior PDF via a set of random samples with associated weights. Similar to KF and other Bayesian inference methods, the PF process contains two steps: (1) prediction: updated posterior PDF of the model parameters at the previous step are used to calculate the system states at the current time through underlying physical models; and (2) update: predicted model parameters and system states, (i.e. particles and their weights) are corrected based on the likelihood function combined with condition monitoring data.

Along the evolution history of PF, sequential importance sampling (SIS) with weights forms the basis for other variants. A common problem with SIS is however the degeneracy phenomenon where after a few iterations all but few particles will have negligible weight. This degeneracy implies that a large computational effort is devoted to updating particles whose contribution to the approximation of the posterior PDF is almost zero [39]. A potential solution to the degeneracy problem derives the second representative of PF: sequential importance resampling (SIR). The basic idea is to eliminate particles that have small weights and to concentrate on particles with large weights as shown in Fig. 7.8b. After the resampling step, the particles are no longer uniformly generated over the search range, but concentrate on the positions with relatively large possibilities [40]. It is however important to realise that the resampling process can result in many repeated particles: those corresponding to the largest likelihoods. This leads to a loss of diversity among the particles. Wang et al. [41] proposed a local search particle filter, which employs the particles that are intentionally inherited from previous iteration to explore a wide range of prior distributions based on the estimation result from last iteration as depicted in Fig. 7.8c.

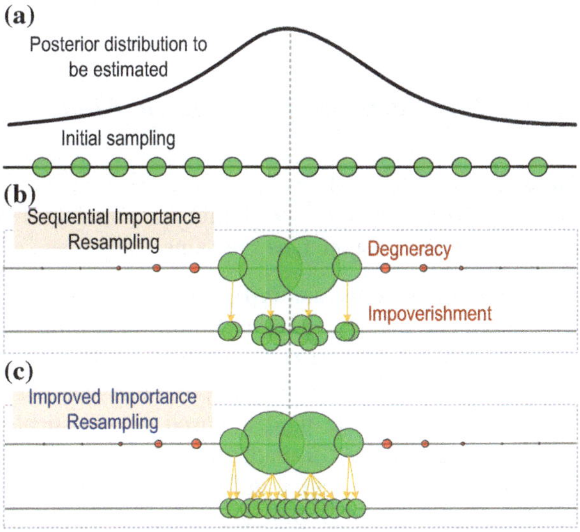


**Fig. 7.8** **a** Initial sampling by PF; **b** sequential importance resampling and associated impoverishment problem; **c** improved importance resampling proposed in Ref. [41]

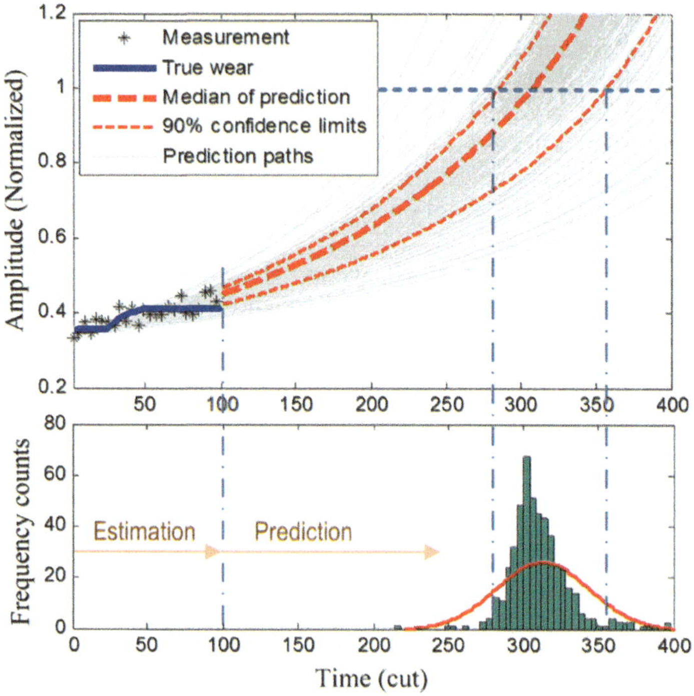


**Fig. 7.9** RUL prediction and uncertainty estimation by PF, adapted from [42]

Compared to KF, which is mature and established, PF is still fast evolving in the field. Wang et al. [41, 42] developed a PF-based framework for precise RUL estimation through a case study on tool wear, as shown in Fig. 7.9.

### *7.3.5 Comparison of Prognostic Models*

The strengths and weaknesses of prognostic methods are individually summarised in the previous sections. It should be noted that all the prognostic techniques presented in the chapter face the same challenges, as listed below:

- Prognostic models are typically developed for specific type of machine or component under certain operating conditions. As a result, it is difficult to generalise a model to be universally employable;
- Uncertainties involved in the prognosis process are difficult to address. Even though probabilistic models based on statistical data-driven and model-based approaches provide a mathematical framework for tracking the evaluation of sensor observation, they are not able to handle uncertainties caused by modelling errors or other sources.

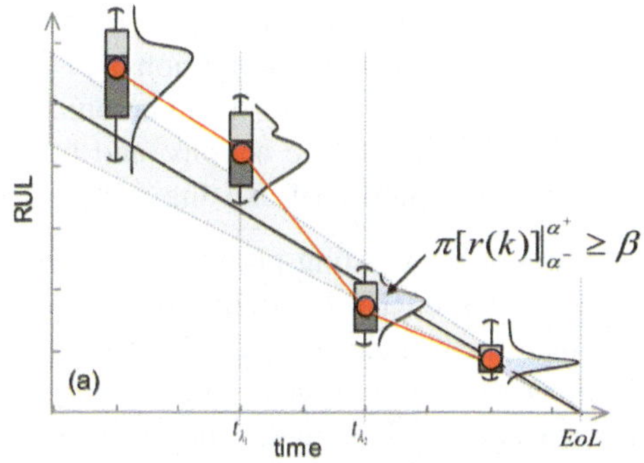


**Fig. 7.10** a-λ accuracy with the accuracy cone shrinking with time on RUL [43]

Practically, evaluation metrics will be needed both quantitatively and qualitatively for selecting the most appropriate method for a specific application. Saxena et al. [43] summarised metrics for prognosis performance evaluation in three groups: algorithm performance, computational performance, and cost benefit. Under algorithm performance, accuracy, robustness, precision and convergence are included. Different methods can be applied to quantify accuracy, such as a-λ accuracy (see Fig. 7.10) and relative accuracy. One indicator of computational performance is the computational complexity, which is especially important for applications where data needs to be monitored in real time to make safety-critical decisions. Other indicators are qualitative in nature, such as robustness within algorithm performance (e.g. practicability of model requirements, sensitivity) and cost benefit.

One major factor that needs to be considered for selecting an appropriate prognostic method is the required information input and assumptions for prognostic models. As described in Sects. 7.3.1 and 7.3.4, both physics-based and model-based prognostic models require good understanding of the physical principles related to machines and the mechanism of fault deterioration. However, the characteristics of and relationships among the various components in a physical system are always too complicated to be modelled effectively. A trade-off between prognosis accuracy and computational cost needs to be carefully considered to be practically meaningful and acceptable.

Another factor for evaluating the prognostic methods is the quantification of uncertainty involved in the prognosis process [44]. The source of this uncertainty can be classified as [20]:

- Modelling error: The failure model that degradation follows should be first determined for prognosis. Various failure models have different triggers to initiate failure and to model failure propagation [45]. Uncertainty in physics-based prognosis models comes from assumptions and simplifications of model structures. Incomplete coverage of data for training empirical models introduces additional uncertainty in data-driven approaches [46].
- Data quality: The selection of condition monitoring features can directly determine the performance of a prognosis system [47], and affect the nonlinear

relationship between features and actual machine health and the sensitivity of features to operating conditions [21].
- Randomness in future degradation: Events, such as changes in operating conditions, maintenance actions, and new failure occurrence, may change the deterioration modes of existing failures [48].

Most of the existing prognostic techniques predefine a threshold for the feature to estimate the RUL by assuming the failure takes place at the instant in time when the increased or decreased feature reaches the predetermined threshold. Practical applications of prognosis systems may commonly yield false-negative and false-positive alarms under the effect of uncertainties discussed above. This problem, caused by an insufficient understanding of prognosis, highlights a future direction for research.

## 7.4 Prognosis-as-a-Service in Cloud Manufacturing

Motivated by the potential of cloud computing (CC) [49, 50] and cloud manufacturing (CM) [51, 52], cloud-enabled prognosis or prognosis-as-a-service represents a new type of service-oriented technology to support multiple enterprises in deploying and managing prognostic services over the Internet. Here, the "cloud" refers to the Internet as a communication network for distributed storage and delivery of computational services. CC brings new opportunity in accelerating the acceptance of advanced manufacturing technologies such as CM. Prognosis, as an integral component of manufacturing, can benefit significantly from CC and CM. The architecture of cloud-enabled prognosis is illustrated in Fig. 7.11.

### *7.4.1 Benefits of Cloud-Enabled Prognosis*

Various assets, such as sensor networks, embedded systems, RFID, and GPS, are integrated in the CM where manufacturing resources (machines, robots, etc.) can be sensed intelligently and connected to the Internet, as well as monitored, controlled, and managed remotely. This creates the Internet of Things (IoT), which is essential to CM. First, machine condition monitoring realised by sensors and data acquisition systems gather data remotely and dynamically on the shop floor. Based on these measurements, remote data analysis and degradation root-cause diagnosis and prognosis are performed. For this purpose, collaborative engineering teams can provide expert knowledge in the cloud, which forms the knowledge base that can be referenced by users through the Internet. The results of prognostic services form the basis for predictive maintenance planning, which can be remotely and dynamically materialised on the factory floor [53].

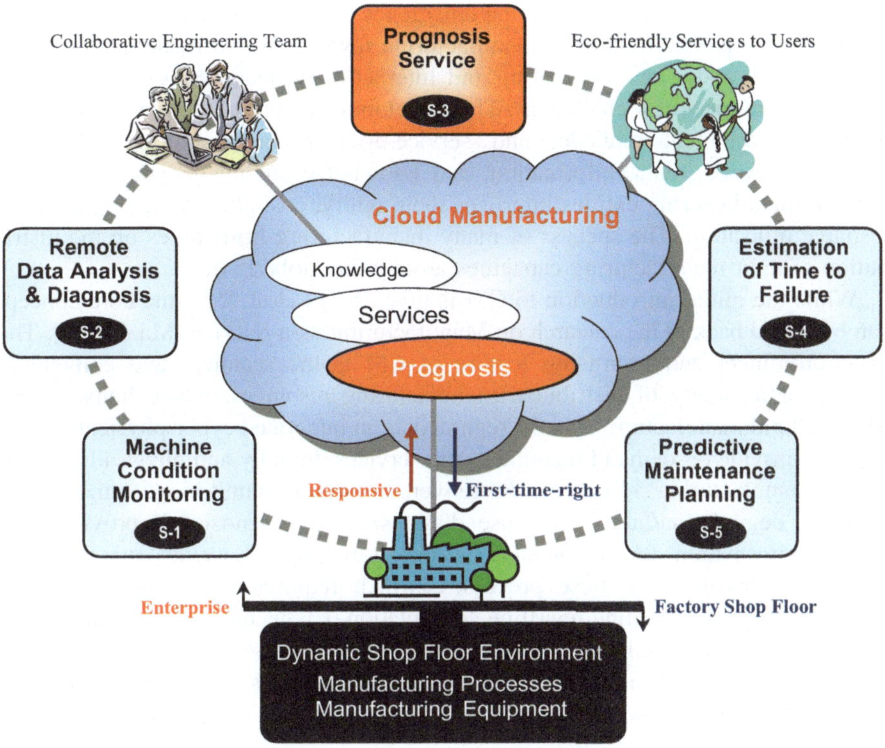


**Fig. 7.11** Prognosis-as-a-Service in cloud manufacturing

Comparing to the state-of-the-art prognostic methods, cloud-enabled prognosis has the following benefits:

- Improved accessibility and robustness: By offering an integrated solution to modular and configurable prognostic services, cloud can increase the robustness of prognosis in manufacturing. Pay-as-you-go prognostic services and varying maintenance options can also be selected from the cloud when necessary or applicable, leading to improved accessibility from customers.
- Improved computational efficiency: Cloud-enabled computation provides efficient computing cycles for complex calculations, due to the higher speed (parallel computing) and lower communication overhead. The characteristics of distributed data storage and computing are essential for cloud.
- Collaboration and distribution: The cloud enables treating machine prognosis as remote services instead of a centralised capability. Through information sharing and fusion realised by crowdsourcing, cloud enables more efficient and effective selection of prognosis models as well as data interpretation, with better interoperability and security.

As stated in Chap. 1, similar to the emergence of cloud-enabled prognosis, there is an ongoing paradigm shift in manufacturing towards global manufacturing networks, which adopt new computing and Internet-based technology such as cloud computing, to meet new challenges. This development leads to the flexible usage of globally distributed, scalable, and service-oriented manufacturing resources. Sharing resources, data, information, and knowledge among geographically distributed manufacturing entities improves their agility, cost-effectiveness, and better resource utilisation. The success of many manufacturing firms relies on the distribution of their manufacturing capacities around the globe [54].

While the initial introduction to CM is given by Li et al. [55], the core concept can be traced back to the research on Manufacturing-as-a-Service (MaaS) [56]. The most prominent and promising feature of CM is the seamless and convenient sharing of a variety of distributed manufacturing resources, which helps realise MaaS. Cloud manufacturing can be regarded as an integrated cyber-physical system that can provide on-demand manufacturing services digitally and physically to best utilise manufacturing resources. Moreover, condition monitoring, remote data analysis, degradation/fault root-cause diagnosis and prognosis all provide supporting information for maintenance decision-making. However, massive data analysis is involved in these processes, which requires significant computing resources to perform online real-time computation. CC techniques can make these tasks more efficient by leveraging infrastructure-oriented services in the cloud for data storage and analysis, while software-oriented services can be performed in a distributive fashion as web-based programmes to interface with manufacturers and consumers.

As illustrated in Fig. 7.12, CM supported by CC in the core encompasses the entire manufacturing process chain within a cloud-enabled environment, from order placement and product design (in a manufacturing system) to machining and asset management (e.g. diagnosis, prognosis and maintenance) [51, 57], where cloud computing represents the core competence of a CM. Based on this concept, more companies in the future would obtain various manufacturing services, including prognosis-as-a-service, through the cloud as conveniently as obtaining utilities in daily operations. Around the outer circle, the conditions of a manufacturing system can be monitored in real-time for diagnosis and prognosis before any well-informed predictive maintenance actions take place.

Within the context, prognosis can support the prediction of resource availability worldwide in addition to predicting machine status and facility performance, which helps determine the most effective and efficient means to manufacture a particular product. Furthermore, cloud-enabled prognosis shares information from similar machines at different stages of service and utilises the power of cloud computing to more efficiently execute the prognostic models in the distributed cloud environment for enhanced decision-making. Challenges in accomplishing this goal include network bandwidth and data transmission speed, security, privacy, reliability, and robustness, which are discussed later in this chapter and in Chap. 3.

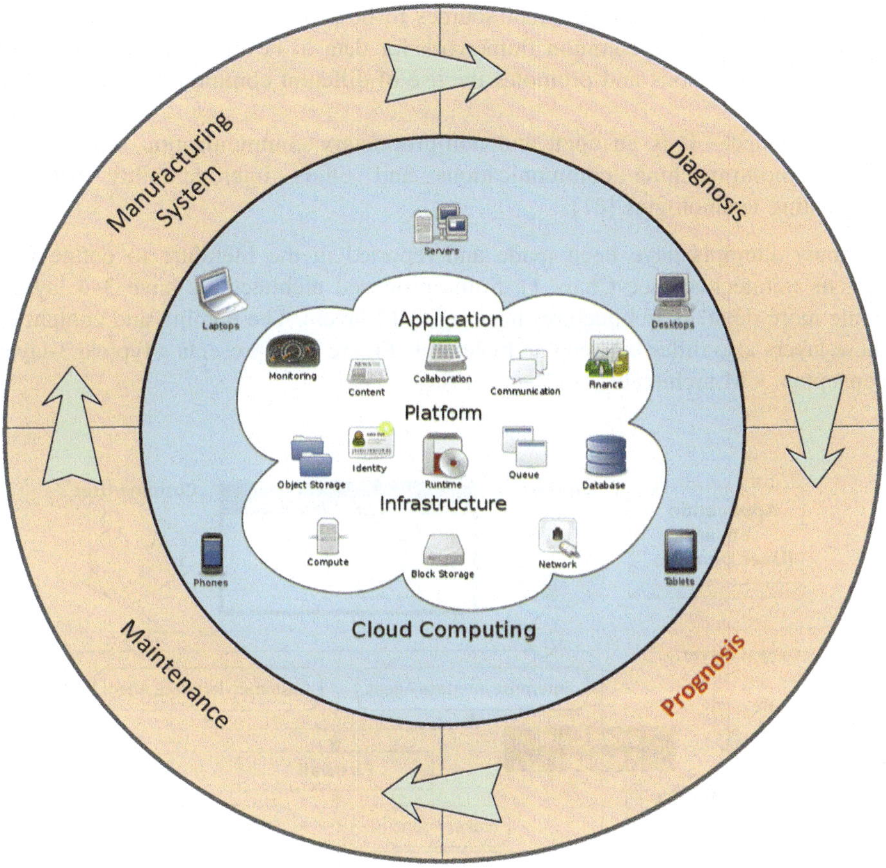


**Fig. 7.12** Cloud-enabled prognosis for closed-loop maintenance support

## 7.4.2 *Supporting Technologies*

The supporting technologies to implement cloud-enabled prognosis include:

- Internet of Things (IoT)—IoT integrates and connects physical assets (e.g. machines, sensors) into an information network, which enables device interoperability and universal manufacturing resource availability and accessibility [58]. IoT is quickly growing with RFID and other sensor technologies, which promotes interconnection between *things*.
- Embedded Systems—The rapid development of embedded systems with IoT enables convenient access to manufacturing resources for status retrieval and control [59].
- Semantic Web—The semantic web facilitates knowledge-based intelligent computation and enables users to search and share data and information easily

by allowing data from different sources to be processed directly by machines [60]. It provides a common framework for data to be represented and reused across applications and promotes the use of different common formats for data exchange.
- MTConnect—It is an open and non-proprietary communication standard for machine-to-machine communications and offers interoperability between existing technologies [61].

Many attempts have been made and reported in the literature to define CM system architecture (see Chap. 1). Some proposed architectures have 3–4 layers, while more detailed architectures have up to 12 layers. The naming and content of these layers also differ between architectures. Figure 7.13 presents a typical 3-layer conceptual CM architecture [62]:

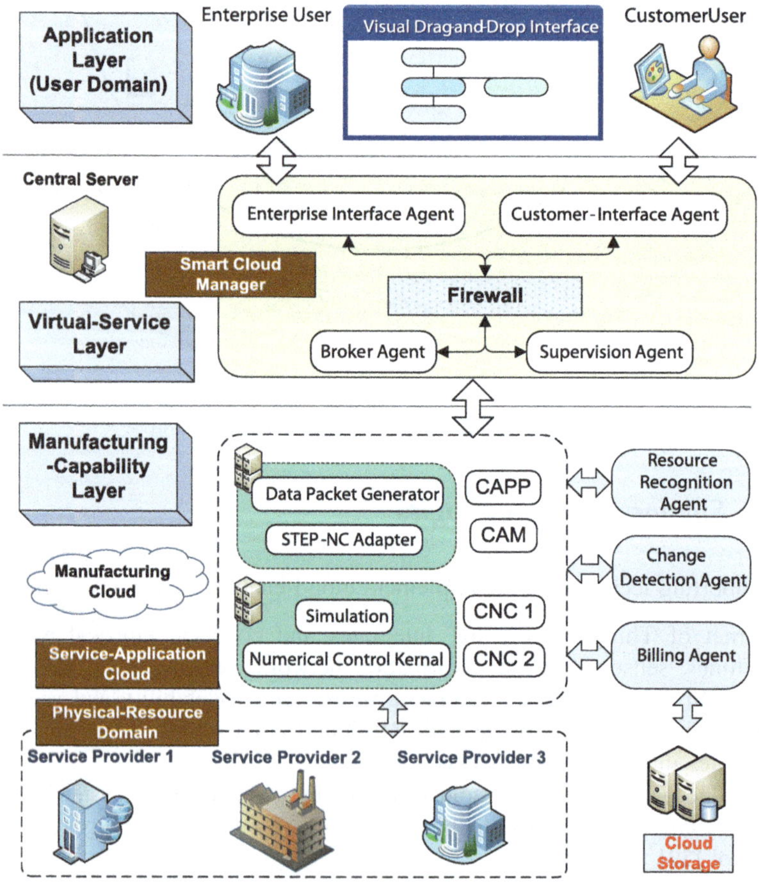


**Fig. 7.13** A 3-layer CM architecture, adapted from [62]

- Manufacturing capability layer: This layer contains the core manufacturing services such as computer-aided process planning (CAPP), computer-aided manufacturing (CAM), computer numerical control (CNC), etc. in a service application cloud. The services and user data can be safely stored in a cloud storage. Physical manufacturing resources are connected to this layer for on-demand access and service realisation.
- Virtual service layer: A central server is placed in this layer for cloud management. Virtual services are matched and mapped to the real services and physical resources based on their availability and capability.
- Application layer: This layer concerns the end users (business and private users) of the cloud services. Comprehensive user interfaces and convenient access to the cloud is the key. User friendliness, thin-client user interface design and timely information presentation are dealt with at this layer.

Despite the difference in architectures, there is an agreement that a CM system has three types of participants: (1) resource/service provider, (2) resource/service consumer, and (3) cloud operator (see more details in Chap. 1).

When implementing prognosis-as-a service in a CM system, security is a major concern. Corporate information often contains sensitive data about operations, trade secrets, and intellectual property. Securing sensitive machine condition data and the ubiquitous availability of requested applications in the cloud are a must for potential users of cloud services. Manifestations of these concerns regularly appear in many existing CC services as a profound unwillingness and anxiety in letting sensitive and important data escape outside the boundaries of the physical company premises. The service models (IasS, PaaS, and SaaS) require different levels of security in a cloud environment. IaaS is the base of all CC and CM services, with PaaS built upon it and SaaS in turn built upon PaaS. Just as capabilities are inherited, so are the cloud security issues and risks [63]. Today, most SaaS business and manufacturing applications that vendors offer are hosted in ISO 27001 and Statement on Auditing (SAS) 70 Type II certified data-centres with service-level agreements offered for applications of 99% and above [64]. More information about CC and CM can be found in [49, 52, 65, 66].

### 7.4.3 Implementing Prognosis in the Cloud

The emerging cloud infrastructure benefits the adoption of prognostic techniques through enhanced computational capability, which not only improves the execution efficiency of prognosis models but also enables more robust decision-making due to information and knowledge sharing by new techniques such as crowdsourcing.

Besides elasticity and cost-effectiveness, another major advantage of cloud-based technologies is the enhanced capability in data storage and computation, which results from the availability of distributed resources. This is especially beneficial in applications involving big data. A typical scenario is depicted in

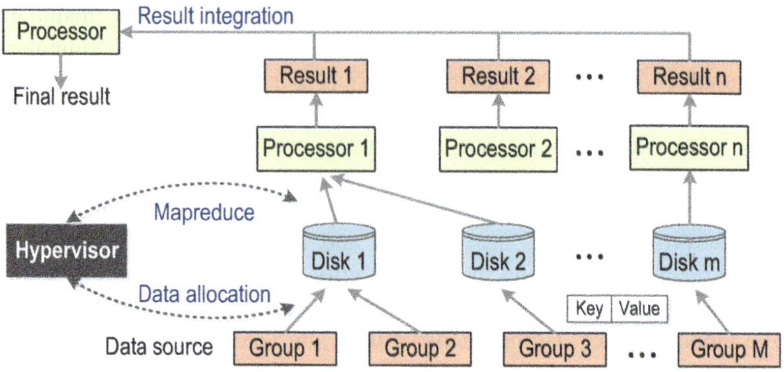


**Fig. 7.14** Structure of distributed data storage and computing [1]

Fig. 7.14, where the hypervisor running in the driver domain is responsible for assigning storage/processing resources and managing the uploaded data [67]. Assuming either the data set can be partitioned and processed independently, or the prognostic programme can be partitioned into sub-tasks via programming models such as Mapreduce, distributed data storage and computation can be performed [49]. Data collected from each component within a machine will be uploaded on to different storage disks and routed into different processors. The prognosis algorithms executed by the processors can vary according to the specific data types and physics of the monitored components. The prognostic results of the various components can be fused to represent the health status of the entire system, which can then be utilised for maintenance decision-making. Such a distributed and parallel computing mode can greatly improve the computing efficiency, to realise real-time condition monitoring and prognosis.

Due to the limitation in network bandwidth, it is impractical to directly transmit raw data from individual machines to the cloud. A cost-effective approach is to have data collected on the shop floor pre-processed, during which failure features or signatures are extracted and subsequently transmitted.

Introducing sensors and networked communication into the shop floor can facilitate smart in-process diagnosis and prognosis, as well as efficient human intervention that improve the robustness and adaptability of processes and systems. Figure 7.15 illustrates the trend of developing cloud-enabled and knowledge-based tools for dispersed engineering teams to perform machine state identification [68], condition-based monitoring, prognosis, and maintenance actions collaboratively, using services enabled by the cloud. All functions of monitoring, prognosis and maintenance are delivered as cloud-based services accessible via web browsers. No expensive software packages are needed for local installation and maintenance.

In recent years, technology advancement in sensing and diagnosis has made an appreciable impact on condition-based monitoring. It can be predicted that research in prognosis and predictive maintenance will continue in the direction of supporting an Internet-based environment. Such a trend is consistent with recent development

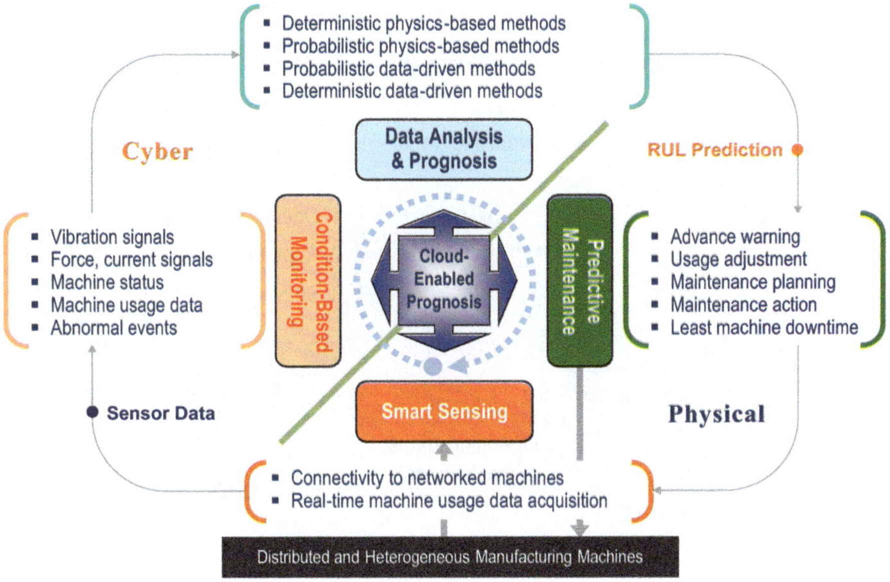


**Fig. 7.15** Relationship between cloud and prognosis

in manufacturing enterprises. For example, the concept of "Industrial Internet" proposed by General Electric enables industries to gather and analyse data from physical objects via IoT, manage operations, and provide value-added services such as predictive maintenance [69]. To support such a collaborative and distributive platform, there is a need for developing the ability to share process and machine data between different applications at different locations seamlessly and collaboratively. Service-oriented cloud manufacturing is a clear path for prognosis and predictive maintenance in the future.

### *7.4.4 Prognosis Applications*

In the area of remote condition monitoring and diagnosis, Teti et al. [70] provided an extensive list of industrial efforts. As an example, the company DMG MORI SEIKI has developed a remote machine monitoring system termed "Mori Net", with the structure shown in Fig. 7.16 [71]. Both machine tool data and corresponding customer information are collected and stored in unified databases located at service centres, where maintenance services such as fault diagnosis and system update are performed. Data collection and communication among machines are based on MTConnect protocol, which enables unified communication interface for different devices and machines by defining a standard set of data types for data collection and storage.

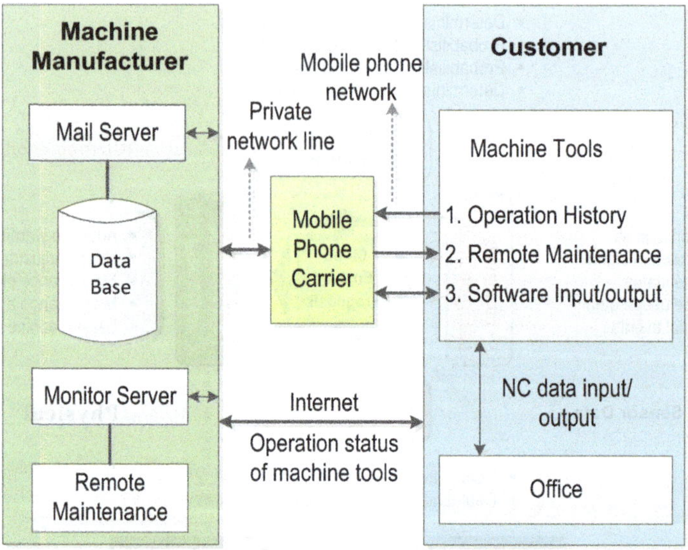


**Fig. 7.16** Remote monitoring and maintenance system developed by DMG MORI SEIKI, adapted from [71]

Currently, the development of condition monitoring, diagnosis, and prognosis in enterprises remains focused on remote monitoring. While not strictly defined as cloud-enabled monitoring and prognosis, remote monitoring employs IoT techniques for data acquisition and network techniques for data and information interaction. The achievement of cloud-enabled monitoring in enterprises still needs to address several challenges, including those presented by communication, information security, and interoperability concerns.

The most important two attributes for maintenance are cost-effectiveness and accuracy, which is a comprehensive factor that includes reliability and probability. A significant advantage of prognosis-enabled condition-based maintenance (CBM) over traditional scheduled maintenance is its effectiveness in reducing maintenance cost. Studies have shown that predictive maintenance can reduce maintenance costs up to 30% and eliminate breakdowns up to 75% relative to scheduled maintenance [72].

The approach of remote monitoring and predictive maintenance has been widely adopted by industries. For example, SANY Heavy Industry has built a remote equipment management system that collects and analyses the real-time data of their equipment sold all over the world, to provide efficient after-sale services including the information support for equipment maintenance [73]. However, predictive maintenance requires a better understanding of the nature of maintenance policies in a mathematical way and incorporation of diagnosis and prognosis results into maintenance rules (i.e. the adaption of maintenance polices). The overall objective of formulating or selecting maintenance rules is to minimise the total maintenance cost, including the hidden cost of risk and reliability. Niu et al. [74] proposed a

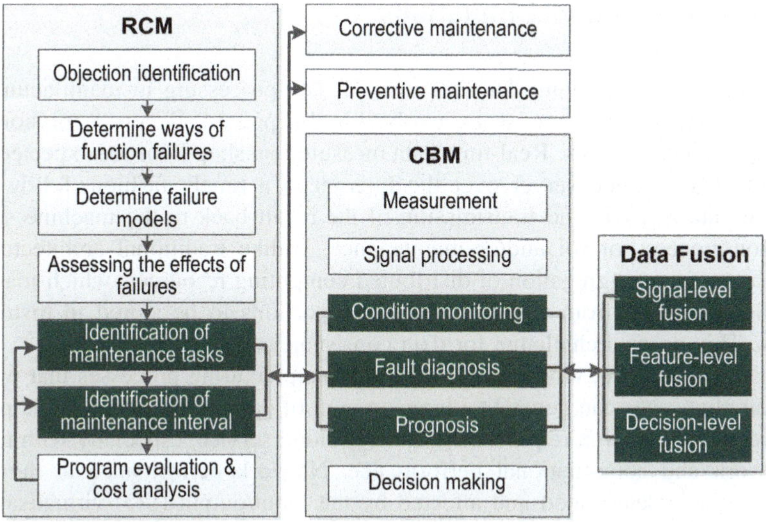


**Fig. 7.17** Condition-based maintenance based on data fusion and reliability-centred maintenance, introduced in [74]

CBM system that employed a reliability-centred maintenance mechanism to optimise maintenance cost, as shown in Fig. 7.17.

For predictive maintenance, the maintenance decision rules should be incorporated with the information obtained from online measurement, data processing (diagnosis and prognosis), or data fusion, which makes sense especially when equipment works in a complex situation and undergoes a different deterioration rate. The maintenance rules should be adapted after the change points where the transitions of mode of system deterioration are assumed to occur. Grall et al. [75] proposed a maintenance strategy with sequential inspection times taking into account the current system state for the choice of the next inspection, as shown in Fig. 7.18. The system deterioration is modelled as a Gamma process, and the system is considered failed if its condition jumps above a pre-set failure level.

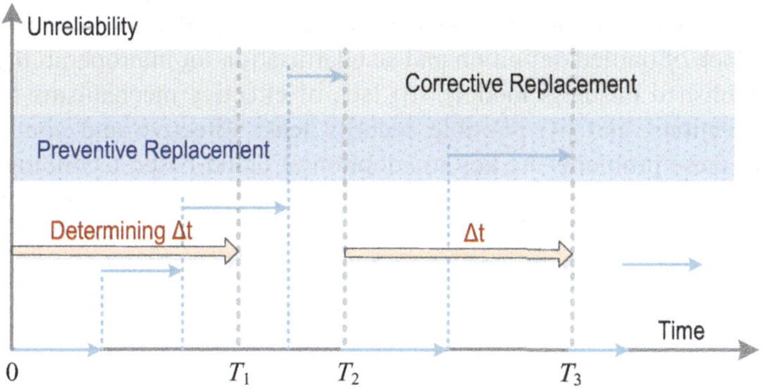


**Fig. 7.18** CBM-based decision-making of inspection time, adapted from [75]

## 7.5 Challenges and Limitations

One of the most important characteristics of data processing in manufacturing is that it is real time with three issues involved in this process: data transmission, data storage, and data analysis. Real-time data measured on shop floor are expected to be transmitted to a cloud server over the Internet in a timely fashion, followed by effective data analysis and transmission of the result back to the machine site for operation/process control and/or maintenance. Unlike traditional architectures, a cloud server is an aggregation of distributed computing resources, which may split data files uploaded from clients into several portions to be stored in distributed servers. This poses a challenge for data consistency.

Sensors (e.g. force, vibration) monitoring manufacturing processes that work at high sampling rates can generate a large amount of data within a short time period. The specific application requires high quality cloud service, especially with respect to network and computational performance. Network performance in the cloud environment is determined and affected by the input/output (I/O) virtualisation—network bandwidth is shared by multiple virtual machines (VMs). Recent research has indicated that the most important issue affecting I/O virtualisation performance is communication between VMs and virtual machine monitor (VMM), which is responsible for assigning storage/computing resources. 30–40% of execution time for a network to transmit or receive an operation is spent in VMM to remap addresses contained in the transmitted data package. It has been demonstrated that the overhead of central processing units (CPUs) and latency increase with the transmitted package rate due to increased communication between the server (VMM) and client (VMs) domains. Especially when dealing with small packets but high packet rate, the throughput is even lower since the software stack does not have enough CPU resources to process.

An important issue determining the virtualisation performance and consequent network and computation performance is dynamic resource management. The most popular option for resource allocation among current cloud-oriented services is to seek trade-off execution quality by the assigned resources via a load balancing mechanism or high availability mechanism.

Today, cloud manufacturing and cloud-enabled monitoring techniques have not been widely accepted in industry practices yet, primarily due to potential problems such as (1) lack of unified definition and standardisation for interoperability, (2) lack of well-established business models, (3) lack of effective mechanisms for privacy and IP protections, and (4) possible security leak. Effective and comprehensive solutions to these problems are key to adoption of cloud-based technologies by the industry, and represent one of the future research directions in manufacturing.

An important goal of IoT in cloud manufacturing is to leverage machine-to-machine (M2M) communication to collect and contextualise data from sources across the manufacturing enterprise. Data analytics may then be used to assess the data and generate information to support different goals, such as prognosis. The difficulty is that M2M communication in a manufacturing environment can be challenging due to considerations related to interoperability and cybersecurity.

Data collection is also a significant challenge because manufacturing equipment is usually old and low in computational power. Many facilities also use a variety of machine-tool types and each may require an interface to communicate with other machines. Every networked device relies on one of several communications protocols (e.g. Modbus, Fieldbus, or Profibus). These interfaces and protocols can grow rapidly if without the appropriate standards that allow for "out-of-the-box" communication. The lack of commonly adopted interfaces and protocols increases the knowledge and resources needed for implementation, which can be substantial given the significant training and setup time required even if expertise is available.

The architecture needed for M2M communication must also enable data and information exchange within one and across several levels of the manufacturing hierarchy (i.e. process to enterprise). It should be scalable for large data volumes and capable of dealing with different time scales (microseconds to days) present in manufacturing data and decision-making. While these characteristics increase the complexity of data collection and analysis, they enable automated monitoring that can support autonomous manufacturing systems where machines identify patterns or disturbances using a cumulated set of knowledge and experiences. These machines can then work with other machines to respond to the disturbance and ensure the continued performance of the manufacturing system.

Cybersecurity remains a significant concern hindering cloud manufacturing applications and services. Chapter 3 highlights some of the issues related to the protection of IP and sensitive information, but the threat to the security of networked devices and assets may be the more important concern for cloud manufacturing. Existing infrastructure, such as supervisory control and data acquisition (SCADA) networks, can be a significant vulnerability given its designed function. Stuxnet is one example of a cyberattack that exploited SCADA networks. Developed to target Iranian efforts to enrich uranium, Stuxnet exploits the SCADA's dynamic-link library (DLL), through which SCADA receives information about the system being controlled [76]. Through the DLL, Stuxnet reprograms the programmable logic controller (PLC) so that the system (i.e. Iranian enrichment reactors) operates as the attacker intends.

Through M2M communication, opportunities exist to target any part of the product lifecycle and its supply chain if these machines are connected through the cloud. Potential attacks could include altering design files, toolpaths, or quality control. Furthermore, the safety of operators and consumers may also be threatened if an attacker can control these systems. Ultimately, the risks must be understood and acknowledged so that technologies can be developed to address them.

## 7.6 Conclusions

Cloud-enabled prognosis can leverage advanced manufacturing by using data and information from across the manufacturing hierarchy to improve efficiency, productivity, and profitability. Recent advances in cloud manufacturing have increased the accessibility to many technologies, such as M2M communications, IoT, and semantic web, and now provide an opportunity to transfer prognosis models and techniques from research labs to industry. Much of the current technological development has focused on providing the infrastructure and architecture to implement prognosis models and techniques. For example, a variety of cloud initiatives and platforms have been suggested to offer different services (e.g. IaaS, PaaS, or SaaS) to manufacturers, and interoperability standards have been proposed for data integration, such as MTConnect. Hardware and software vendors have also started to provide cloud-enabled diagnosis and prognosis solutions, such as remote monitoring and diagnosis of machine tools and shop floor equipment.

The key challenges for cloud-enabled prognosis will be in data collection and management. Standards will be needed for data interfaces, collection, transmission, and interoperability. Methods to anonymise and remove sensitive information from data and to synthesise data streams from multiple and varied sources will be critical in dealing with the large data volume that may be collected from across the manufacturing hierarchy. Cybersecurity must protect IP sensitive information and the security of networked devices and assets to deploy much of this technology in industry. If these issues are resolved, the potential exists to exploit many aspects of the cloud, such as crowdsourcing, to improve manufacturing efficiently and effectively by providing knowledge and value to actors throughout the product lifecycle, which would drive innovation beyond manufacturing.

Currently, most of the research activities related to prognosis are confined within controlled laboratory conditions, due to the fact that prognosis models are application specific. For example, the parameters involved in the Paris' formula for tool wear prediction vary with the type of tools used. Crowdsourcing, if integrated with cloud-based techniques, presents an opportunity for prognosis in an industrial setting. A challenge, as well as an opportunity, in crowdsourcing is the feasibility and interoperability of data for the purpose of fusion given the variety of data (e.g. condition monitoring data and features). Establishing guidelines for designing a prognosis system in a cloud environment, including sensor selection, data transmission, database creation, prognostic method selection, and cooperative, and intelligent decisions, would have a significant impact on advancing the state of prognosis in the context of cloud.

As with cloud-enabled prognosis and its computing capability, dynamic resource allocation can be another research direction, especially in the context of big data analytics. Typically, sensor outputs, after pre-processing by local agents, are transmitted to computing resources in the cloud. Challenges and opportunities lie in how to allocate efficiently these data and fuse analytical results to ensure remote yet online and real-time manufacturing process and equipment monitoring and prognosis. Also of interest is effective and efficient M2M communication, including data collection, sharing, and transmission, to minimise the bottleneck of current cloud-based techniques and maximise cloud resource utilisation.

Cloud-enabled prognosis benefits from both advanced computing capability and information sharing for intelligent decision-making. Cloud-enabled prognosis, as well as cloud-enabled design and manufacturing services allocation, is part of cloud manufacturing, which requires an association of distributed manufacturing service providers for information and resource sharing. Significant challenges exist in the creation of mechanisms or standards for information and resource sharing, to maximise the benefit and minimise the potential hazards for industries. Another challenge is effective communication between clients and encapsulated service providers. It is expected that specific service requirements can be intelligently and automatically assigned to one or several industries associated with the cloud with minimal human intervention. Overcoming these challenges will make cloud-enabled prognosis an effective tool for the widespread adoption of cloud manufacturing.

## References


1. R. Gao, L. Wang, R. Teti, D. Dornfeld, S. Kumara, M. Mori, M. Helu, Cloud-enabled prognosis for manufacturing. CIRP Ann. Manuf. Technol. **64**(2), 749–772 (2015)
2. C.I. Henschke, D.F. Yankelevitz, D.M. Libby, M.W. Pasmantier, J.P. Smith, O.S. Miettinen, Survival of patients with stage 1 lung cancer detected on CT screening. N. Engl. J. Med. **355** (17), 1763–1771 (2006)
3. M.E. Newman, Spread of epidemic disease on networks. Phys. Rev. E **66**(1), 016128 (2002)
4. T. Gneiting, A.E. Raftery, Weather forecasting with ensemble methods. Science **310**(5746), 248–249 (2005)
5. E.F. Fama, Market efficiency, long-term returns and behavioral finance. J. Financ. Econ. **49** (3), 283–306 (1998)
6. D.J. Inman, C.R. Farrar, J.V. Lopes, J.V. Steffen, *Damage Prognosis for Aerospace Civil and Mechanical Systems* (John Wiley & Sons Ltd, Chichester, UK, 2005)
7. G. Vachtsevanos, F. Lewis, M. Roemer, A. Hess, B. Wu, *Intelligent Fault Diagnosis and Prognosis for Engineering Systems* (John Wiley & Sons, inc., Hoboken, New Jersey, 2006)
8. A.K.S. Jardine, D. Lin, D. Banjevic, A review on machinery diagnostics and prognostics implementing condition-based maintenance. Mech. Syst. Signal Process. **20**(7), 1483–1510 (2006)
9. H.K. Tonshoff, J.P. Wulfsberg, H.J.J. Kals, W. Konig, C.A. van Luttervelt, Development and trends in monitoring and control of machining process. CIRP Ann. Manuf. Technol. **37**(2), 611–622 (1988)
10. K. Jemielniak, Commercial tool condition monitoring systems. Int. J. Adv. Manuf. Technol. **15**(10), 711–721 (1999)
11. Y. Peng, M. Dong, M.J. Zuo, Current status of machine prognostics in condition-based maintenance: a review. Int. J. Adv. Manuf. Technol. **50**, 297–313 (2010)
12. A. Elwany, N. Gebraeel, Sensor driven prognostic models for equipment replacement and sparse parts inventory. IEEE Trans. Ind. Electron. **40**(7), 629–639 (2008)
13. P.W. Marksberry, I.S. Jawahir, A comprehensive tool-wear/tool-life performance model in the evaluation of NDM (near dry machining) for sustainable manufacturing. Int. J. Mach. Tools Manuf. **48**(7–8), 878–886 (2008)

14. R. Kothamasu, S. Huang, W. VerDuin, System health monitoring and prognostics-a review of current paradigms and practices. Int. J. Adv. Manuf. Technol. **28**, 1012–1024 (2006)
15. Z.D. Zhou, Y.P. Chen, J.Y.H. Fuh, A.Y.C. Nee, Integrated condition monitoring and fault diagnosis for modern manufacturing systems. CIRP Ann. Manuf. Technol. **49**(1), 387–390 (2000)
16. R. Gouriveau, N. Zerhouni, Connexionist-systems-based long term prediction approaches for prognostics. IEEE Trans. Reliab. **61**(4), 909–920 (2012)
17. B. Sun, S. Zeng, R. Kang, M.G. Pecht, Benefits and challenges of system prognostics. IEEE Trans. Reliab. **61**(2), 323–335 (2012)
18. X.S. Si, W. Wang, C.H. Hu, D.H. Zhou, Remaining useful life estimation–a review on the statistical data driven approaches. Eur. J. Oper. Res. **213**(1), 1–14 (2011)
19. J.Z. Sikorska, M. Hodkiewicz, L. Ma, Prognostic modelling options for remaining useful life estimation by industry. Mech. Syst. Signal Process. **25**(5), 1803–1836 (2011)
20. P. Baraldi, F. Mangili, E. Zio, Investigation of uncertainty treatment capability of model-based and data-driven prognostic methods using simulated data. Reliab. Eng. Syst. Saf. **112**, 94–108 (2013)
21. A. Heng, S. Zhang, A.C. Tan, J. Mathew, Rotating machinery prognostics: state of the art, challenges and opportunities. Mech. Syst. Signal Process. **23**(3), 724–739 (2009)
22. K. Tracht, G. Goch, P. Schuh, M. Sorg, J.F. Westerkamp, Failure probability prediction based on condition monitoring data of wind energy systems for spare parts supply. CIRP Ann. Manuf. Technol. **62**(1), 127–130 (2013)
23. B. Mills, A. Redford, *Machinability of Engineering Materials* (Applied Science, London, 1983)
24. H.A. Kishawy, S. Kannan, M. Balazinski, Analytical modeling of tool wear progression during turning particulate reinforced metal matrix composites. CIRP Ann. Manuf. Technol. **54** (1), 55–58 (2005)
25. P.C. Paris, M.P. Gomez, W.E. Anderson, A rational analytic theory of fatigue. The Trend in Engineering **13**, 9–14 (1961)
26. K.C. Fan, H.M. Chen, T.H. Kuo, Prediction of machining accuracy degradation of machine tools. Precision Engineering **36**, 288–298 (2012)
27. R. Teti, S.R.T. Kumara, Intelligent computing methods for manufacturing systems. CIRP Ann. Manuf. Technol. **46**(2), 629–652 (1997)
28. H.R. DePold, F.D. Gass, The application of expert systems and neural networks to gas turbine prognostics and diagnostics. J. Eng. Gas Turbines Power **121**(4), 607–612 (1999)
29. J. Dong, K.V.R. Subrahmanyam, Y.S. Wong, G.S. Hong, A.R. Mohanty, Bayesian-inference–based neural networks for tool wear estimation. Int. J. Manuf. Technol. **30**, 797–807 (2006)
30. J.M. Van-Noortwijk, A survey of the application of gamma processes in maintenance. Reliab. Eng. Syst. Saf. **94**(1), 2–21 (2009)
31. L.R. Rabiner, A tutorial on hidden Markov models and selected applications in speech recognition. Proc. IEEE **77**(2), 257–286 (1988)
32. F. Camci, R.B. Chinnam, Health-state estimation and prognostics in machining processes. IEEE Trans. Autom. Sci. Eng. **7**(3), 581–597 (2010)
33. P. Baruah, R. Chinnam, HMMs for diagnostics and prognostics in machining processes. Int. J. Prod. Res. **43**(6), 1275–1293 (2005)
34. J. Yan, C. Guo, X. Wang, A dynamic multi-scale Markov model based methodology for remaining life prediction. Mech. Syst. Signal Process. **25**(4), 1364–1376 (2011)
35. W. Caesarendra, A. Widodo, P.H. Thom, B.S. Yang, J.D. Setiawan, Combined probability approach and indirect data-driven method for bearing degradation prognostics. IEEE Trans. Reliab. **60**(1), 14–20 (2011)
36. S.J. Engel, B.J. Gilmartin, K. Bongort, A. Hess, Prognostics, the real issues involved with predicting life remaining. IEEE Aerosp. Conf. Proc. **6**, 457–470 (2000)
37. D. An, J.H. Choi, T.L. Schmitz, N.H. Kim, In situ monitoring and prediction of progressive joint wear using Bayesian statistics. Wear **270**, 828–838 (2011)

38. G. Welch, G. Bishop, An Introduction to the Kalman Filter (1995)
39. M.S. Arulampalam, S. Maskell, N. Gordon, T. Clapp, A tutorial on particle filters for online nonlinear/non-Gaussian Bayesian tracking. IEEE Trans. Signal Process. **50**(2), 174–188 (2002)
40. J. Sun, H. Zuo, W. Wang, M.G. Pecht, Application of a state space modeling technique to system prognostics based on a health index for condition-based maintenance. Mech. Syst. Signal Process. **28**, 585–596 (2012)
41. P. Wang, R. Gao, Z. Fan, Switching local search particle filtering applied to heat exchanger system. in *Proceedings of 2015 IEEE International Instrumentation and Measurement Technology Conference*, Pisa, Italy, May 2015
42. J. Wang, P. Wang, R. Gao, Particle filter for tool wear prediction. in *Proceedings of 42nd North American Manufacturing Research Conference*, Detroit, Michigan, June 2014
43. A. Saxena, J. Celaya, B. Saha, S. Saha, K. Goeble, Metrics for offline evaluation of prognostic performance. Int. J. Prognostics Health Manage. **1**, 4 (2010)
44. F. Zhao, Z. Tian, Y. Zeng, Uncertainty quantification in gear remaining useful life prediction through an integrated prognostics method. IEEE Trans. Reliab. **62**(1), 146–159 (2013)
45. J. Son, Q. Zhou, S. Zhou, X. Mao, M. Salman, Evaluation and comparison of mixed effects model based prognosis for hard failure. IEEE Trans. Reliab. **62**(2), 379–394 (2013)
46. C. Hu, B.D. Youn, P. Wang, J. Taek-Yoon, Ensemble of data-driven prognostic algorithms for robust prediction of remaining useful life. Reliab. Eng. Syst. Saf. **103**, 120–135 (2012)
47. J. Sun, H. Zuo, W. Wang, M.G. Pecht, Prognostics uncertainty reduction by fusing on-line monitoring data based on a state-space-based degradation model. Mech. Syst. Signal Process. **45**(2), 396–407 (2014)
48. R. Moghaddass, M.J. Zuo, An integrated framework for online diagnostic and prognostic health monitoring using a multistate deterioration process. Reliab. Eng. Syst. Saf. **124**, 92–104 (2014)
49. Q. Zhang, L. Cheng, R. Boutaba, Cloud computing: state-of-the-art and research challenges. J. Internet Serv. Appl. **1**(1), 7–18 (2010)
50. M. Armbrust, A. Fox, R. Griffith, A.D. Joseph, R. Katz, A. Konwinski, M. Zaharia, A view of cloud computing. Commun. ACM **53**(4), 50–58 (2010)
51. L. Ren, L. Zhang, L. Wang, F. Tao, X. Chai, Cloud manufacturing: key characteristics and applications. Int. J. Comput. Integr. Manuf. **30**(6), 501–515 (2017)
52. D. Wu, M.J. Greer, D.W. Rosen, D. Schaefer, Cloud manufacturing: strategic vision and state-of-the-art. J. Manuf. Syst. **32**(4), 564–579 (2013)
53. Y. Yang, R. Gao, Z. Fan, J. Wang, L. Wang, Cloud-based prognosis: perspective and challenge. in *Proceedings of the ASME 2014 International Manufacturing Science and Engineering Conference*, Detroit, Michigan, USA, June 2014
54. O.F. Valilai, M. Houshmand, A collaborative and integrated platform to support distributed manufacturing system using a service-oriented approach based on cloud computing paradigm. Robot. Comput. Integr. Manuf. **29**, 110–127 (2013)
55. B. Li, L. Zhang, S. Wang, F. Tao, J. Cao, X. Jiang, X. Song, X. Chai, Cloud manufacturing: a new service-oriented networked manufacturing model. Comput. Integr. Manufact. Syst. **16**(1), 1–7 (2010)
56. J.D. Goldhar, M. Jelinek, Manufacturing as a service business: CIM in the 21st century. Comput. Ind. **14**(1), 225–245 (1990)
57. T. Masood, R. Roy, A. Harrison, Y. Xu, S. Gregson, C. Reeve, Integrating through-life engineering service knowledge with product design and manufacture. Int. J. Comput. Integr. Manuf. **28**(1), 59–74 (2015)
58. F. Xiang, Y. Hu, in *Cloud Manufacturing Resource Access System Based on Internet of Things*, ed. by D. Sun et al. Applied Mechanics and Materials, vol. 121–126 (2012), pp. 2421–2425
59. S. Heath, *Embedded Systems Design*, 2nd edn. (Butterworth-Heinemann, Oxford, UK, 2003)

60. D. Martin, M. Burstein, D. Mcdermott, S. Mcilraith, M. Paolucci, K. Sycara, D.L. Mcguinness, E. Sirin, N. Srinivasan, Bringing semantics to web services with OWL-S. World Wide Web **10**(3), 243–277 (2007)
61. MTConnect Institute (2014), http://www.mtconnect.org/downloads/standard.aspx
62. X.V. Wang, X.W. Xu, An interoperable solution for cloud manufacturing. Robot. Comput. Integr. Manuf. **29**(4), 232–247 (2013)
63. X. Xu, From cloud computing to cloud manufacturing. Robot. Comput. Integr. Manuf. **28**, 75–86 (2012)
64. B. Adiseshan, Manufacturing in the cloud (2012), http://www.automation.com/pdf_articles/inkumo/inkumo_white_paper_manufacturing_in_the_cloud.pdf
65. W. Venters, E.A. Whitley, A critical review of cloud computing: researching desires and realities. J. Info. Technol. **27**, 179–197 (2012)
66. G. Adamson, L. Wang, M. Holm, P. Moore, Cloud manufacturing—a critical review of recent development and future trends. Int. J. Comput. Integr. Manuf. **30**(4–5), 347–380 (2017)
67. S. Abiteboul, I. Manolescu, P. Rigaux, M.C Rousset, P. Senellart, Web data management, (Cambridge University Press, 2011)
68. L. Wang, Machine availability monitoring and machining process planning towards cloud manufacturing. CIRP J. Manufact. Sci. Technol. **6**(4), 263–273 (2013)
69. General Electric, Industrial Internet Insights Report (2015), https://www.gesoftware.com/
70. R. Teti, K. Jemielniak, G. O'Donnell, D. Dornfeld, Advanced monitoring of machining operations. CIRP Ann. Manuf. Technol. **59**(2), 717–739 (2010)
71. M. Mori, M. Fujishima, M. Komatsu, B. Zhao, Y. Liu, Development of remote monitoring and maintenance system for machine tools. CIRP Ann. Manuf. Technol. **57**(1), 433–436 (2008)
72. U.S. Department of Energy, Operations & Maintenance Best Practices: A Guide to Achieving Operational Efficiency (2010)
73. SANY, Heavy Industry (2015), http://www.sanygroup.com/
74. G. Niu, B.S. Yang, M. Pecht, Development of an optimized condition-based maintenance system by data fusion and reliability-centered maintenance. Reliab. Eng. Syst. Saf. **95**(7), 786–796 (2010)
75. L. Dieulle, C. Berenguer, A. Grall, M. Roussignol, Sequential condition-based maintenance scheduling for a deteriorating system. Eur. J. Oper. Res. **150**, 451–461 (2003)
76. R. Langner, Stuxnet: dissecting a cyberwarfare weapon. IEEE Secur. Priv. **9**(3), 49–51 (2011)

# Part III
# Sustainable Robotic Assembly in CPS Settings

# Chapter 8
# Resource Efficiency Calculation as a Cloud Service

## 8.1 Introduction

The environmental impact of energy consumption became an increasingly important matter in industry. The optimisation of energy consumption needs to be performed in order to reserve natural resources and lower the production cost. Therefore, different companies nowadays are focusing on improving energy efficiency and sustain high product quality and production throughput at the same time. However, finding a practical yet industrially feasible solution is still a challenge. The aim of this chapter is to discover how to minimise the energy consumption of the assembly robots used in today's factories with minimum investment.

This chapter first presents a novel approach to minimise the energy consumption of a robot. This approach in particular determines the most energy-efficient joint configurations of the robot when a predefined assembly task is given. The approach is then evaluated using two case studies to compare its results with both commercial software and real robot measurements. In the next section, a summary of the related work is presented to prepare readers for the right context.

## 8.2 Related Work

The significance of minimising energy consumption has been understood by several researchers and equipment manufacturers. Related efforts are numerous. Okwudire and Rodgers [1] presented an approach to control a hybrid feed drive for energy-efficient NC machining, their results showed several improvements in energy consumption and performance over the traditional approaches. Other researchers [2, 3] focused on planning a collision-free trajectory of the robot. On the other hand, some researchers focused on minimising the energy consumption for robots with relatively complex kinematics such as the work presented in [4, 5], which consider



L. Wang and X. V. Wang, *Cloud-Based Cyber-Physical Systems in Manufacturing*,
https://doi.org/10.1007/978-3-319-67693-7_8

the energy consumption of a hexapod robot and a hopping robot, respectively. Another research [6] presented a thorough analysis of energy consumed during the production of a single product. The approach reported many improvements focused on product production and its design, these improvements reduced the energy consumption up to 50%. Another method presented by Weinert et al. [7] focused on the reduction of energy consumption and managed to reduce the energy consumption by describing the production operations as a set of energy blocks and then determining the energy consumed in each block. Several researchers, on the other hand, examined the possibilities of minimising the energy consumption of machine tools. For Example, the work presented in [8] which proposed a cloud-based framework to provide adaptive process planning based on the availability and capability of machine tools. Another example is the work reported by Mori et al. [9] who demonstrated the ability to reduce the energy consumption by adjusting specific cutting conditions as well as controlling the acceleration to maintain a proper synchronisation between the spindle acceleration and its feed system. This approach provides a useful tool for changeable machining operations. Furthermore, the work presented by Vijayaraghavana and Dornfeld [10] highlighted the impact of monitoring the energy consumption of machine tools by associating consumed energy with performed machining tasks.

Behrendta et al. [11] investigated the nature of energy consumption in different machine tools and evaluated their efficiencies, accompanied by energy demand modelling [12]. Within the assembly domain, many research teams focused their work on studying the energy efficiency of industrial robots. Several tools have been developed to calculate and analyse the energy consumption of the robots. For example, the work presented by [13] analysed the energy consumption of an ABB IRB140 robot and suggested an optimisation module to efficiently reduce energy consumption in robot-related applications. Other researchers [14] presented a method to determine the schedule that minimises the energy consumption of a robotic production line. The work reported in [15] showed the importance of minimising the energy consumption of a 6-axis robot in industrial environment.

Other research groups looked at the robot energy efficiency from the time perspective. As an example, the work reported by [16] stated that the optimal time-energy path for the robot can be determined with a few seconds with the help of the modern computing units. By taking the robot smooth movement into account, a similar approach described in [17] performed an energy optimisation on a defined smooth robot trajectory. By defining the energy consumption as a cost function, the approach was tested on a SCARA robot and proved that it can save the energy consumption of the robot in the long run. Another approach reported in [18] examined the impact of the robot operating parameters such as the payload and the velocity on the energy consumption by studying the dynamic behaviour of a 6-degree-of-freedom robot. Nevertheless, others like [19] highlighted the difficulty of simulating the dynamic model of the robot since it depends on the accurate information about the mass properties of the robot.

At the same time, several researchers [20, 21] focused on constructing the mathematical model of the mechatronic components of the robot to be able to analyse the robotic energy consumption and find methods to minimise it. It is clear that many research projects have focused on the planning of energy-efficient trajectories for robots. One of these projects is the work reported [22], which defined holonomic constraints to move a robot on a prescribed path with the existence of obstacles. Another project [23] focused on the effect of payload on the energy consumption of a robotic system. Furthermore, researchers like [24] focused on optimising the velocity of the robot to minimise the electromechanical losses of the robot's motors. In addition, a research group [25] developed an approach to generating an energy-efficient trajectory using a cost function and implemented it on an industrial robot.

Building the dynamic model of a robot is the key to analyse its energy consumption. However, building that model accurately requires the understanding of several losses (mechanical, electrical, etc.) involved in the calculation of the energy consumption [26]. Despite the significant effort made toward energy-efficient machines and machining processes, successful use of energy during robotic assembly remains a challenge and requires further investigations. This is due to the fact that kinematic and dynamic features of the robots are governed by robot controllers instead of shop-floor operators.

## 8.3 System Overview

In order to fulfil the research objective, an optimisation module has been developed using MATLAB®. As shown in Fig. 8.1, this module aims to accomplish five tasks: ❶ to affiliate the predefined trajectory of a robot's TCP (tool centre point) with its velocity and acceleration; ❷ to solve the inverse kinematics for the trajectory and determine the robot's joint configurations; ❸ to calculate both the forward and backward recursions of the robot model, the results of which are used for solving the inverse dynamics of the robot trajectory; ❹ to determine the energy consumption for each joint configuration; and ❺ to select the robot's optimal joint configuration based on the calculated values of energy consumption. Here, each joint configuration represents one set of joint angles to place the TCP in the defined position and orientation along the trajectory.

## 8.4 Methodology and Implementation

The optimisation module is based on the mathematical model of a robot, which can describe the kinematic and dynamic behaviours of the robot in particular. After analysing the force and torque on each joint, the most energy-efficient joint configuration of the robot is determined. An ABB IRB 1600 industrial robot is selected to demonstrate the feasibility of this approach.

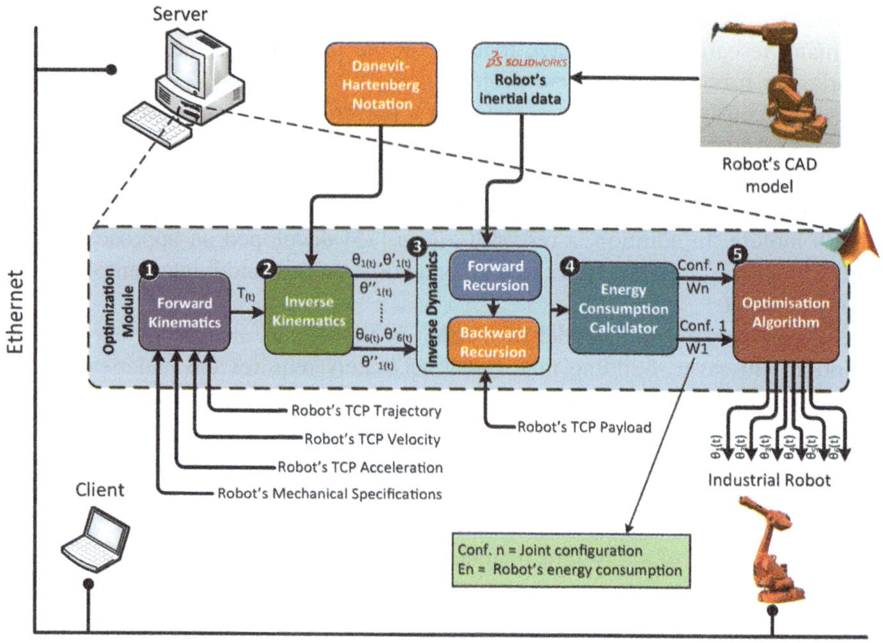


**Fig. 8.1** System configuration

## 8.4.1 Denavit-Hartenberg (D-H) Notation

By implementing the D-H notation [27], the kinematic coefficients are identified and the joint frames of the robot are defined. The notation assigns frames at the joints of the robot from the base to the end-effector. The Z-axes of these frames are aligned with the joints' rotational axes, as shown in Fig. 8.2. For simplification, frames 4–6 are placed on the robot wrist with the same origin.

## 8.4.2 Forward Kinematics

The forward kinematics of the robot is calculated by multiplying the transformation matrices of the robot joints as clarified in Eq. (8.1) to define the position and orientation of the end-effector of the robot with respect to its base.

$$ {}^{0}_{TCP}T(\theta_1 \ldots \theta_6) = {}^{0}_{1}T(\theta_1).{}^{1}_{2}T(\theta_2).{}^{2}_{3}T(\theta_3).{}^{3}_{4}T(\theta_4).{}^{4}_{5}T(\theta_5).{}^{5}_{6}T(\theta_6).{}^{6}_{TCP}T = \begin{bmatrix} {}^{0}_{TCP}R & P \\ 0 & 1 \end{bmatrix} \tag{8.1} $$

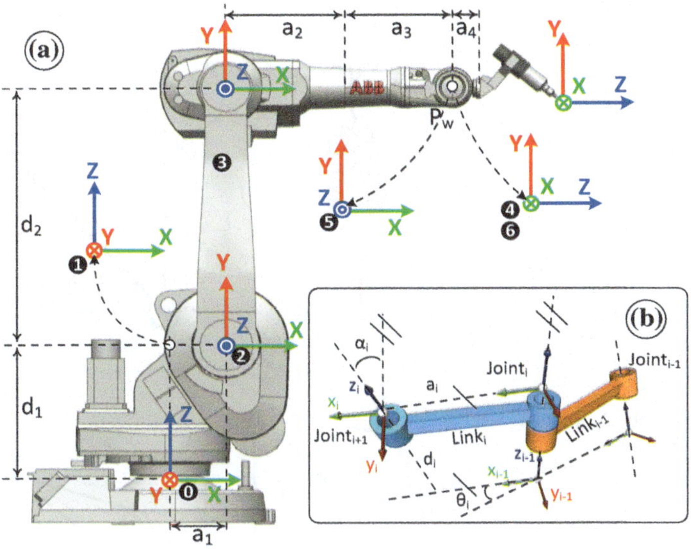


Fig. 8.2 (A) Robot's frame assignments, (B) D-H parameters

where, ${}^{i}_{j}T$ is the transformation matrix between link i and j, ${}^{6}_{TCP}T$ is that between joint 6 and TCP, ${}^{0}_{TCP}R$ and P are the rotation matrix and translation vector, respectively.

### 8.4.3 *Inverse Kinematics*

Based on the kinematic features of the robot, the first three joints control the end-effector's position and the last three joints control its orientation. The process is started first by solving the configuration of the first joint $\theta_1$ in Eq. (8.2). It is accomplished by considering that $\theta_1$ changes the position of robot's wrist in the X-Y plane as illustrated in Fig. 8.3.

$$\theta_1 = \begin{cases} atan2\left(P_{yw}, P_{xw}\right) \\ atan2\left(-P_{yw}, -P_{xw}\right) \end{cases} \tag{8.2}$$

The calculation continues by determining the values of the next two joints. It is achieved by using $\text{Link}_2$ and $\text{Link}_3$ to form XY0-Z0 plane as shown in Fig. 8.4. Equation (8.3) shows the calculation for joint3 value, resulting in two possible results.

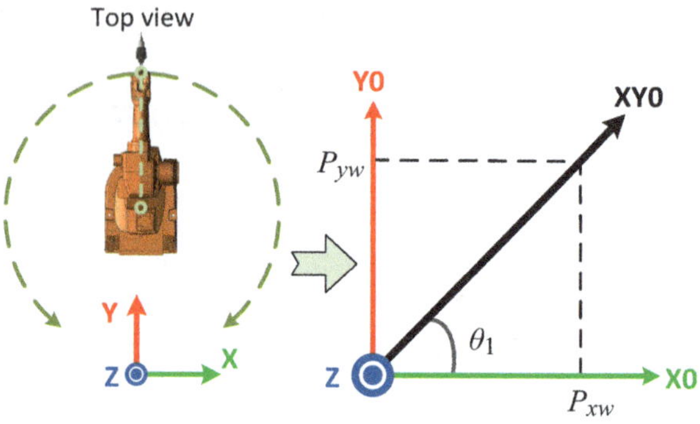


**Fig. 8.3** The robot's joint 1 projected on X-Y plane

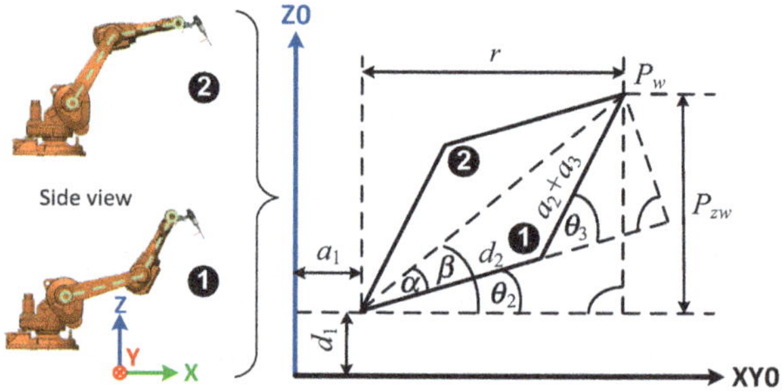


**Fig. 8.4** The robot's joints 1 and 2 projection

$$\theta_3 = \begin{cases} atan2\left(+\sqrt{1-cos^2\theta_3}, cos\theta_3\right) \\ atan2\left(-\sqrt{1-cos^2\theta_3}, cos\theta_3\right) \end{cases} \tag{8.3}$$

The calculations then continue by finding joint2 value $\theta_2$ using Eq. (8.4).

$$\theta_2 = atan\frac{P_{zw}-d_1}{r} - atan\frac{(a_2+a_3)sin\theta_3}{d_2+(a_2+a_3)cos\theta_3} \tag{8.4}$$

The orientation of the robot wrist against the base is then determined by using Eq. (8.5). The rotation matrix of the rest of the joints can be calculated by Eq. (8.6).

$$^{0}_{w}\mathbf{R} = {}^{0}_{3}\mathbf{R}\ (\theta_1, \theta_2, \theta_3).{}^{3}_{4}\mathbf{R}(\theta_4 = 0) \tag{8.5}$$

$$R_{zxz}(\theta_4, \theta_5, \theta_6) = {}^{0}_{W}R.{}^{0}_{TCP}R.{}^{6}_{TCP}R^T = \begin{bmatrix} r_{11} & r_{12} & r_{13} \\ r_{21} & r_{22} & r_{23} \\ r_{31} & r_{32} & r_{33} \end{bmatrix} \tag{8.6}$$

The configurations of the last three joints can be computed in Eqs. (8.7)–(8.10) using Euler angles $\alpha$, $\beta$, and $\gamma$.

$$\theta_4 = \alpha, \theta_5 = -\beta, \theta_6 = \gamma \tag{8.7}$$

$$\alpha = atan2\left(\frac{r_{13}}{sin\beta}, -\frac{r_{23}}{sin\beta}\right) \tag{8.8}$$

$$\beta = \begin{cases} atan2\left(+\sqrt{r_{31}^2 + r_{32}^2}, r_{33}\right) \\ atan2\left(-\sqrt{r_{31}^2 + r_{32}^2}, r_{33}\right) \end{cases} \tag{8.9}$$

$$\gamma = atan2\left(\frac{r_{13}}{sin\beta}, -\frac{r_{23}}{sin\beta}\right) \tag{8.10}$$

## 8.4.4 *Inverse Dynamics*

Recursive Newton-Euler Algorithm (RNEA) [28] is adopted in this chapter for its reliable results. The first step is to calculate the inertial tensor matrices of the robot using the 3D model of the robot together with SolidWorks® software.

The procedure of solving the inverse dynamics is divided into two steps: forward and backward recursions as explained below.

### 8.4.4.1 Forward Recursion

Starting from the first robot link to the last one, the algorithm determines the linear and angular motions of each link of the robot. Since the robot will start from a standstill state, the initial values of velocities and accelerations are set to zeros. Consequently, the angular velocity $\omega_i$ and acceleration $\alpha_i$ of link $i$ are calculated together with the linear accelerations $a_i$ and $a_{ci}$ of link $i$, using Eq. (8.11).

$$\begin{aligned} \omega_i &= {}^{i-1}_{i}\mathbf{R}^T.\omega_{i-1} + z_i.\dot{\theta}_i \\ \alpha_i &= {}^{i-1}_{i}\mathbf{R}^T.\alpha_{i-1} + z_i.\ddot{\theta}_i + \omega_i \times z_i.\dot{\theta}_i \\ a_i &= {}^{i-1}_{i}\mathbf{R}^T.a_{i-1} + \dot{\omega}_i + r_{i-1,i} + \omega_i \times \left(\omega_i \times r_{i-1,i}\right) \\ a_{ci} &= {}^{i-1}_{i}\mathbf{R}.a_{i-1} + \dot{\omega}_i \times r_{i-1,ci} + \omega_i\left(\omega_i \times r_{i-1,ci}\right) \end{aligned} \tag{8.11}$$

### 8.4.4.2 Backward Recursion

The process continues by obtaining the force and torque that affect the movement of each joint. The calculation begins with the last link and ends with the base of the robot. Using the angular velocities and accelerations calculated in the previous step, the gravity vector $g_0$ is represented in a frame for each link in Eq. (8.12).

$$g_i = {}^0_i\mathbf{R}^T . g_0 \tag{8.12}$$

The force $f_{i+1}$ and torque $\tau_i$ of link $i$ are calculated using Eqs. (8.13)–(8.14). At the same time, the external force $f_{N+1}$ and torque $\tau_{N+1}$ applied to the robot's end-effector are considered implicitly. The mechanical losses are modelled through Coulomb and viscous friction with respective coefficients $\tau_{ci}$ and $\tau_{vi}$.

$$f_i = {}^i_{i+1}\mathbf{R} \cdot f_{i+1} + m_i(a_{ci} - g_i) \tag{8.13}$$

$$\begin{aligned}\tau_i = {}^i_{i+1}\mathbf{R} \cdot \tau_{i+1} - f_i \times r_{i-1,ci} + {}^i_{i+1}\mathbf{R} \cdot f_{i+1} \times r_{i,ci} + \omega_i \times (I_i \cdot \omega_i) + I_i \cdot a_i \\ + \tau_{ci} sign\left(\dot{\theta}_i\right) . z_i + \tau_{vi}\dot{\theta}_i . z_i\end{aligned} \tag{8.14}$$

### 8.4.5 Energy Consumption

At the beginning, the power consumption at each joint in a certain time interval $k$ is calculated in Eq. (8.15), based on joint velocity obtained from inverse kinematics and required torque from backward recursion step of inverse dynamics. Consequently, the power consumptions of all joints are accumulated to get the total power consumption of the robot, as described in Eq. (8.16).

$$P_i(k) = \left(\tau_i(k) \cdot \dot{\theta}_i(k)\right) \tag{8.15}$$

$$P(k) = \sum_{i=1}^{n} P_i(k) \tag{8.16}$$

The process continues by computing the energy consumption in Eq. (8.17), where $dt_k$ is the time duration of the robot path.

$$E = \int_{t_0}^{t_M} P(t)dt \cong \sum_{k=0}^{M} P(k).dt_k \tag{8.17}$$

### 8.4.6 Energy Optimisation

Energy optimisation is performed as a final step to select the most energy-efficient robot configuration from the calculated configurations to perform the assembly tasks.

By looking at Eqs. (8.2)–(8.10), it is possible to see that there are two solutions for first joint $\theta_1$, each solution is used to calculate two solutions for both second joint $\theta_2$ using Eq. (8.4) and third joint $\theta_3$ using Eq. (8.3), This leads to four solutions for the first three joints. The process continues by using Eqs. (8.7) and (8.9) to find two solutions for the fifth joint $\theta_5$, each solution for $\theta_5$ is then used to calculate $\theta_4$ and $\theta_6$ using Eqs. (8.7), (8.8) and (8.10), which generates a single solution for each case. Therefore, the solutions for the last three joints of the robot are two. The total number of solutions for the robot is then equal to eight, calculated by combining the solutions of the first three joints with the ones from the last three joints. Without a doubt some of the calculated solutions are unfeasible among the whole path, in that case they are omitted and the total number of solutions can be less than eight. Having this limited number of solutions for the robot makes the optimisation a simple and straightforward step. Finding the suitable configuration can be performed quickly by selecting the one that has the lowest energy consumption.

## 8.5 Case Studies

Two case studies have been examined to evaluate the capability of the proposed energy optimisation approach as described in the following subsections.

### *8.5.1 Energy Map of Robot Workspace*

In this case study, a square shape path was examined at different locations within the robot's workspace. Energy consumption is optimised at each location in the workspace. Figure 8.5 depicts a 3D energy map of the robot, where the energy consumptions are represented in colours from green to red. Using the energy map as a guide, an engineer can place a workpiece in a location of low energy level in a robotic assembly cell. It is obvious that the areas in red or close to red should be avoided for energy saving.

### *8.5.2 Energy Measurement in Predefined Paths*

A case study emulates collision-free assembly operations which require moving the robot in a certain hypothetical trajectory is introduced. As shown in Fig. 8.6, this path is tested in two different locations to examine the correctness of using the energy map. Two identical paths placed respectively in "red" and "green" zones are considered.

Table 8.1 describes the robot's joints values of the first target in the first and second path for two possible joint configurations, respectively.

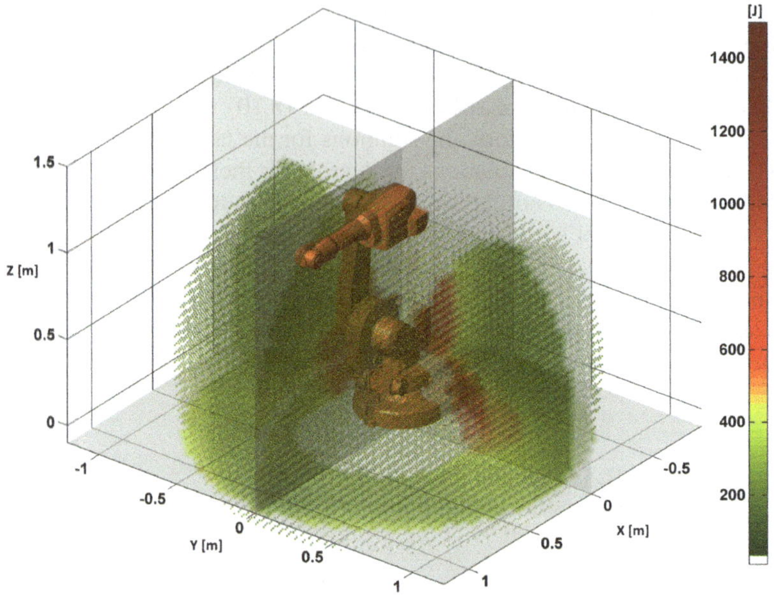


Fig. 8.5 An energy map in the workspace of an ABB IRB 1600 robot

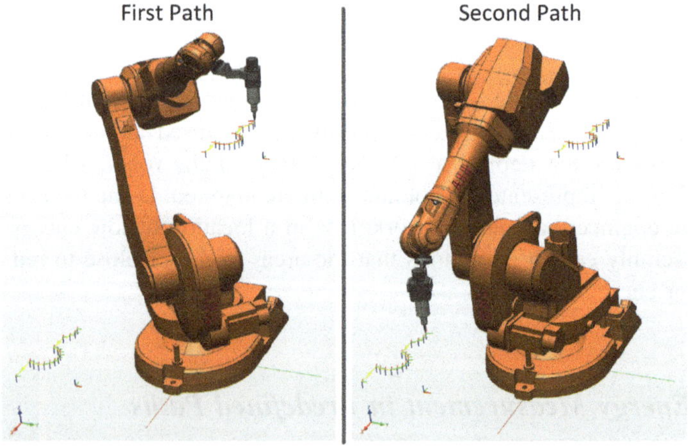


Fig. 8.6 The hypothetical paths

The case study is introduced to the optimisation module as well as to ABB RobotStudio® and the real robot. The results of energy-optimised path-following are illustrated in Fig. 8.7.

The case study is conducted in RobotStudio® and the results are recorded for later comparison. Furthermore, the energy consumptions of the hypothetical paths

**Table 8.1** The joint values (deg) of the experiment paths with corresponding simulated energy consumption

| Path | | First Joint | Second Joint | Third Joint | Fourth Joint | Fifth Joint | Sixth Joint | Energy [J] |
|---|---|---|---|---|---|---|---|---|
| First | Conf. A | 53 | 38 | -13 | 30 | -91 | -139 | 463 |
| | Conf. B | -126 | 68 | -2 | 90 | 149 | 131 | 434 |
| Second | Conf. A | -6 | -9 | -29 | 45 | -22 | -153 | 201 |
| | Conf. B | 173 | 110 | 1 | 18 | 121 | 79 | 451 |

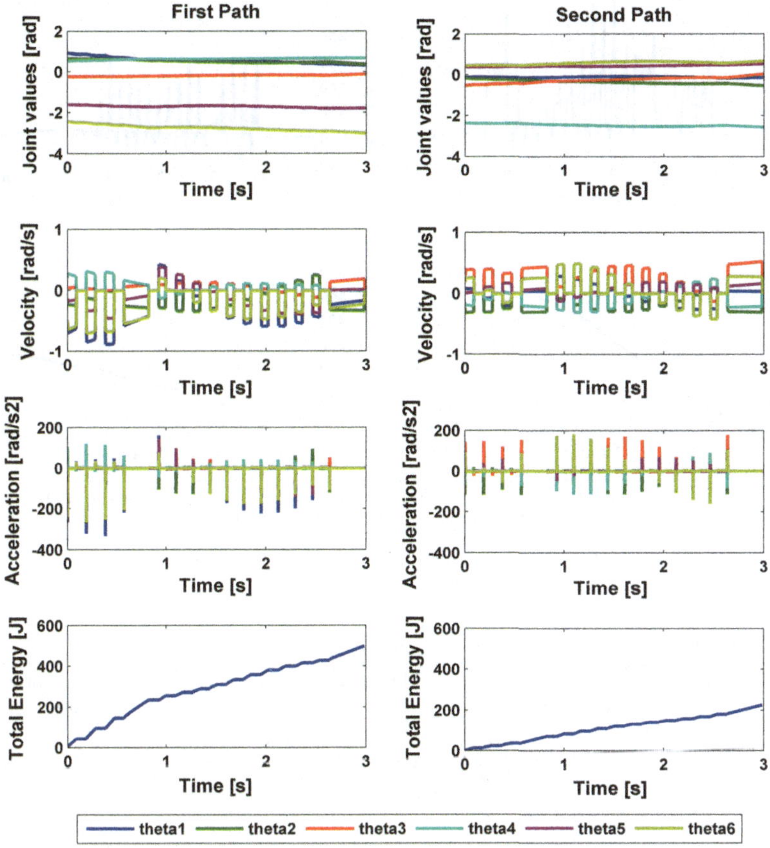


**Fig. 8.7** The output of the energy optimisation for the hypothetical paths

are measured on a real robot (ABB IRB1600) to evaluate the accuracy of the optimisation. The results are illustrated in Fig. 8.8.

The measurements are conducted using a 3-phase voltage module (NI 9244) and a current module (NI 9238) from National Instrument™. Table 8.2 summarises the results of the optimisation module introduced in this chapter and the ones from

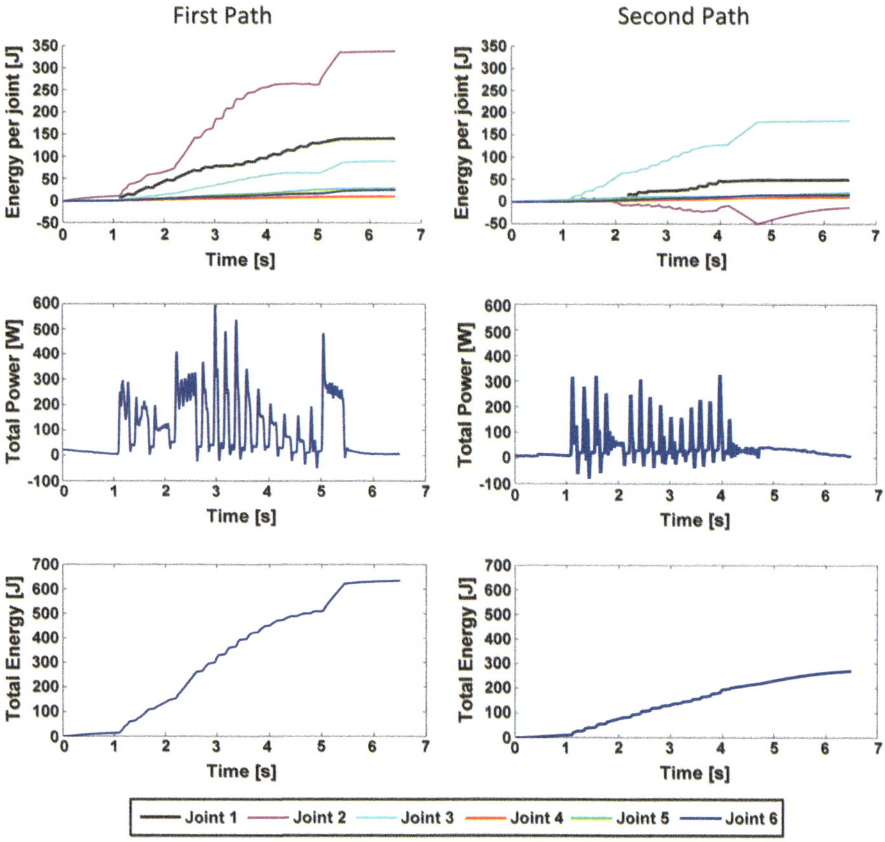


Fig. 8.8 The measurements on the real robot of the hypothetical paths

**Table 8.2** Comparison for the energy consumption results

| Path | Energy consumption [J] | | |
|---|---|---|---|
| | Measured | RobotStudio® | Optimisation module |
| First | 635 | 170 | 463 |
| Second | 275 | 60 | 201 |

RobotStudio®, respectively. Due to different placements of the path, a difference in energy consumption is clear from the results. The results showed that the robot can have energy harvesting during its movement, such as the energy in Joint 2 illustrated in Fig. 8.8.

The results illustrated in Fig. 8.7 and Fig. 8.8 for the case study show that the energy map can be used as a tool to study the robot envelop and design the robotic cell based on that. Furthermore, the results show that the energy map can identify how sustainable the robot is from the energy efficiency perspective.

## 8.6 Conclusions

Energy consumption of machines and robots is an important factor in a production line, from both environmental and economical aspects. Robots heavily used in assembly lines can contribute to the reduction of energy consumption by selecting energy-efficient configurations together with the wise usage of the robot workspace in the low-energy area. The optimisation method introduced in this chapter uses this strategy to select the robot's joint configurations that consume the minimum energy. Given that the energy consumption also relies on the position and orientation of a workpiece in the robot workspace, the concept of energy map is proposed to advise shop floor engineers for workpiece placement towards low energy consumption. On the other hand, energy consumption in robotic assembly is influenced by robot kinematics, dynamics, task requirements and the technical features of robots in terms of design. Except the last one, the energy behaviours of robots are modelled mathematically, which is useful for energy-efficient robot control.

With few modifications, the presented energy module can be used for a wide range of industrial robots. These modifications are needed to identify the robot and the payload properties (kinematic model, mass properties, payload on the end-effector, etc.). Several other robots have the same kinematic structure like the one presented in the chapter, only parameters of kinematics and dynamics models need to be identified and applied to integrate those robots with the presented energy module. On the other hand, integrating robots that have different kinematic structure needs new calculation for the inverse kinematics as well.

The experimental results presented previously showed a noticeable difference between the measured energy consumption and the simulated one in RobotStudio®. This indicates that the mathematical model of the robot in RobotStudio® is not fully identical to the real robot. Addressing the main reasons for the difference is not possible at the moment because RobotStudio® is proprietary software which means that its source code is not publicly available. This difference can be clarified if the robot manufacturer decides in the future to open their software to developers.

The model used in the reported optimisation can be improved further, so that the approach can have better identification of the robot's dynamic specifications. The technical features of robots such as energy loss in motors, gears and couplings are the remaining challenges of energy modelling as future work. Further validation of the energy model and the optimisation approach deserves more attentions from the research community.

## References

1. C. Okwudire, J. Rodgers, Design and control of a novel hybrid feed drive for high performance and energy efficient machining. CIRP Ann. - Manuf. Technol. **62**(1), 391–394 (2013)
2. Z. Liu, W. Bu, J. Tan, Motion navigation for arc welding robots based on feature mapping in a simulation environment. Robot. Comput. Integr. Manuf. **26**(2), 137–144 (2010)

3. R. Saravanan, S. Ramabalan, C. Balamurugan, Evolutionary multi-criteria trajectory modeling of industrial robots in the presence of obstacles. Eng. Appl. Artif. Intell. **22**(2), 329–342 (2009)
4. P.G. de Santos, E. Garcia, R. Ponticelli, M. Armada, Minimizing energy consumption in hexapod robots. Adv. Robot. **23**(May), 681–704 (2009)
5. B. Vanderborght, N.G. Tsagarakis, R. Van Ham, I. Thorson, D.G. Caldwell, MACCEPA 2.0: Compliant actuator used for energy efficient hopping robot Chobino1D. Auton. Robots **31**(1), 55–65 (2011)
6. S. Rahimifard, Y. Seow, T. Childs, Minimising embodied product energy to support energy efficient manufacturing. CIRP Ann. - Manuf. Technol. **59**(1), 25–28 (2010)
7. N. Weinert, S. Chiotellis, G. Seliger, Methodology for planning and operating energy-efficient production systems. CIRP Ann. - Manuf. Technol. **60**(1), 41–44 (2011)
8. D. Mourtzis, E. Vlachou, N. Xanthopoulos, M. Givehchi, L. Wang, Cloud-based adaptive process planning considering availability and capabilities of machine tools. J. Manuf. Syst. **39**, 1–8 (2016)
9. M. Mori, M. Fujishima, Y. Inamasu, Y. Oda, A study on energy efficiency improvement for machine tools. CIRP Ann. - Manuf. Technol. **60**(1), 145–148 (2011)
10. A. Vijayaraghavan, D. Dornfeld, Automated energy monitoring of machine tools. CIRP Ann. - Manuf. Technol. **59**(1), 21–24 (2010)
11. T. Behrendt, A. Zein, S. Min, Development of an energy consumption monitoring procedure for machine tools. CIRP Ann. - Manuf. Technol. **61**(1), 43–46 (2012)
12. T. Peng, X. Xu, L. Wang, A novel energy demand modelling approach for CNC machining based on function blocks. J. Manuf. Syst. **33**(1), 196–208 (2014)
13. A. Mohammed, B. Schmidt, L. Wang, L. Gao, Minimizing Energy Consumption for Robot Arm Movement. Procedia CIRP **25**, 400–405 (2014)
14. A. Vergnano et al., Modeling and optimization of energy consumption in cooperative multi-robot systems. Tase **9**(2), 423–428 (2012)
15. M. Pellicciari, G. Berselli, F. Leali, A. Vergnano, Mechatronic s A method for reducing the energy consumption of pick-and-place industrial robots. Mechatronics **23**(3), 326–334 (2013)
16. D. Verscheure, B. Demeulenaere, J. Swevers, J. De Schutter, and M. Diehl, Time-Optimal Path Tracking for Robots: A Convex Optimization Approach, 54 (10), 2318–2327 (2009)
17. J.F. Gauthier, J. Angeles, and S. Nokleby, Optimization of a test trajectory for SCARA systems. Adv. Robot Kinemat. Anal. Des. 225–234 (2006)
18. M. Brossog, J. Kohl, J. Merhof, S. Spreng, J. Franke, Energy consumption and dynamic behavior analysis of a six-axis industrial robot in an assembly system. Procedia CIRP **23**, 131–136 (2014)
19. P. Ystgaard, T.B. Gjerstad, T.K. Lien, and P.A. Nyen, "Mapping Energy Consumption for Industrial Robots, 2012
20. R. Chhabra, M. Reza Emami, Holistic system modeling in mechatronics. Mechatronics **21**(1), 166–175 (2011)
21. D. Gielen, M. Taylor, Modelling industrial energy use: The IEAs Energy Technology Perspectives. Energy Econ. **29**(4), 889–912 (2007)
22. J. Gregory, A. Olivares, E. Staffetti, Energy-optimal trajectory planning for robot manipulators with holonomic constraints. Syst. Control Lett. **61**(2), 279–291 (2012)
23. R. Saravanan, S. Ramabalan, C. Balamurugan, Evolutionary optimal trajectory planning for industrial robot with payload constraints. Int. J. Adv. Manuf. Technol. **38**(11–12), 1213–1226 (2008)
24. E.S. Sergaki, G.S. Stavrakakis, A.D. Pouliezos, Optimal robot speed trajectory by minimization of the actuator motor electromechanical losses. J. Intell. Robot. Syst. Theory Appl. **33**(2), 187–207 (2002)
25. C. Hansen, J. Oltjen, D. Meike, and T. Ortmaier, Enhanced Approach for Energy-Efficient Trajectory Generation of Industrial Robots, pp. 1–7, 2012

26. P. Matthias and B. Martin, Reducing the energy consumption of industrial robots in manufacturing systems, pp. 1315–1328, 2015
27. J. Denavit, R.S. Hartenberg, A kinematic notation for lower-pair mechanisms based on matrices. J. Appl. Mech. **22**, 215–221 (1955)
28. J.Y.S. Luh, M.W. Walker, R.P.C. Paul, On-Line computational scheme for mechanical manipulators. J. Dyn. Syst. Meas. Control **102**(2), 69–76 (1980)

# Chapter 9
# Safety in Human-Robot Collaborative Assembly

## 9.1 Introduction

A human-robot collaborative system requires the coexistence of both humans and robots. The consistent safety of human in such environment is paramount, including both passive collision detection and active collision avoidance by monitoring the operators and controlling the robots, respectively, at runtime.

Several approaches for human-robot collaborations have been reported, recently. Agravante et al. [1] and Monje et al. [2] introduced a control system for a humanoid robot to carry out a joint operation with an operator. Takata and Hirano [3] presented a solution that adaptively allocates human operators and industrial robots in a shared assembly environment. Chen et al. [4] revealed an optimisation process with multiple objectives based on simulation for assigning and strategy generation of human-robot assembly subtasks. Krüger et al. [5] highlighted the merits and available technologies of human-robot collaborative assembly cells. Using a human-robot shared approach can offer both the reliability of robots and the adaptability of humans. On the other hand, however, such a system can provoke additional stress to human operators if implemented in poorly designed assembly lines. Therefore, Arai et al. [6] measured an operator's mental strain caused by the location and speed of a robot with respect to the operator, intending to establish a beneficial hybrid assembly environment. Furthermore, Kuli and Croft [7] used robot motion as a stimulus to estimate the human effective state in real time; the developed system analysed human biological indicators like heart pulse, perspiration level and facial expression.

Several recent approaches attempted to detect and protect operators in locations shared by humans and robots. Two methods were widely considered: (1) using a vision system to perform 3D inspection [8] through 3D models as well as skin colour detection to perform 3D tracking of human body in a robotic cell, and (2) inertial sensor-based approach [9] using geometry representation of operators through a special suit for motion capturing. Real-world experiments indicate that



L. Wang and X. V. Wang, *Cloud-Based Cyber-Physical Systems in Manufacturing*,
https://doi.org/10.1007/978-3-319-67693-7_9

the latter approach may not be considered as realistic solutions as it relies on the existence of a particular uniform with sensing devices and the inadequacy of capturing the movement around the person wearing the uniform, leaving the neighbouring objects unsupervised. This can create a safety leak, as there may be a possibility of collision between a moving object and a standing-still operator. More details of varying sensing methods can be found in a 2010 literature survey [10].

Among vision-based methods, the efficiency of collision detection has been the motivation for many researchers. For example, Gecks and Henrich [11] implemented a multi-camera collision detection system, whereas a high-speed emergency stop was utilised in [12] to avoid a collision using a specialised vision chip for tracking. A projector-camera based approach was presented in [13], which consists of defining a protected zone around the robot by projecting the boundary of the zone. The approach is able to dynamically and visually detect any safety interruption. In [14], a triple stereovision system was reported for capturing the motion of a seated operator (upper-body only) by wearing colour markers. Nonetheless, relying on the colour consistency may not be suitable in uneven environmental lighting conditions. In addition, the tracking markers of mobile operators may not appear clearly in the monitored area. Instead of markers, a ToF (time-of-flight) camera was adopted in [15] for collision detection, and an approach using 3D depth information was proposed in [16] for the same purpose. Using laser scanners in these approaches offers suitable resolution but requires longer computational time, since each pixel or row of the captured scene is processed independently. On the other hand, ToF cameras provide high performance solution for depth images acquisition, but with insufficient level of pixel resolution (capable of reaching $200 \times 200$) and with rather high expense. Lately, Rybski et al. [17] acquired data from 3D imaging sensors to construct a three-dimensional grid for locating foreign objects and identifying human operators, robots and background. More recently, an integrated approach for collision avoidance using depth information from single Kinect sensor was reported in [18].

In addition, other researchers focused on combining different sensing techniques to track the human and the robot on the shop floor like the work presented in [19] which used both the ultrasonic and the infrared proximity sensor to establish a collision free robotic environment. Meanwhile, other researchers like Cherubini et al. [20] incorporated both the force/torque sensors and the vision systems into a hybrid assembly environment to provide a direct contact between the human and the robot.

Among commercial systems of safety protection solutions, SafetyEYE® [21] is a popular choice. It computes 2½D data of a monitored region using a single stereo image and detects violation of predefined safety zones. Accessing into any of the safety zones will trigger an emergency stop of the monitored environment. However, these safety zones cannot be updated during operation.

To maintain high productivity in human-robot collaboration, there is a necessity to introduce an inexpensive and reliable online protection system for assembly lines where onsite operators share the tasks with industrial robots in a fenceless environment. Although there have been advancements in safety protection in the past decade, the surveyed methods and systems are either too expensive or excessively limited in handling real-world applications. Aiming to solve this problem, this chapter presents a novel approach to providing a safe and protected environment for human operators to work with robots alongside. Its novelty consists of: (1) effective detection of any collision between 3D models of robots and depth camera images of humans in an augmented reality environment, and (2) active avoidance of any possible collision through online robot control.

During the recent years, researchers have developed various tools to programme, monitor and control industrial robots. The aim is to reduce possible robot downtime and avoid collisions caused by inaccurate programming, through simulation and/or augmented reality [22]. However, these tools require pre-knowledge about a robotic system. Introducing unknown objects to the robotic system may produce unrealistic solutions that cause a breakdown to the physical robotic system due to no-longer valid robot programmes.

Laser scanners and vision cameras are common techniques to convert unknown objects to virtual 3D models. Modelling objects using stereo vision cameras was a main focus for research [23–25], whereas others including [26] adopted a predefined library of 3D models to match the real desired objects. However, the stereo vision camera-based approach suffers from two drawbacks: (1) it requires expensive and less compact equipment, and (2) it lacks the ability to capture and model complex shapes from a fixed single viewpoint due to limited visibility.

2D vision systems can also be applied to model unknown objects. By taking a number of snapshots of an object from different viewports, the object can be modelled based on analysing the captured silhouette in each snapshot. For example, Petit et al. [27], and Atsushi et al. [28] focused on modelling the object in high accuracy and with details.

Despite the fact that these approaches were successful in their reported applications, they are unable to model multiple objects in a single run, besides their lack of ability to model objects remotely.

In this chapter, we introduce a new approach for constructing 3D models of multiple arbitrary objects, simultaneously, based on a set of snapshots taken for the objects from different angles. This approach is implemented through a system that analyses the captured silhouettes of the objects and constructs 3D representations for the objects in a web-based environment, allowing an operator to perform assembly operations from a distance.

## 9.2 Human Robot Collaboration

Based on Wise-ShopFloor architecture [29–31], an active collision avoidance solution is illustrated in Fig. 9.1. Due to their high performance and flexibility, both C++ and Java are used for developing the system. An industrial robot, ABB IRB 140, is employed to construct a physical human-robot collaborative assembly cell for testing and verification. A local server responsible for collision avoidance is

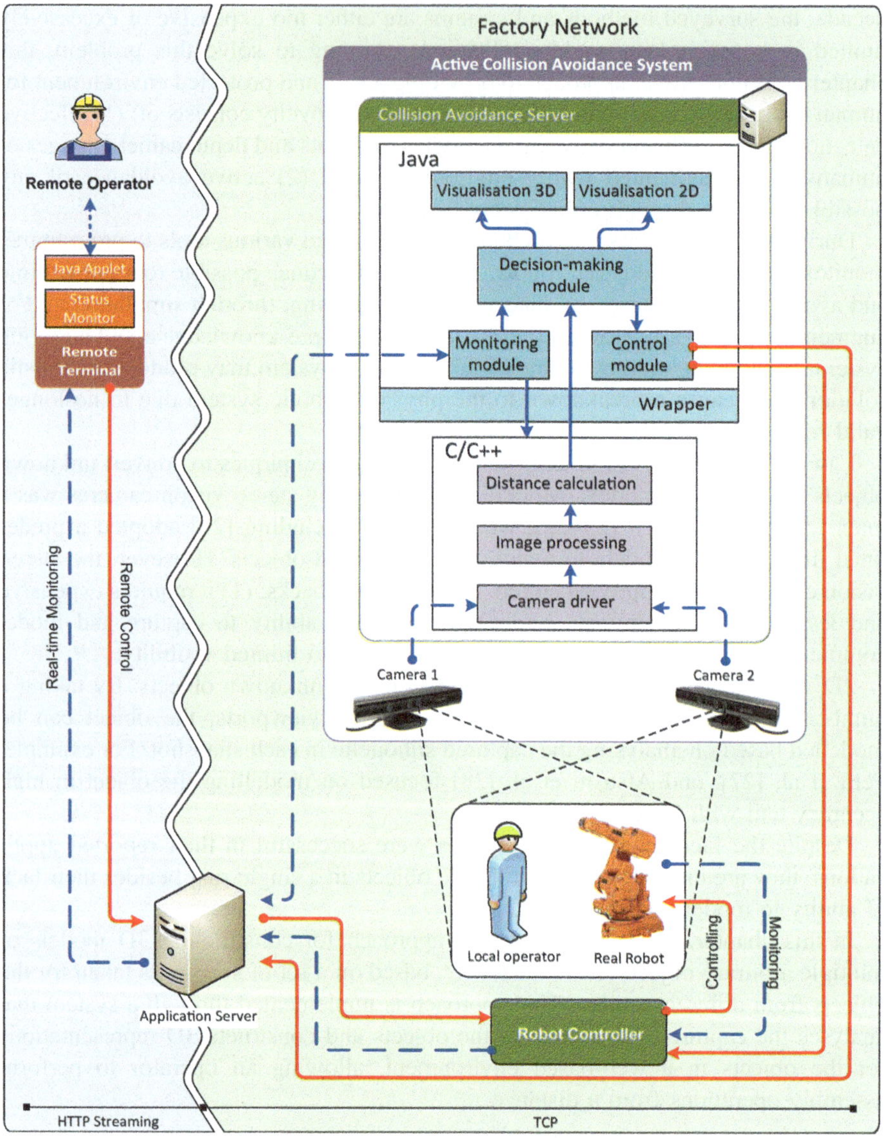


**Fig. 9.1** System design for active collision avoidance

configured in a PC of Intel 2.7 GHz Core i7 CPU with 12 GB RAM, and running a 64-bit Windows operating system. With the help of Java 3D, this collision avoidance server is used for image processing and establishing a collision-free environment. Furthermore, two Microsoft Kinect sensors (depth cameras) are installed to obtain the depth images of operators in the robotic cell.

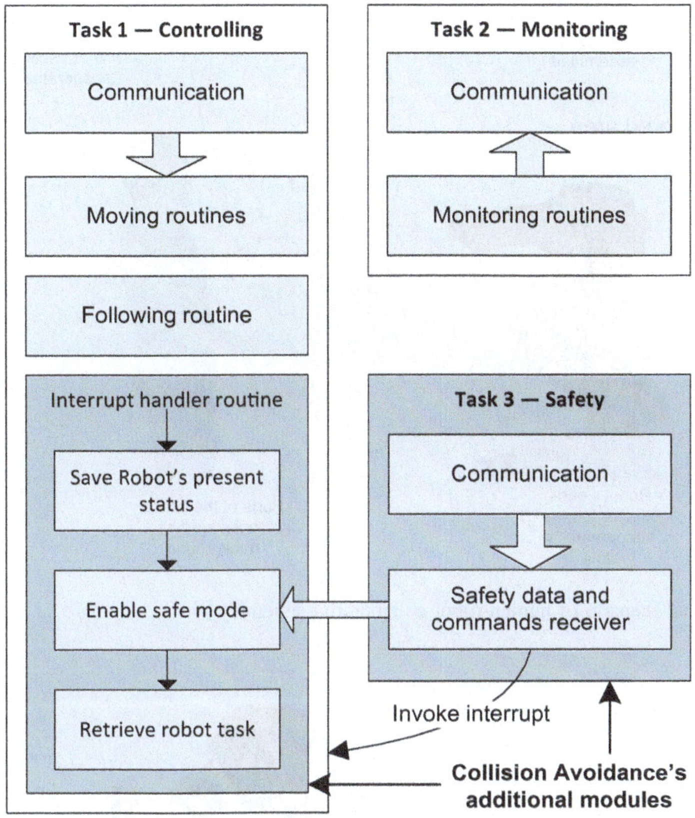


**Fig. 9.2** Task configurations of robot controller

Initially in Wise-ShopFloor, the robot controller has two tasks running simultaneously to control and monitor the robot. For the purpose of active collision avoidance, a third task is added to the robot controller as shown in Fig. 9.2.

The collaboration scenario shown in Fig. 9.3 is typical where the robot follows the operator's hand to deliver needed assistance during a shared assembly operation. Another scenario could be that the robot keeps a safety distance from the operator during assembly. For example, in Fig. 9.4, the robot is used to assemble a shaft and a washer, and insert the assembled parts in an output magazine. The operator's responsibility is to take out the assembled parts from the output magazine and fill fresh parts into an input magazine. The active collision avoidance is activated when the robot moves to/from the two magazines. Avoiding collision is restricted only to control the robot at the time of picking and delivering parts. For seamless human-robot collaboration, switching between the two behaviours of the robot (i.e. hand following and collision avoiding) can be implemented by a simple button press or through a voice command.

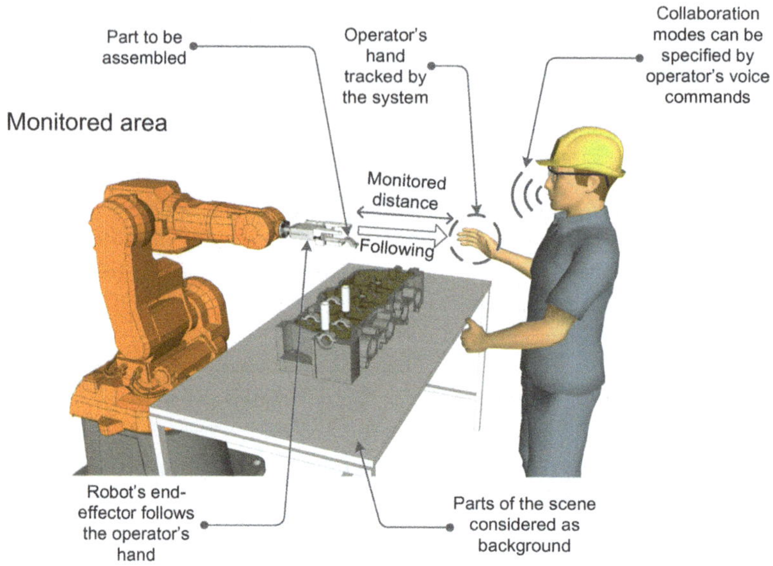


**Fig. 9.3** One scenario of human-robot collaborative assembly

**Fig. 9.4** A mini robotic assembly cell for testing

## 9.3 Depth Sensor-Driven Active Collision Avoidance

The present approach starts by calibrating the Kinect sensors, followed by acquiring the depth information from them. The process continues by determining the closest distance between the robot and obstacles (including operators), and active collision avoidance is then performed. The velocity of the approaching operator is also calculated to improve the system responsiveness. The following sections describe in detail the mechanism of this approach.

### *9.3.1 Kinect Sensors Calibration*

The depth vision systems selected in this system are Microsoft Kinect sensors (depth cameras) equipped with spatial resolution of 640 × 480, 11-bit depth, and a field of view of 58 × 40°, which can measure the depth information from 0.8 to 3.5 m, as illustrated in Fig. 9.5. A calibration of the Kinect sensors is needed to calculate the distance values. It is accomplished by measuring a particular surface from different distances. To improve the accuracy in the distance calculation, Eq. (9.1) is implemented for optimising the sensors' parameters.

$$d = k_3 \tan\left(\frac{n}{k_2} + k_1\right) - k_4 \tag{9.1}$$

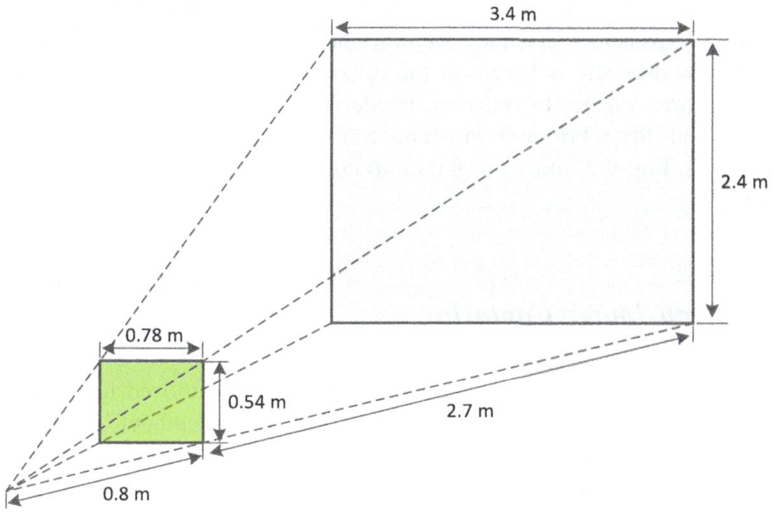


**Fig. 9.5** Kinect sensor's field of view

**Table 9.1** Kinect sensors' parameters after calibration

| Parameter | Measured value | | Unit |
|---|---|---|---|
| | Kinect sensor 1 | Kinect sensor 2 | |
| $k_1$ | 1.1873 | 1.18721 | rad |
| $k_2$ | 2844.7 | 2844.1 | 1/rad |
| $k_3$ | 0.12423 | 0.1242 | m |
| $k_4$ | 0.0045 | 0.01 | m |

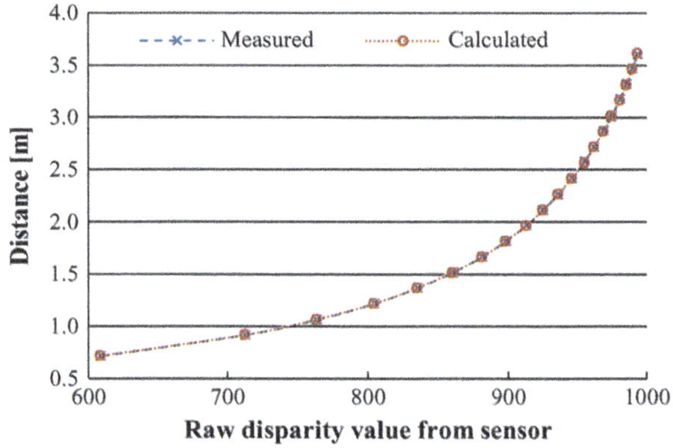


**Fig. 9.6** Outcomes of distance calculation after calibrating one sensor

where $n$ represents the raw 11-bit to describe the distance from one Kinect sensor and $k_1$–$k_4$ are the calibration parameters. The parameters are determined after optimisation with their values described in Table 9.1.

Taking one of the sensors as an example, its calibration outcomes of distance calculation versus sensor accuracy are described in Figs. 9.6 and 9.7, respectively, where the raw disparity is based on the difference between the projected and the recorded infrared image. In order to decide a suitable threshold value for safety protection, both the robot environmental conditions and the measurement accuracies, shown in Fig. 9.7, must be taken into consideration so as not to compromise the safety.

## 9.3.2 *Depth Image Capturing*

The *libfreenect userspace* driver of Microsoft Kinect is utilised to communicate with the depth sensors and access to the raw data (colour, depth, etc.). The first step is to initialise the driver so as to detect all connected Kinect sensors, followed by the second step of reading depth streams through a call-back function. Retrieving the depth frame in the third step and processing the depth data pixel by pixel in the fourth step complete the whole process as shown in Fig. 9.8. A call-back function is

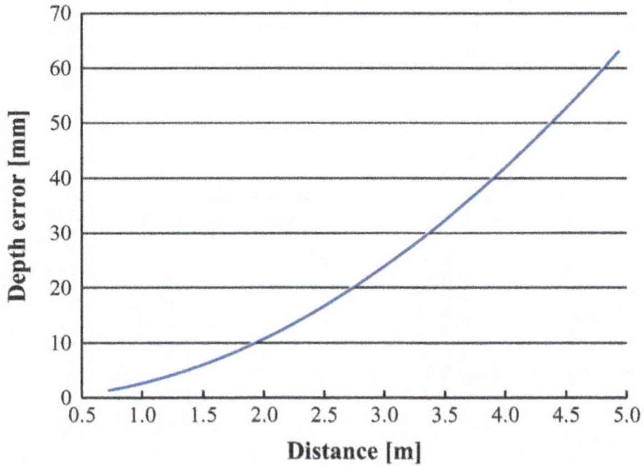


**Fig. 9.7** Depth error versus distance of one Kinect sensor

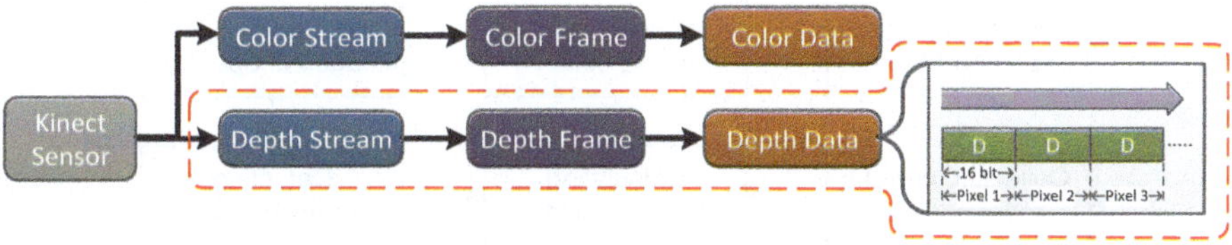


**Fig. 9.8** Kinect sensor data interface

triggered when a new depth frame is available to copy the data, pre-process the data and pass the data to the processing stage.

### *9.3.3 Depth Image Processing*

In this section, 3D models are introduced to represent a well-defined shop-floor environment. Physical motion sensors are linked to the collision avoidance system to drive the behaviour of the 3D models and monitor the shop-floor in real time. By reading the joint values of a robot from its controller, the present pose of the robot can be retrieved to the basic human-robot shared environment for visualising the 3D model of the robot. At the same time the human operator can be represented as point cloud with the help of the depth images from the Kinect sensors. Two Kinect sensors are employed for surveillance of unstructured foreign objects in the robotic cell, including mobile operators who lack the representation in the 3D space. The concept is depicted in Fig. 9.9. To sustain the rapid processing, the closest range between the virtual 3D model of the robot and camera information about the operator's location is used to detect any collision in an augmented environment.

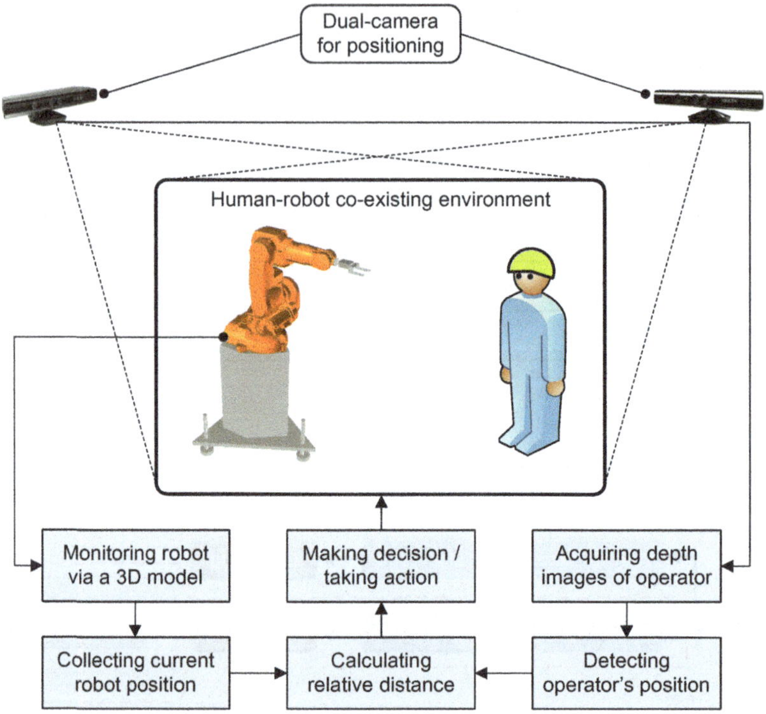


**Fig. 9.9** Concept of collision detection in an augmented environment

Supported by the calculated minimum distance, suitable decision can be made, which leads to efficient prevention of any possible collision.

The detailed procedures of depth images capturing and analysis are given in Fig. 9.10. To maintain the efficiency of the system, the procedure begins by removing the background from the depth images using the background images captured during calibration. Depth information related to the movement of the robot is subtracted as well from the captured depth images by projecting back the robot model to the depth images. Therefore, merely the unknown objects are retained, as shown in Fig. 9.10 where the images in the third row depict a human operator as the subject of interest after employing a *noise-removal* filter and a *connected-component* algorithm. After the background removal, the noise in the captured point cloud is eliminated by adding a statistical *outlier removal* filter.

Identifying the human operator from both cameras is achieved by converting the captured images to point clouds represented in the robot coordinate system, then combining them into one point cloud after registering the images. The captured point cloud of the human operator is superimposed to the 3D model of the robot to generate an augmented environment, allowing the system to calculate the minimum distance between the point cloud and the 3D robot model effectively. Figure 9.10 also illustrates the outcome of the depth images processing at varying stages.

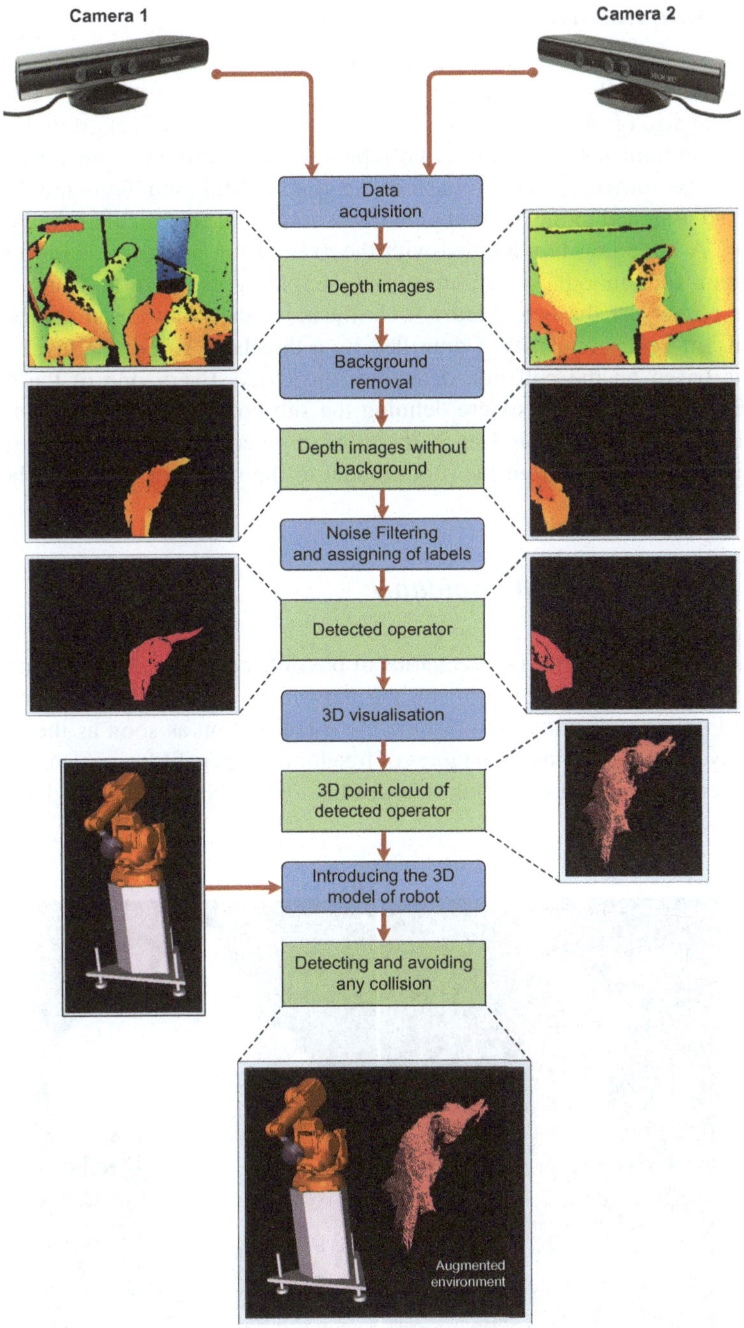


**Fig. 9.10** Procedures and outcomes of depth image processing

### 9.3.4 Minimum Distance Calculation

The point cloud of the human operator contains a large quantity of data, despite the fact that the size of the image is considerably decreased after background removal. Therefore, minimising the point cloud representation becomes a necessity, which leads to the performance improvement of the system. Minimum bounding boxes are chosen and assigned to the 3D point clouds to accelerate the calculation of collision detection. A bounding box aligned with the axes of the 3D space is introduced in the system for smooth visualisation and simple representation using the two opposite corners of the box. Figure 9.11 illustrates the point cloud and bounding boxes in different granularity. Controlled by a threshold value, the level of granularity is based on the collision detection sensitivity. Every one of the boxes is considered as the smallest sphere defining the sub-box that further helps to accelerate the distance calculation. Hence, the problem of collision detection is treated as the distance calculation from the robot model to the centre of every sub-box.

### 9.3.5 Active Collision Avoidance

It is not difficult to identify two scenarios of possible collisions during human-robot collaboration. The first one is to stop the movement of the robot if operators are detected in close proximity, and resume the robot motion as soon as the operators walk away. Tasks with limited degrees of freedom can benefit from such a scenario, for instance an inserting operation in assembly. The second one is to dynamically

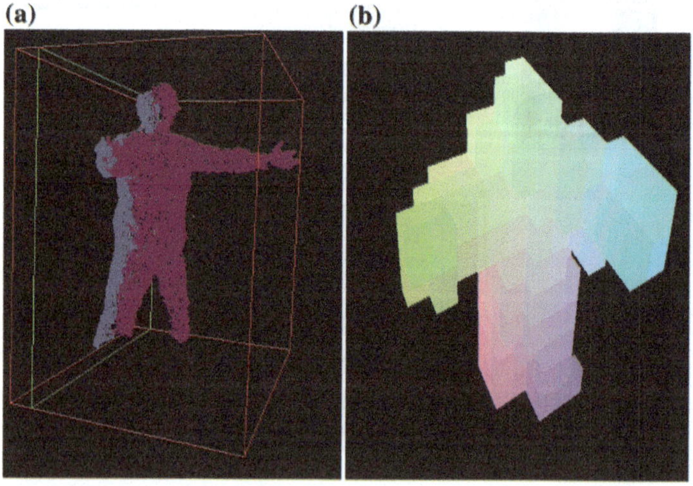


**Fig. 9.11** Simple representation of point cloud: **a** a minimum box bounding the captured point cloud, and **b** a matrix of sub-boxes

modify a robot's trajectory to avoid collision with any foreign obstacles (including humans). It is more appropriate to be employed for such applications as material transfer where the path modification can less affect the operations while keeping the humans safe.

Detecting an obstacle in a planned robot path triggers the system to dynamically alter the robot's trajectory. The change in the robot trajectory is driven by the calculated distance to the obstacle. Figure 9.12a shows the vectors needed for dynamic robot path control. Collision vector **c** is determined and represented as the vector from the robot's end-effector to the nearest obstacle. The vector symbolising the robot's direction of movement $\mathbf{v_c}$ can be decomposed into a parallel component $\mathbf{v_c} \| \mathbf{c}$ and a perpendicular component $\mathbf{v_c} \perp \mathbf{c}$ against the collision vector **c**. The parallel component is determined in Eq. (9.2) by calculating the dot product between the movement vector and a collision versor pointing to the direction of the collision vector. The collision versor represents a unit vector and shares the same direction of the collision vector. The perpendicular component is then computed using Eq. (9.3). The parallel component representing the motion approaching the obstacle is thus modified to prevent any possible collision. The modification is supported by the computation of the distance between the obstacle and the robot end-effector in this case, resulting in a vector $\mathbf{v_a} \| \mathbf{c}$. A new vector of modified robot motion can then be generated from $\mathbf{v_c} \perp \mathbf{c}$ and $\mathbf{v_a} \| \mathbf{c}$ as explained in Eq. (9.4).

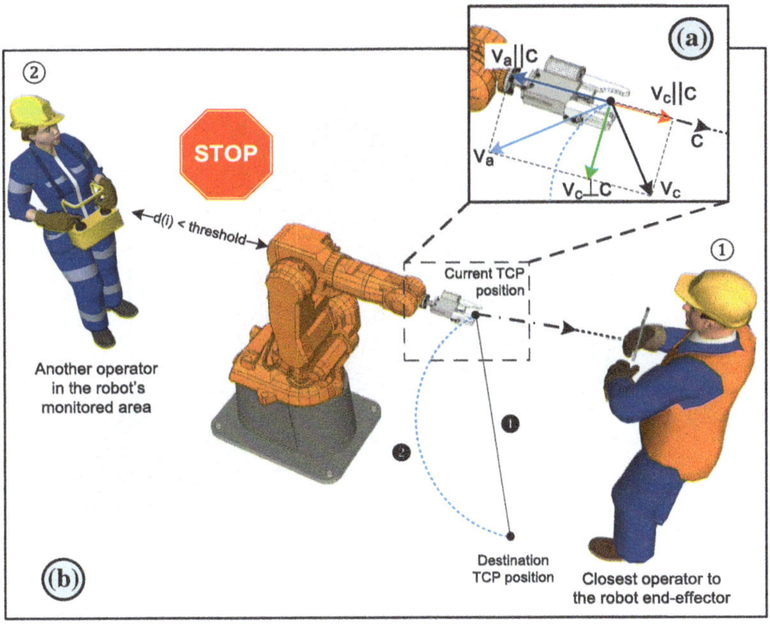


**Fig. 9.12** Robot trajectory modification in real-time (**a**), and collision avoidance in a multi-operator environment (**b**)

$$\mathrm{v_c}||c = (\mathrm{v_c} \cdot \hat{c}) \cdot \hat{c}, \quad \text{where } \hat{c} = \frac{\mathrm{c}}{|\mathrm{c}|} \tag{9.2}$$

$$\mathrm{v_c}\perp c = \mathrm{v_c} - \mathrm{v_c}||c \tag{9.3}$$

$$\mathrm{v_a} = \mathrm{v_a}||c + \mathrm{v_c}\perp c \tag{9.4}$$

Taking distance-to-obstacle $\|\mathbf{c}\|$ into account, Fig. 9.13 shows the variation of the parallel component against the movement vector. The colour spectrum indicates the modified value of the parallel component $\mathbf{v_a}\|\mathbf{c}$ of the collision vector. The line indicated by ❶ is the anticipated movement toward the obstacle, and line ❷ is the anticipated movement in the opposite direction. To modify the parallel component of the movement vector, two threshold values $d_{th1}$ and $d_{th2}$ are defined. The parallel component remains the same as long as the distance-to-obstacle $\|\mathbf{c}\|$ is greater than $d_{th2}$. When $\|\mathbf{c}\|$ is smaller than $d_{th1}$, the parallel component receives a negative defined value, pointing to the opposite direction of the obstacle. In the case that $d_{th1} < \|\mathbf{c}\| < d_{th2}$, the parallel component is modified linearly as shown in Fig. 9.13 to maintain the continuity of the movement.

Based on the calculated shortest distance-to-obstacle, one of four safety strategies is applied to control the robot. Figure 9.14 explains four cases where the four safety strategies can be employed: ❶ an audio warning is fired as soon as an operator walks into the monitored area, and at the same time the speed of the robot is reduced to prepare for a full stop; ❷ a retrievable stop interruption is sent to the

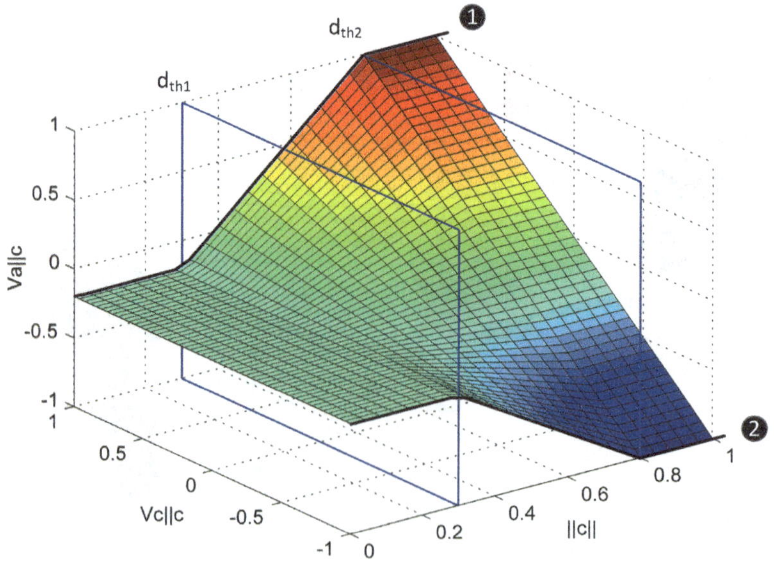


Fig. 9.13 Modification of movement vector

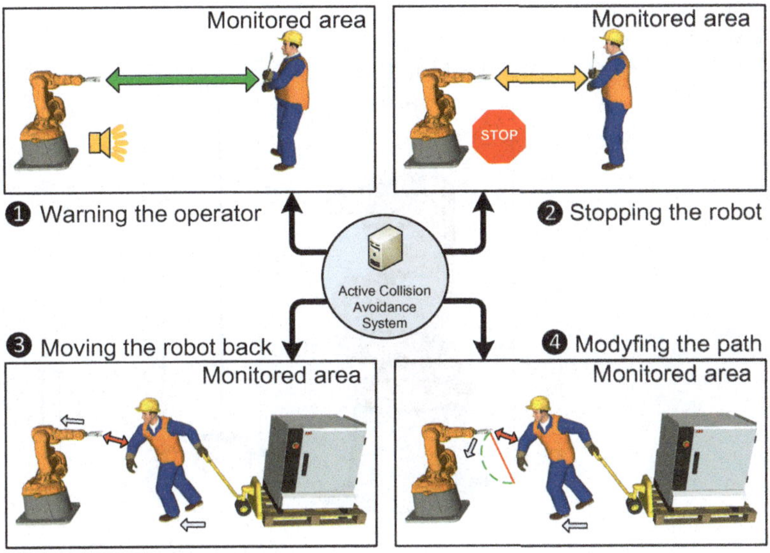


**Fig. 9.14** Four safety strategies for active collision avoidance

robotic system if the human steps into a defined hazard zone; nonetheless; ❸ when the human continues to move towards the robot (e.g. for inspection, etc.), the robot arm will move away automatically to keep a safe distance from the operator for collision avoidance; and ❹ in light of the possibility to change the path, the current robot trajectory to the target is modified dynamically to prevent any collision with the operator while the robot is moving. As soon as the operator exits the monitored area, the robot (in cases ❷ and ❸) will resume the task from where it stopped. Implementing the four safety strategies ensures a minimum downtime on the robot side as it replaces the traditional emergency stops with recoverable temporary interruptions. At the same time, the operator is assured to step in and out the robotic cell freely and safely, despite the fact that the robot is moving, which leads to enhanced productivity on the human side. It is inevitable to notice that the last three safety strategies are especially important for operators to perform tasks sharing the same space with robots. Switching between the three safety strategies is possible at any moment. It is also possible to integrate the strategies into an assembly plan created for human-robot collaborative assembly.

### *9.3.6 Velocity Detection*

Further analyses of system performance have been performed to identify its capability of calculating the velocity of any foreign object in the robotic cell. This

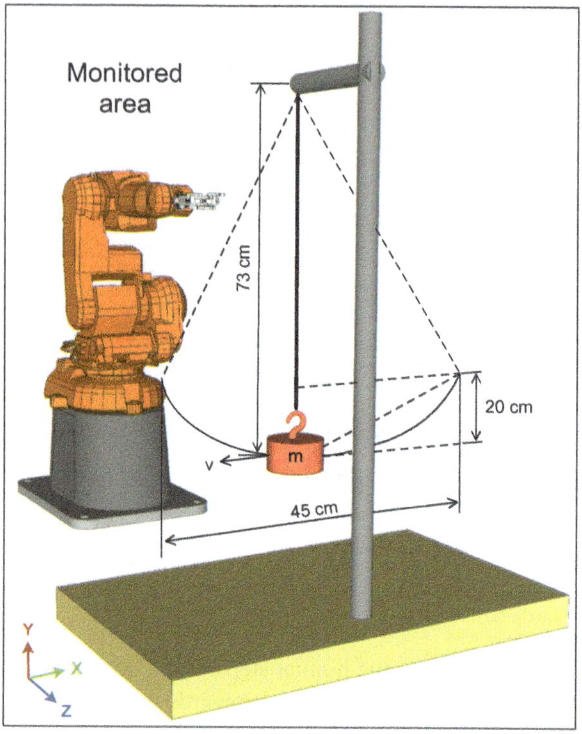


**Fig. 9.15** Experimental setup for velocity measurement

has been achieved by monitoring the object's positions in 3D space with regard to time. To evaluate the efficiency of the velocity calculation, an experimental setup was established by mounting a pendulum in the robotic cell and detecting its movement during the steady state swinging. Equations (9.5) and (9.6) provide the basis for calculating the maximum velocity of the pendulum mathematically and for comparing it with actual measured velocity. Figure 9.15 shows the pendulum setup in the robotic cell, and Fig. 9.16 depicts the measured velocity of the pendulum during the experiment.

$$PE + E_k = \text{const} \Rightarrow mgh + \frac{mv^2}{2} = \text{const} \Rightarrow mgh_{\text{max}} = \frac{mv_{\text{max}}^2}{2} \tag{9.5}$$

$$v_{\text{max}} = \pm\sqrt{2gh_{\text{max}}} \Rightarrow v_{\text{max}} = \pm\sqrt{2 * 9.80665\,\text{m/s}^2 * 0.2\,\text{m}} \Rightarrow v_{\text{max}} = \pm 1.98\,\text{m/s} \tag{9.6}$$

where $PE$ is potential energy, $E_k$ is kinetic energy, $m$ is mass of the pendulum, $g$ is gravity of Earth, $h$ is pendulum position over the reference level, $h_{\text{max}}$ is pendulum position in its highest point, $v$ is pendulum velocity, $v_{\text{max}}$ is maximal velocity of pendulum in its lowest point.

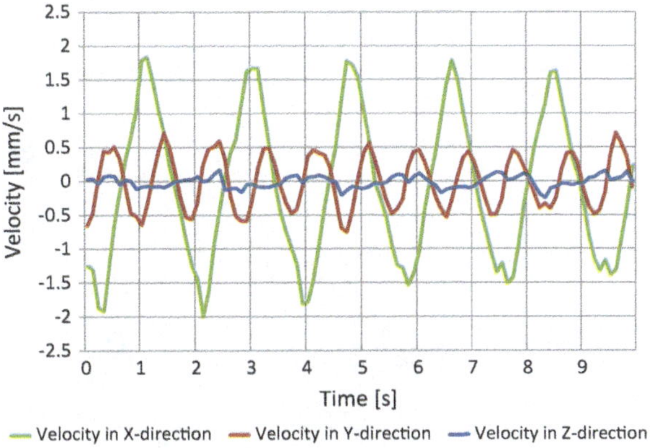


**Fig. 9.16** Pendulum's velocity measured by Kinect sensors

A Kalman filter can be used to reduce the measurement uncertainty and statistical noise captured by Kinect, and it uses a recursive approach to estimate the object velocity based on a series of measurements perceived by Kinect over time. Using this filter gives more accurate results than those based on single measurements.

The results indicate that the system can be used to detect the velocity of any foreign object and control the robot accordingly. Further development is suggested to integrate the results of this section to the collision avoidance system. For instance, detecting the velocity allow the system to predict the position of the obstacle which can help in the presence of processing delays and fast movement of humans.

## 9.4 System Verification

A verified example of the robot's TCP (tool centre point) path during active collision avoidance is illustrated in Fig. 9.17. The first target of the robot path is marked by ⓪, followed by the targets: ①, ②, and then back to ①. In the absence of identified obstacle, the robot follows the linear movement ❶ between the planned targets. Once the system detects an obstacle within a certain distance to the end-effector, the present path of the robot is modified adaptively to prevent any possible collision during the movement to the following position. Locations along the modified path are indicated by circles ❷. These locations are taken into account during adaptive trajectory planning and are indicated by crosses ❸; they represent the nearest locations to the robot end-effector. Lines ❹ describe the distances between the end-effector and the detected nearby obstacle along the robot's path.

The verified outcomes of the robot's end-effector when following an operator are plotted in Fig. 9.18. The identified locations of the operator's hand are indicated by red crosses. These locations are concentrated in areas indicated by ①, ② and ③

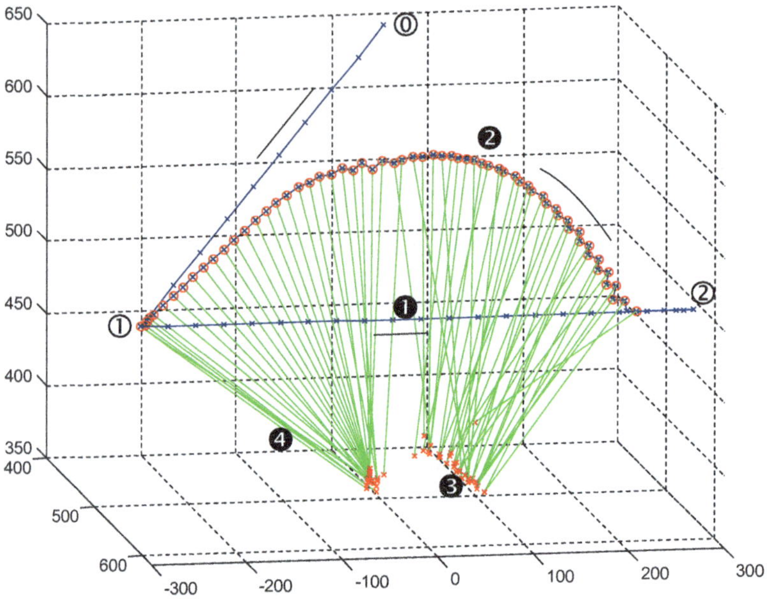


**Fig. 9.17** Recorded path of robot TCP for collision avoidance

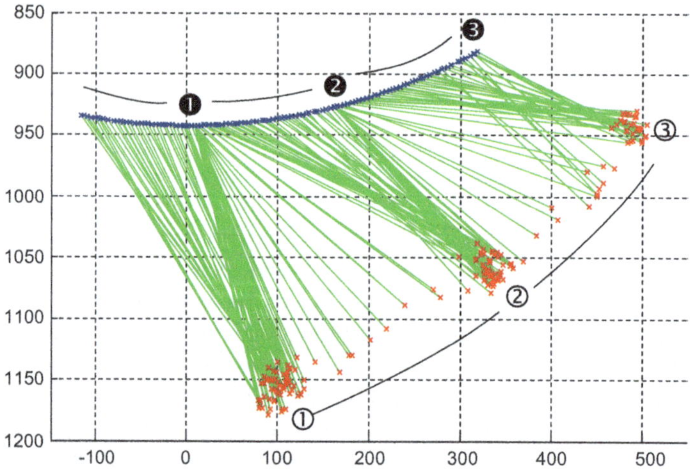


**Fig. 9.18** Recorded path of robot TCP for operator following

which represent the locations where the hand of the operator stayed for longer periods of time. The blue crosses indicate the locations of the robot's end-effector, where ❶, ❷ and ❸ are the robot end-effector's ponding locations corresponding to the operator's hand.

## 9.5 A Remote Assembly Application

As mentioned before, human safety can be assured by active collision avoidance in a shared human-robot collaborative environment. Although less critical, human safety remains as an important issue in remote human-robot collaborations. One example is remote assembly via human-robot collaboration as explained in this section.

Instead of video image streaming, 3D models can be used to guide an offsite operator during remote assembly to meet real-time constraint over the Internet. The 3D models of the parts to be assembled by a robot can be constructed based on a sequence of images captured in varying poses by a robot-mounted camera. The camera is then turned off during robotic assembly to save network bandwidth for better performance. In this context, the robot is treated as a manipulator, which mimics the human's operations but from distance. To safeguard people around the remote robot, the aforementioned active collision avoidance system can be applied.

### *9.5.1 System Configuration*

The remote assembly system is configured with four modules as shown in Fig. 9.19: (1) an application server for image processing and 3D modelling, (2) a real robot for physical assembly, (3) a vision camera for capturing unknown objects, and (4) a user interface to a remote operator for monitoring and control of the entire system. Note that the safety module is not discussed here. This system is capable of identifying and modelling incoming parts of unknown geometries to be assembled. The new parts are then merged into a virtual environment, *Wise-ShopFloor* [29], with existing 3D models of the robotic cell for 3D model-driven remote assembly.

In the system implementation, the camera is placed near the robot's end-effector to capture objects freely. The process starts by moving the camera facing the objects from above when taking the first snapshot. The system creates the initial models of the objects by converting the silhouettes in the top-view snapshot to a set of vertical pillars of a given height. Moving the camera, the system then takes a set of new snapshots of the objects from different angles. Projecting the silhouettes of the snapshots back to the 3D space can trim the pillars (3D models) to approximate the objects. Figure 9.20 shows one scenario of 2D trimming, where the outer bounding polygon approximates the inner actual object.

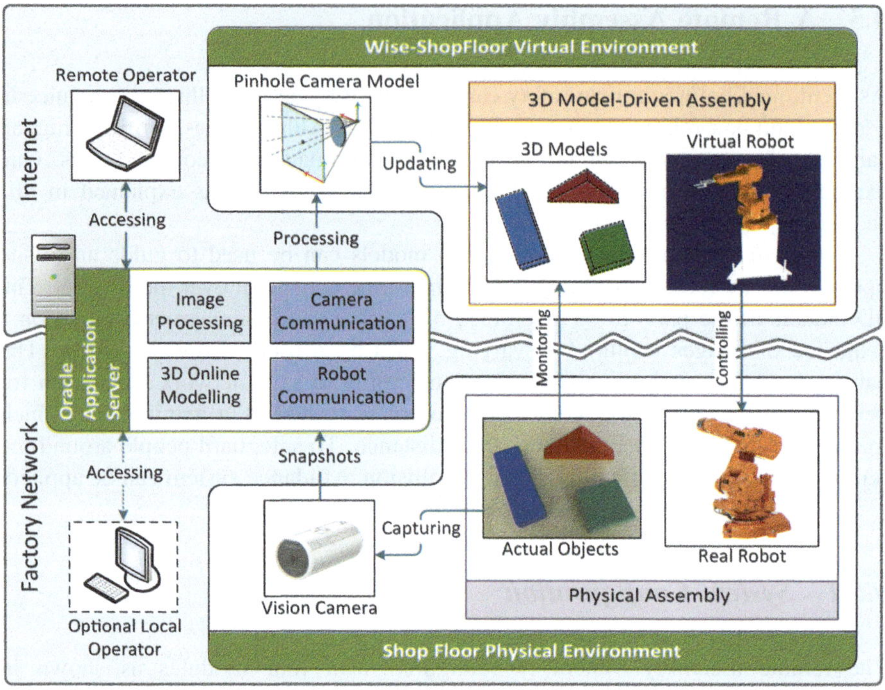


**Fig. 9.19** System configuration

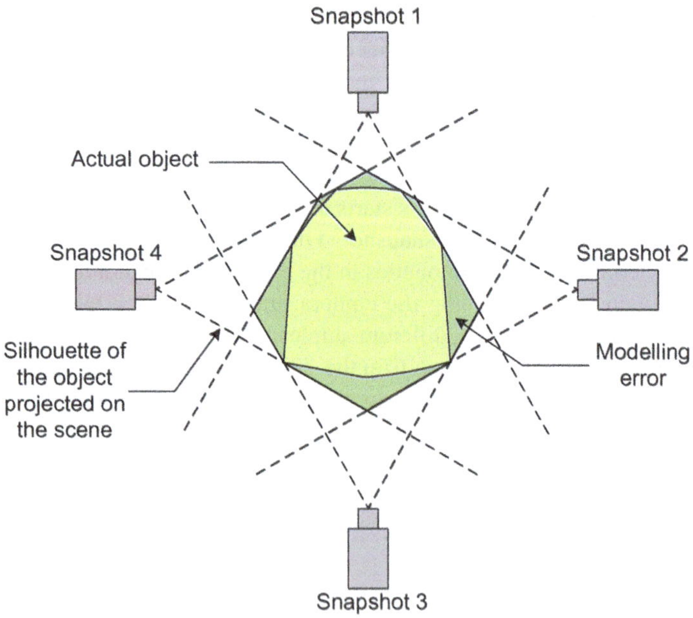


**Fig. 9.20** Shape approximation by trimming of a 2D example

## 9.5.2 System Implementation

### 9.5.2.1 Image Processing

The aim of image processing is to extract the silhouettes of the captured objects. The processing details are explained below.

*Converting to grayscale*—To minimise the computational load, the captured colour images are converted to grayscale by taking the average RGB value of each pixel in the images.

*Gaussian filtering*—Noise in an image affects the accuracy of the silhouette. A zero-mean Gaussian filter given in Eq. (9.7) is applied to each pixel at $(i, j)$ in the image matrix of $(m \times n)$ to remove the noise and to smooth the image. The output image $H(i, j)$ is the convolution of an input image $f(i, j)$ and Gaussian mask $g(\alpha, \beta)$.

$$\begin{aligned} H(i,j) &= f(i,j)g(\alpha,\beta) \\ &= \sum_{\alpha=-[(n-1)/2]}^{(n-1)/2} \sum_{\beta=-[(m-1)/2]}^{(m-1)/2} f(i-\alpha, j-\beta)g(\alpha,\beta) \end{aligned} \tag{9.7}$$

*Image thresholding*—This process extracts the silhouette pixels from the image. It is achieved by scanning the image pixel by pixel while comparing its intensity value with a threshold value. Based on the comparison, the pixel is converted to white or black.

*Silhouettes labelling*—This process identifies the silhouettes in an image by assigning a unique label to each of them, using the component labelling algorithm [32]. The process scans the image pixel by pixel to find a match to one of the silhouettes, followed by checking its neighbouring pixels. If one or more neighbouring pixels hold a label, the algorithm assigns the lowest label to the pixel; otherwise, a new label is assigned. The result of this process is a two-dimensional array in which each element represents a pixel, and each silhouette is represented by a unique label. The background holds a zero value.

### 9.5.2.2 3D Modelling

Constructing 3D models of the captured objects is based on the silhouettes retrieved from the captured images, as follows:

- *Calibration of camera*

A pinhole camera model [33] is adopted for camera calibration. Constructing 3D models precisely requires calibrating the camera to determine its parameters and identify its physical location, e.g. the camera's optical centre, image centre, focal coefficients $f_x$ and $f_y$ (Fig. 9.21Ⓐ), and radial and tangential distortion coefficients (not shown). A 2D coordinate system U-V is defined on the image plane, specifying

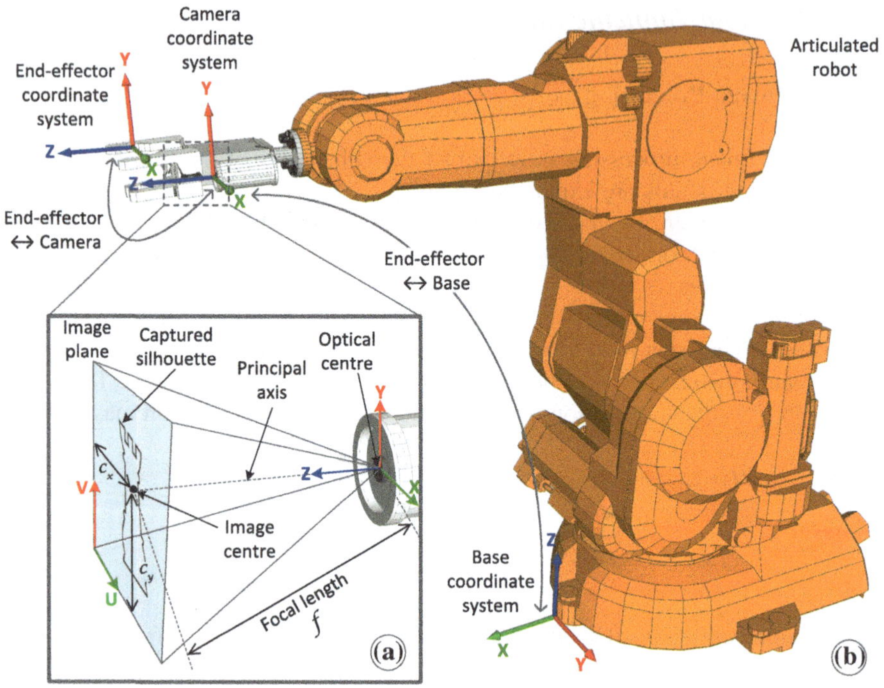


Fig. 9.21 Parameters and coordinate systems for camera calibration

pixel locations in a captured image. Moreover, the camera with respect to the robot's end-effector is specified by a transformation matrix, and its relationship to the robot base is defined as shown in Fig. 9.21Ⓑ. The calibration needs to be performed only once when the camera is mounted on the robot for the first time, with minor adjustments at regular service intervals to minimise any deviations.

- *Construction of pillars*

The first snapshot providing the top view of the objects is used to construct initial 3D models, using the extracted silhouettes. These models are represented by a set of pillars of pixel diameter in 3D space. Figure 9.22 depicts the construction of the initial pillars.

An initial value is assigned as the height to all pillars. This value is the maximum possible height of the objects. The construction of the initial pillars is accomplished by applying Tsai's pinhole camera model [33], as shown in Fig. 9.21Ⓐ. Given a 3D point $(x, y, z)$, its projected 2D point $(u, v)$ on the U-V plane is described as

$$u = f_x \times x'' + c_x, \quad v = f_y \times y'' + c_y \tag{9.8}$$

$$x'' = x'(1 + k_1 r^2 + k_2 r^4 + k_3 r^6) + 2p_1 x' y' + p_2(r^2 + 2x'^2) \tag{9.9}$$

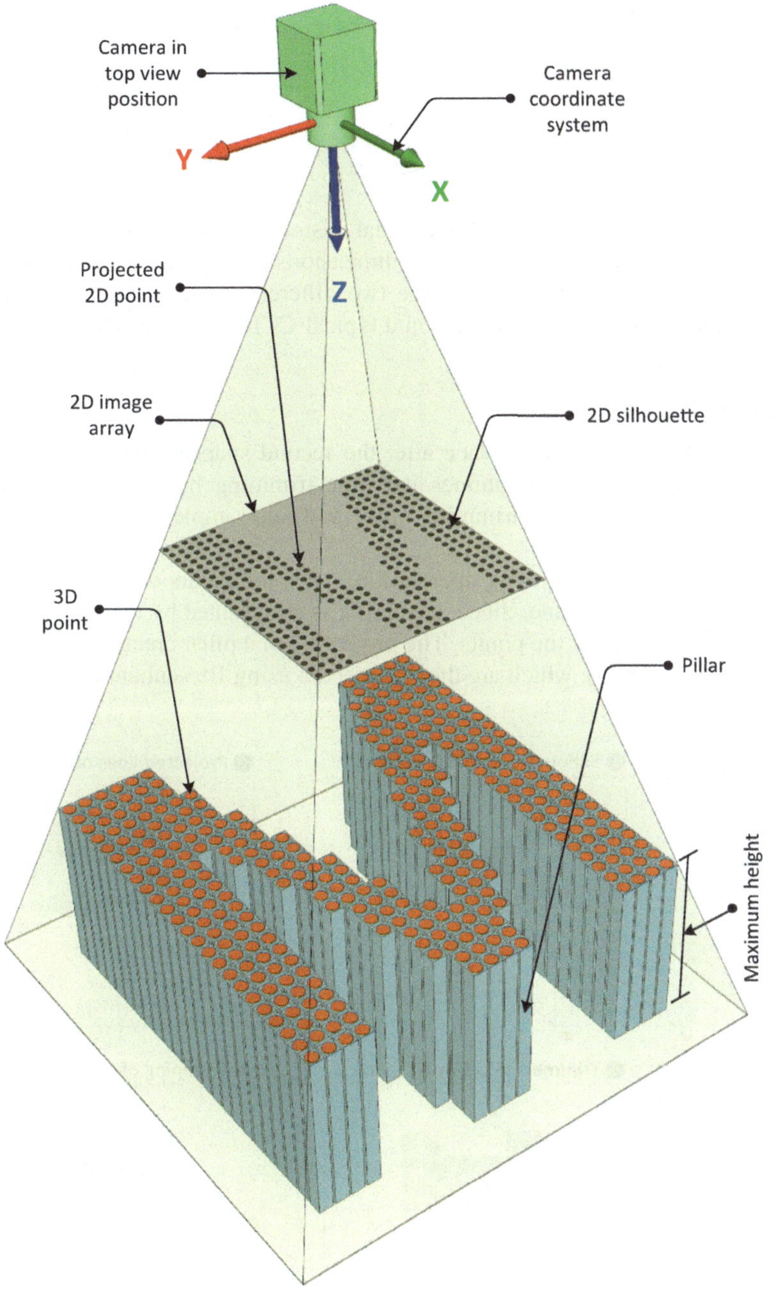


**Fig. 9.22** Construction of initial pillars

$$y'' = y'(1 + k_1 r^2 + k_2 r^4 + k_3 r^6) + 2p_2 x' y' + p_1 (r^2 + 2y'^2) \tag{9.10}$$

$$x' = \frac{x}{z}, \quad y' = \frac{y}{z}, \quad r^2 = x'^2 + y'^2 \tag{9.11}$$

$$f_x = f \times s_x, \quad f_y = f \times s_y \tag{9.12}$$

where $k_1$, $k_2$, $k_3$ and $p_1$, $p_2$ are the radial distortion coefficients and tangential distortion coefficients, respectively. The dimensions of a pixel are defined by $s_x$ and $s_y$. Equations (9.8) and (9.12) introduce two different focal coefficients: $f_x$ and $f_y$. This is because the individual pixels on a typical CCD image sensor are rectangles in shape.

- *Trimming of pillars*

The trimming operation takes place after the second snapshot has been processed (silhouette extracted), and continues until the trimming by the last silhouette is done. Figure 9.23 shows the trimming process of two sample pillars with reference to one silhouette.

In other words, the trimming operation projects the pillars one by one from the 3D space to the image plane. Since each pillar is represented by two 3D end points, the projection is only for the points. The projection of a pillar creates two 2D points, calculated by Eq. (9.8), which are then connected using Bresenham algorithm [34].

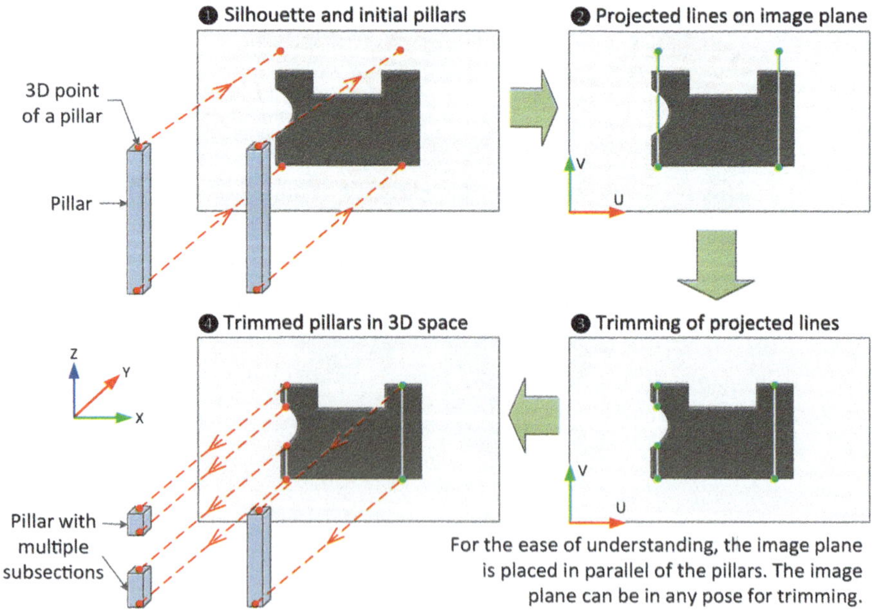


**Fig. 9.23** Example of pillar-trimming process

Extracting the pixels shared by the projected line and the silhouette reveals a trimmed line. Finally, the trimmed 2D line is projected back to the 3D space, resulting in a trimmed pillar. The trimming process is repeated for all pillars and all snapshots. For more complex shape, the concavity of the object results in pillars with multiple subsections.

- *Solid prism representation*

Despite the fact that the aforementioned processes can trim the pillars as closely as possible to mimic the real objects, the pillars alone are neither intuitive nor computationally efficient for 3D visualisation due to the fact of non-solid geometry. Moreover, the modelled shapes need to be compatible with the robot 3D model in *Wise-ShopFloor* [29]. This, however, can be achieved by creating a solid prism representation for the trimmed pillars.

Two objectives are considered here: (1) to localise the process, and (2) to create a uniform representation of a given object. The pillars are first divided into groups of three according to their immediate neighbouring connectivity. The prism creation is then divided into three steps to construct: (1) the top surface, (2) the bottom surface, and (3) the three sides of each prism. The order of the end points is crucial when building a surface patch of each prism as its surface normal affects its visibility. As shown in Fig. 9.24, three end points in counter-clockwise order are used to create a surface patch with an outer visibility. Moreover, three cases of pillar division (cut) caused by the subsections of pillars are also considered during prism creation, as illustrated in Fig. 9.24.

### *9.5.3 Case Study*

As shown in Fig. 9.25, three simple parts are chosen for a proof-of-concept case study to validate the functionality of the 3D model-driven remote assembly system. Once the 3D models of the randomly placed parts are generated and integrated with the 3D model of the robotic assembly cell, the camera is switched off to save network bandwidth, leaving a low-volume data connection with the robot controller alive. A remote operator assembles the 'parts' (3D models) using the 3D robot model in the cyber world, whereas the real robot mimics the virtual robot and assembles the actual parts simultaneously in the physical world—a typical cyber-physical system. During remote assembly, only the robot control commands are transmitted from the virtual robot to the real one instantly and automatically, without extra robot programming. It is worth mentioning that image processing can also identify the geometric centres and orientations of the parts, leading to semi-automate pick-and-place and grasping operations during remote assembly. Figure 9.26 depicts the results of 3D model creation of the parts, as well as those of 3D model-driven remote assembly.

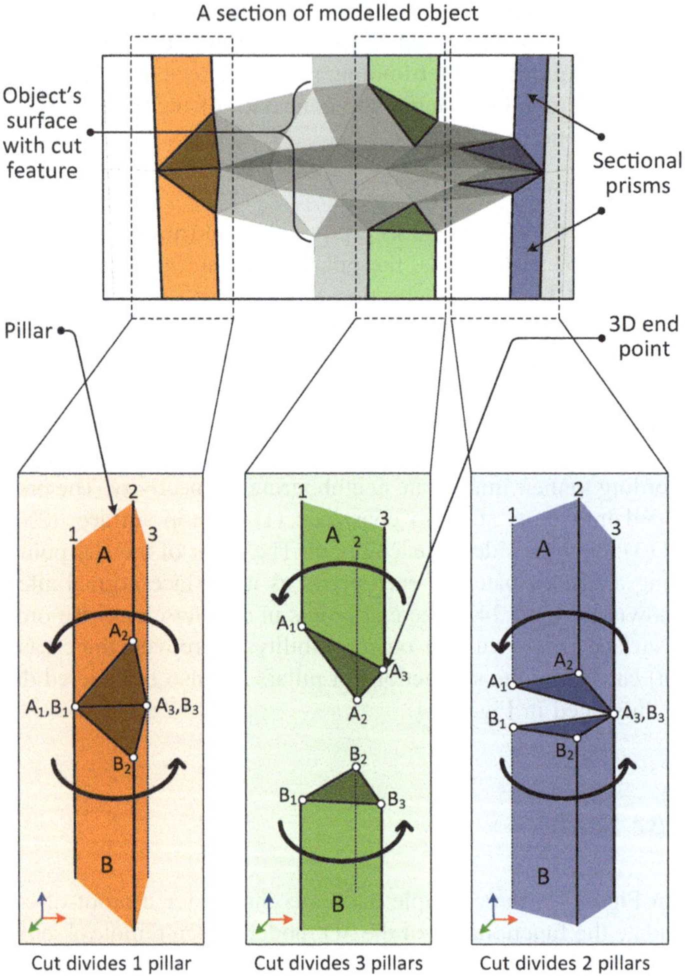


**Fig. 9.24** Prism creation with different cut cases

In this case study, seven snapshots taken from different angles are used to model the parts. More snapshots would improve the quality of the 3D models, although the processing time would be longer. A performance analysis is therefore conducted under the following specifications to understand its relationship: Intel Core i5 processor of 2.6 GHz, graphics card of GeForce GT 240, a 4 GB RAM, and running under the operating system of Windows Vista.

The image-based 3D model generation has been tested for ten times and the average computation time for each processing step was calculated and recorded as illustrated in Fig. 9.27, with error bars indicating deviations in the recorded

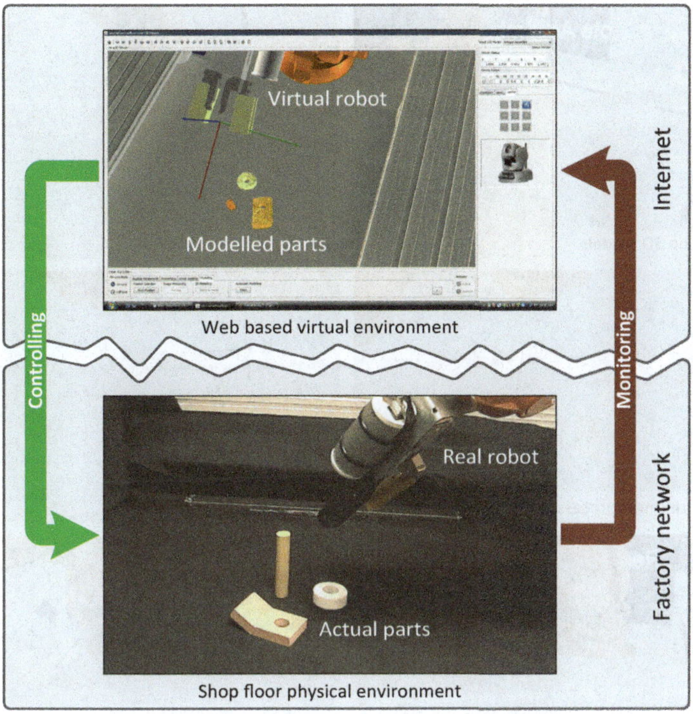


**Fig. 9.25** 3D model-driven remote assembly

processing times. It is found that the silhouette labelling process consumes in average 1.75 s in processing time, with the highest deviation. The reason is due to the fact that the labelling algorithm employed examines all the neighbouring pixels when spotting a non-zero pixel in an image during the pixel-by-pixel scanning. It consumes a high percentage of processing time that varies from one test to another. Despite this fact, the system can process one image in about 4.8 s.

The results of the pillar trimming process for each snapshot are also recorded. Figure 9.28 manifests the accuracy by comparing the actual height of a real object and the trimmed pillar height of its 3D model after processing each snapshot excluding the first top-view image. As can be seen in the figure, the accuracy of pillar trimming is reasonably high after processing the 7th snapshot as the error has converged quickly to a small value in 22 s. with a modelling accuracy of less than 1 mm. In terms of the efficiency of remote assembly, the real robot lags behind the virtual robot by 30 ms over the Internet.

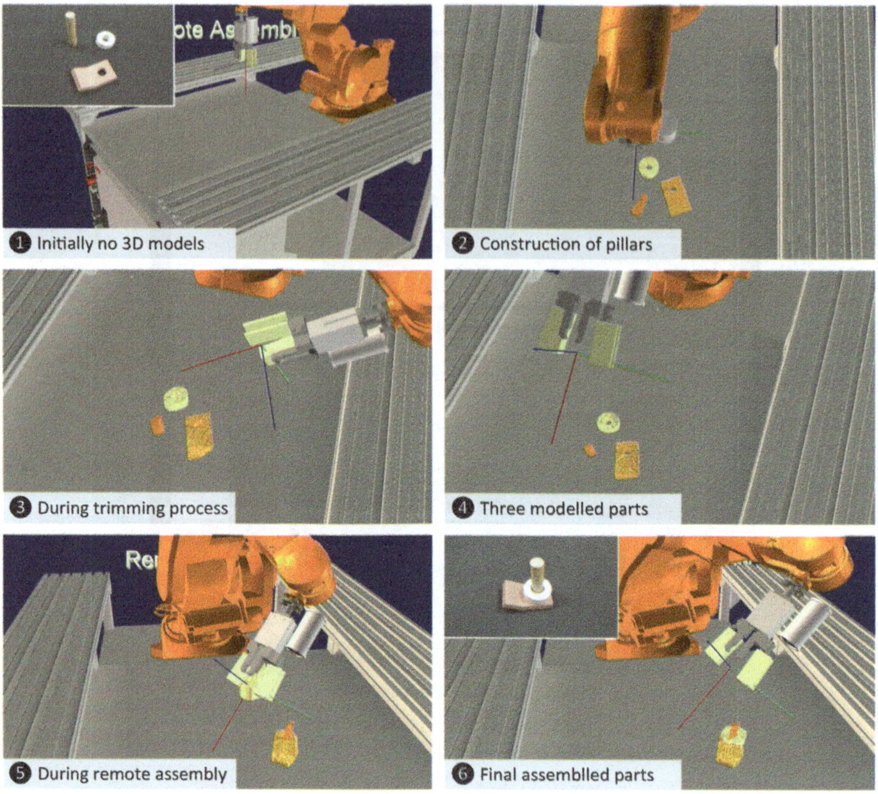


**Fig. 9.26** Results of case study for 3D modelling and remote assembly

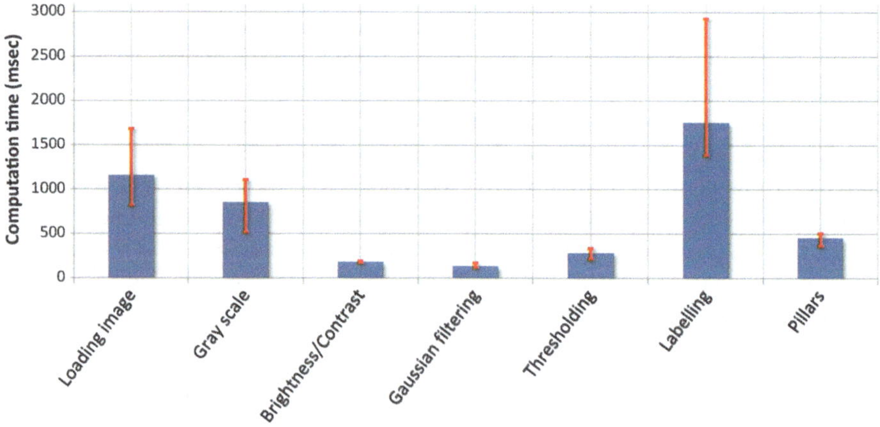


**Fig. 9.27** Comparison of computation time of different processing steps

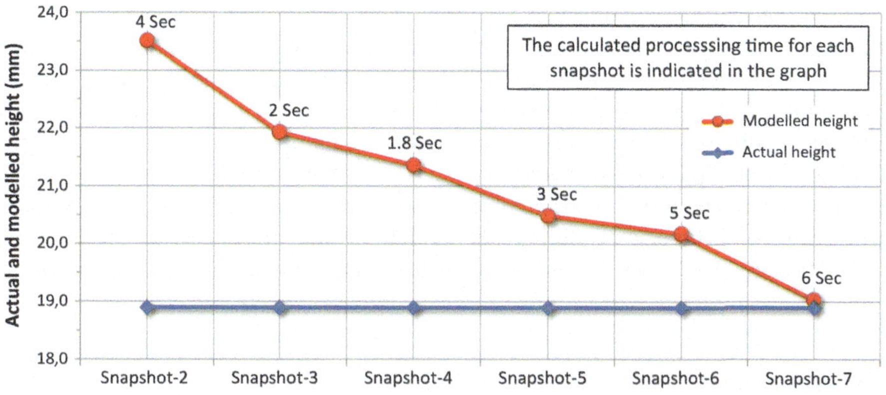


**Fig. 9.28** Modelling errors versus number of snapshots processed

## 9.6 Conclusions

Human safety is an important necessity in any human-robot collaborative systems. Targeting this necessity, this chapter introduced a cost-effective and time-efficient solution for actively avoiding collisions that can provide consistent and reliable safety protection in a human-robot collaborative environment. The aim of this solution is to improve the overall robotic system performance by associating robot's virtual 3D models with human operators to a series of depth and vision sensing units for online collision detection and avoidance in an augmented environment. Instead of using the traditional emergency stops that do not allow human-robot coexistence but increase the downtime of a robotic cell, the introduced approach detects in real-time any possible collision and actively control the robot via four safety modes: alarming a human operator, stopping a robot, moving the robot away from the approaching operator via recoverable interruptions, or modifying the robot trajectory at runtime. The approach provides better flexibility and productivity. Furthermore, a human-robot collaborative scenario has been verified for the purpose of enabling the robot to track the operator to facilitate a shared assembly task. Introducing the four safety strategies to assembly planning is the future aim along this direction so that the behaviour of a robot fits the nature of a required task, leading to more advanced human-robot collaborative assembly.

A 3D model-driven robot-in-the-loop approach is presented in the second half of this chapter for remote assembly, where an off-site operator can manipulate a real robot instantly via cyber robot control. The 3D models of the parts to be assembled are generated based on a set of snapshots of the parts captured by a robot-mounted camera. The generated 3D models are then integrated with the 3D model of a real robotic cell. The advantages are: (1) elimination of video-image streaming during remote assembly, and (2) robot programming-free to users. The needed robot control commands are generated automatically and transmitted from the virtual robot to the real one for physical assembly. This method can generate a set of 3D models in 22 s from seven snapshots. The efficiency can be improved by performing image processing in parallel with moving the camera to the next position. For complex geometries, more snapshots can be used to improve the modelling accuracy. This remote assembly system is for manual assembly where the robot is treated as a manipulator. No matter simple or complex, assembly sequence is up to the assembler. The future work includes more comprehensive feature identification, e.g. centre of a hole, etc., during 3D modelling and more tests of realistic and complex parts assembly. This contributes to remote human-robot collaboration.

## References


1. D.J. Agravante, A. Cherubini, A. Kheddar, Using vision and haptic sensing for human-humanoid joint actions, in *Proceedings of the IEEE Conference on Robotics, Automation and Mechatronics, RAM* (2013), pp. 13–18
2. C.A. Monje, P. Pierro, C. Balaguer, A new approach on human–robot collaboration with humanoid robot RH-2. Robotica **29**(6), 949–957 (2011)
3. S. Takata, T. Hirano, Human and robot allocation method for hybrid assembly systems. CIRP Ann.-Manuf. Technol. **60**(1), 9–12 (2011)
4. F. Chen, K. Sekiyama, H. Sasaki, J. Huang, B. Sun, T. Fukuda, An assembly strategy scheduling method for human and robot coordinated cell manufacturing, in *IEEE International Conference on Intelligent Robots and Systems* (2011), pp. 4670–4675
5. J. Krüger, T.K. Lien, A. Verl, Cooperation of human and machines in assembly lines. CIRP Ann.-Manuf. Technol. **58**(2), 628–646 (2009)
6. T. Arai, R. Kato, M. Fujita, Assessment of operator stress induced by robot collaboration in assembly. CIRP Ann.-Manuf. Technol. **59**(1), 5–8 (2010)
7. D. Kuli, E.A. Croft, Affective state estimation for human–robot interaction. IEEE Trans. Robot. **23**(5), pp. 991–1000 (2007)
8. J. Krüger, B. Nickolay, P. Heyer, G. Seliger, Image based 3D surveillance for flexible man-robot-cooperation. CIRP Ann.-Manuf. Technol. **54**(1), 19–22 (2005)
9. J.A. Corrales, F.A. Candelas, F. Torres, Safe human-robot interaction based on dynamic sphere-swept line bounding volumes. Robot. Comput. Integr. Manuf. **27**(1), 177–185 (2011)
10. Z.M. Bi, L. Wang, Advances in 3D data acquisition and processing for industrial applications. Robot. Comput. Integr. Manuf. **26**(5), 403–413 (2010)
11. T. Gecks, D. Henrich, Human-robot cooperation: safe pick-and-place operations, in *Proceedings of the IEEE International on Robot and Human Interactive Communication*, vol. 2005 (2005), pp. 549–554
12. D. Ebert, T. Komuro, A. Namiki, M. Ishikawa, Safe human-robot-coexistence: emergency-stop using a high-speed vision-chip, in *2005 IEEE/RSJ International Conference on Intelligent Robots and Systems, IROS* (2005), pp. 1821–1826
13. C. Vogel, C. Walter, N. Elkmann, A projection-based sensor system for safe physical human-robot collaboration, in *IEEE International Conference on Intelligent Robots and Systems* (2013), pp. 5359–5364
14. J.T.C. Tan, T. Arai, Triple stereo vision system for safety monitoring of human-robot collaboration in cellular manufacturing, in *Proceedings of the 2011 IEEE International Symposium on Assembly and Manufacturing, ISAM* (2011), pp. 1–6
15. R. Schiavi, A. Bicchi, F. Flacco, Integration of active and passive compliance control for safe human-robot coexistence, in *Proceedings of the IEEE International Conference on Robotics and Automation* (2009), pp. 259–264

16. M. Fischer, D. Henrich, 3D collision detection for industrial robots and unknown obstacles using multiple depth images, in *Advances in Robotics Research* (2009), pp. 111–122
17. P. Rybski, P. Anderson-Sprecher, D. Huber, C. Niessl, R. Simmons, Sensor fusion for human safety in industrial workcells, in *IEEE International Conference on Intelligent Robots and Systems* (2012), pp. 3612–3619
18. F. Flacco, T. Kroeger, A. De Luca, O. Khatib, A depth space approach for evaluating distance to objects: with application to human-robot collision avoidance. J. Intell. Robot. Syst. Theory Appl. **80**, 7–22 (2015)
19. B. Dániel, P. Korondi, T. Thomessen, Joint level collision avoidance for industrial robots. IFAC Proc. **45**(22), 655–658 (2012)
20. A. Cherubini, R. Passama, A. Crosnier, A. Lasnier, P. Fraisse, Collaborative manufacturing with physical human-robot interaction. Robot. Comput. Integr. Manuf. **40**, 1–13 (2016)
21. Pilz GmbH & Co. KG (2016), [Online]. Available: http://www.safetyeye.com/
22. A.Y.C. Nee, S.K. Ong, G. Chryssolouris, D. Mourtzis, Augmented reality applications in design and manufacturing. CIRP Ann.-Manuf. Technol. **61**(2), 657–679 (2012)
23. S. Tzafestas, D. Aristos, Simultaneous object recognition and position tracking for robotic applications. in *Proceedings of the 2009 IEEE International Conference on Mechatronics* (2009), pp. 145–169
24. X. Tian, H. Deng, M. Fujishima, K. Yamazaki, Quick 3D modeling of machining environment by means of on-machine stereo vision with digital decomposition. CIRP Ann.-Manuf. Technol. **56**(1), 411–414 (2007)
25. D. Samak, A. Fischer, D. Rittel, 3D reconstruction and visualization of microstructure surfaces from 2D images. CIRP Ann.-Manuf. Technol. **56**(1), 149–152 (2007)
26. Y. Sumi, Y. Kawai, T. Yoshimi, F. Tomita, 3D object recognition in cluttered environments by segment-based stereo vision. Int. J. Comput. Vis. **46**(1), 5–23 (2002)
27. B. Petit et al., Multicamera real-time 3D modeling for telepresence and remote collaboration. Int. J. Digit. Multimedia Broadcast. **2010**, 1–12 (2010)
28. K. Atsushi, H. Sueyasu, Y. Funayama, T. Maekawa, System for reconstruction of three-dimensional micro objects from multiple photographic images. Comput. Aided Des. **43**(8), 1045–1055 (2011)
29. L. Wang, Wise-ShopFloor: an integrated approach for web-based collaborative manufacturing. IEEE Trans. Syst. Man Cybern. Part C Appl. Rev. **38**(4), 562–573 (2008)
30. L. Wang, R. Sams, M. Verner, F. Xi, Integrating Java 3D model and sensor data for remote monitoring and control. Robot. Comput. Integr. Manuf. **19**(1–2), 13–19 (2003)
31. L. Wang, M. Givehchi, G. Adamson, M. Holm, A sensor-driven 3D model-based approach to remote real-time monitoring. CIRP Ann.-Manuf. Technol. **60**(1), 493–496 (2011)
32. M. Jankowski, J.-P. Kuska, Connected components labelling—algorithms in Mathematica, Java, C++ and C#, in *International Mathematica Symposium* (2004)
33. R. Tsai, A versatile camera calibration technique for high-accuracy 3D machine vision metrology using off-the-shelf tv cameras and lenses. IEEE J. Robot. Autom. **3**(4), 323–344 (1987)
34. M.S. Fawad, Adapting Bresenham algorithm. J. Theor. Appl. Inf. Technol. **2**(2), 27–30 (2007)

# Chapter 10
# Cloud Robotics Towards a CPS Assembly System

## 10.1 Introduction

In recent years, *cloud* has become a popular technology which gained huge market success globally. Cloud concept indicates a model for enabling ubiquitous, convenient, on-demand network access to a shared pool of configurable computing resources (e.g. networks, servers, storages, applications, and services) that can be rapidly provisioned and released with minimal management effort or service provider interaction [1]. It provides elastic and flexible supports for service-oriented production models. Instead of investing on costly IT equipment or software licenses as a whole, the cloud users are able to pay for the exact amount of software or hardware usage based on pay-as-you-go principle. It is particularly helpful for small and medium-sized enterprises (SMEs) that are normally short of start-up capitals for new investments on equipment.

It is thus logical and reasonable for manufacturing researchers and stakeholders to adopt cloud into the manufacturing industry so as to improve current production performance. Based on NIST's definition, Xu [2] extended the cloud concept to manufacturing as a model enabling ubiquitous, convenient, on-demand network access to a shared pool of configurable manufacturing resources (e.g. manufacturing software tools, manufacturing equipment, and manufacturing capabilities) that can be rapidly provisioned and released with minimal management effort or service provider interaction. The cloud-based automation is predicted to alter the automation landscape by means of improved perception, faster planning, accurate modelling, lifelong learning, large-scale systems, new ways of interacting with humans, and so forth [3]. The availability of new technologies, e.g. computing cloud, big data management, open platform, and broad bandwidth, and their possible uses in robotics have opened the door to a whole new line of research called Cloud Robotics [4]. However, manufacturing facilities are conducted by different vendors using different standards, platforms, communication protocols, and interfaces. Thus, it forms a heterogeneous environment which experiences difficulties in



L. Wang and X. V. Wang, *Cloud-Based Cyber-Physical Systems in Manufacturing*,
https://doi.org/10.1007/978-3-319-67693-7_10

interaction and integration. Compared with cloud computing systems, one of the biggest challenges cloud manufacturing facing is involving numerous types of physical resources, e.g. machine tools and robots.

As a matter of fact, the proposal of integrating robots with cloud was made 20 years ago. In 1995, Bohus et al. [5] suggested implementing remote robot control over the Internet cloud. In 2010, James Kuner at Google introduced the term Cloud Robotics [6] to describe a new approach to robotics that takes advantage of the Internet as a resource for massively parallel computation and real-time sharing of vast data resources. As a new way of merging robots and ICT, cloud robots are predicted as an evolutionary jump for robots and a transformational change of paradigm [3].

This chapter focuses on merging the technologies from both computing cloud and industrial applications in the manufacturing sector. In the first half of this chapter, relevant cloud manufacturing research works are reviewed. In the second half, a novel cloud manufacturing system is presented to integrate manufacturing applications in the cloud paradigm. The system is introduced from the perspective of system architecture, integration mechanisms and robotic applications.

## 10.2 Cloud Robotics

In the past years, the cloud robotics research has been conducted worldwide. Many approaches are proposed in different sectors. In this section, cloud robotics related research are reviewed and discussed from two perspectives, i.e. robotic systems and applications.

### *10.2.1 Cloud Robotics at System Level*

As an important spirit of cloud, SOA (service-oriented architecture) refers to a system consisting of a collection of loosely coupled services that communicate with each other through standard interfaces and via standard message-exchanging protocols [7]. In the cloud robotic centre, Du et al. [8] proposed a framework following the general cloud computing paradigm, which was suggested to address the current limitations in capacity and versatility of robotic applications. A broker mechanism was deployed to look up the services and applications available in the unit's directory. Chu et al. [9] proposed a platform offering telematics services deployed to form the cloud platform, namely Cargo. This architecture assisted the service providers to establish a service platform in order to include varied developing technologies.

In cloud computing, there are different levels of service deployments from high to low, such as Infrastructure-as-a-Service (IaaS), Platform-as-a-Service (PaaS), Software-as-a-Service (SaaS) [1]. These models are reflected in the robotic clouds as well. Mourandian et al. [10] focused on the IaaS aspects of robotic applications as cloud computing services. They proposed an architecture that enabled cost efficiency through virtualisation and dynamic task delegation to robots, including the robots that might belong to other clouds. Gherardi et al. [11] suggested a PaaS approach for configurable product lines based on cloud robotic applications. It allowed robot developers to relieve end users from the low-level decisions required for configuring the architecture of complex systems distributed on the robot and the cloud. It was predicted that numerous robotic applications could be developed in this area; for instance, REALabs platform was built based on the PaaS model [12]. Mohanarajah et al. [13, 14] presented the design and implementation of Rapyuta. Rapyuta was an open source PaaS framework designed specifically for robotic applications. It helped robots to offload heavy computation by providing secured customisable computing environments in the cloud. Artificial neural network was used for the training of locations [15]. The idea was to establish the communication between the cloud and robot over a large environment and identify the location from the images sent by the robot at the SaaS level.

Chen and Hu [16] discussed Internet of intelligent things and Robot-as-a-Service (RaaS). The idea of achieving RaaS was through autonomous and intelligent mobile physical services or robots to form a local pool of intelligent devices and that could make local decisions without communications with the cloud. Bekris et al. [17] described how solutions from the recent literature could be employed on the cloud during a periodically updated pre-processing phase to efficiently answer manipulation queries on the robot given changes in the workspace. In this setup, interesting trade-offs arose between path quality and computational efficiency. These trade-offs motivated further research on how motion planning should be executed given access to a computing cloud.

To recap, the current cloud robotic systems aim to improve the robot performance with the help of enhanced computing power and data/knowledge management capability from the cloud. Extended from Kato et al. [18], the comparison among conventional, web-based and cloud-based robotics is shown in Fig. 10.1. In a traditional robotic cell, all knowledge and service modules are integrated along with the control system and physical components locally. In contrast, web-based

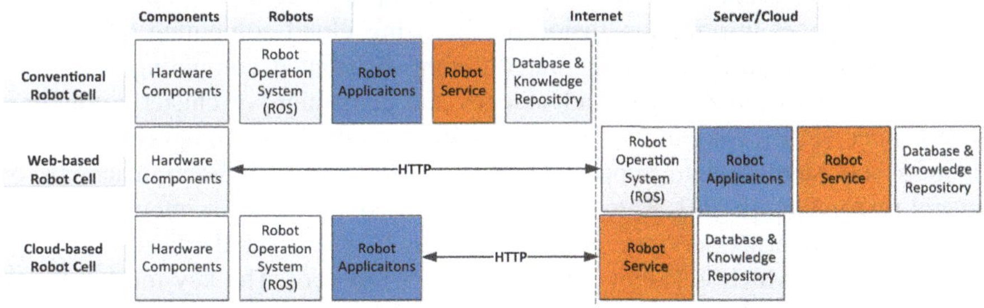


**Fig. 10.1** Comparison of conventional, web-based and cloud-based robotic cells

robotic cell aims to control the physical devices remotely over the network. Thus, the information and operating systems are deployed in remote servers.

However, this system structure is challenged by technical issues, e.g. real-time control, synchronisations and stability risks. As an intermediate solution, cloud-based robotics is able to resolve the conflict between heavy computing needs and local control requirements. Information sharing and data management is able to be implemented remotely in the cloud, while the physical applications are imaged and maintained in terms of virtual cloud services.

### *10.2.2 Cloud Robotics at Application Level*

At the application level, numerous cloud-related robotic applications are developed. Kamei et al. [19] discussed networked robotics connected to the cloud. New fields are predicted, e.g. daily activity and accessible support. Some of the issues were identified as future challenges including multi-robot management, multi-area management, user attribute management, and service coordination management. A Human-Robot Cloud (HRC) was also proposed as an extension to cloud computing across the support of physical human and the cognitive "components" of the cloud, which were neither expected to be experts nor to be engaged with the cloud in full-time [20].

From the machine intelligence's perspective, Ren [21] proposed the concept of an advanced intelligence machine, which was a device that used both natural and artificial intelligence and was capable of effective recognition and generation of effective speech and behaviour. Morariu et al. [22] introduced a classification of virtualised manufacturing execution systems and shop floor devices, which was presented focusing on the virtualisation techniques suitable for each device type, considering the level of distributed intelligence and the virtualisation overhead. The implementation using six Adapt robots and an IBM Cloud-Burst 2.1 private cloud, was described; and virtualisation overhead in terms of event propagation delays was measured and presented in several scenarios of resource workload collocation on physical cloud blades.

During the development of cloud robotics, a number of supportive technologies can also be observed. At data level, Liu et al. [23] presented a comprehensive view on a system level information fusion design using cloud computing technology. A systematic comparison among four different distributed computing paradigms was given, which illustrated the advantages and constraints of cluster computing, peer-to-peer (P2P) computing, grid computing, and cloud computing. Apache Hadoop was suggested to support large data sets across different machines by several researchers [9, 24]. Kato et al. [18] proposed a system for integrating robot services with Internet services. The feature was to use a standardised communication protocol, which is Robot Service Network Protocol. The key mechanism of the proposed framework was a robot assignment function, which discovered distributed robot resources and assigned the requested tasks by end users to suitable robots [25]. Aguero and Veloso [26] developed a transparent multi-robot communication exchange mechanism for executing robot behaviours. Jang and Kim [27] developed a script language-based template for the source code generation and exchange.

To summarise, despite the above-mentioned achievements in cloud-based manufacturing, there is still a lack of research in a cloud system that is able to support manufacturing chain as a whole solution. Thus in this chapter, the Interoperable Cloud Manufacturing System (ICMS) is presented along with the system structure and integration mechanisms.

## 10.3 ICMS: an Example of Cloud Robotics System

The ICMS system architecture is illustrated in Fig. 10.2. Physical production resources are integrated in the system in terms of manufacturing services. The *Cloud* layer works as the service coordinator and supervisor of the whole production system. Cloud users and administrators are able to access the cloud over the network, with the help of standardised Application Programming Interface (API). Inside the Cloud layer, the Service Manager mechanism is the core execution module which interacts with cloud users, and executes the service packages accordingly. The Cloud Database maintains information regarding cloud user, cloud service packages, service histories, and most importantly resource profiles that are utilised to schedule and execute cloud services. These specifications guarantee the capability, availability and feasibility of production facilities at the Physical Resource level.

At the Physical Resource Layer, manufacturing tasks assigned by cloud are taken by control units of production devices, e.g. Robot-as-a-Service (RaaS) unit and Machine-as-a-Service (MaaS) unit. Robot Operating Systems (ROS) and CNC controllers interpret the production documents from cloud into process working

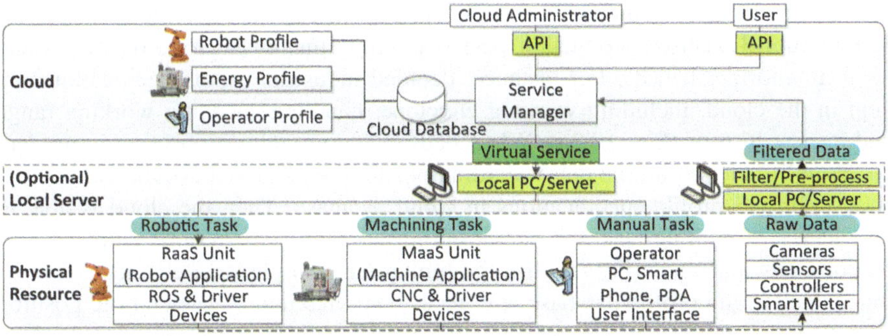


**Fig. 10.2** ICMS system architecture

steps and then controlling signals that directly drive physical devices eventually. In this top-down approach, human operators are also able to interact with the Cloud layer via devices like smart phones, PCs and PDAs. Real performance on shop-floor is monitored by range cameras, sensors, smart meters and device controllers. Monitoring results are fed back to cloud for service supervision and future improvements. Thus, it forms a closed-loop production system.

Between the Cloud layer and Physical Resource Layer, Local Servers are optional due to two main reasons. First, during the monitoring process, shop-floor sensors generate huge amount of data dynamically, e.g. power, current, vibration, and force readings. It is inefficient to stream all raw data to cloud directly, since most contents are not essential but generate heavy network traffics. Thus a local PC or server is necessary in this case to play as the data filter and pre-processor. Raw data is locally filtered and processed by the server, and then uploaded to the cloud. It thus balances bandwidth loads and cloud data management.

Second, in some cases the local server needs to work as an interface between the Cloud layer and the Physical Resource Layer. In practice, many commercial control units (ROS and CNC controller) are designed as a semi-closed system. To some degree it guarantees the robustness and safety of the unit. However, these systems are difficult to interact with the cloud directly. Thus in these cases, a local PC or server is needed to interact with operating systems at low level via user interface on one hand (over local network in most cases), and communicate with the cloud via the Internet on the other hand. It is particularly suitable for the integration of robotic applications and related devices, since most of the legacy devices, monitors, cameras, measuring devices, etc. are not cloud-ready. The Local Server is able to interface with the legacy devices via native Application Programming Interfaces or User-Defined Interfaces at lower level, and interact with the Cloud Layer at higher level directly. It guarantees the connectivity of the production environment.

## *10.3.1 Integration Mechanisms in ICMS*

The communication methods between manufacturing facilities and cloud are shown in Fig. 10.3. Localised production plan is possible thanks to cloud databases and local monitoring devices. For instance, detailed industrial robot specifications are kept in the cloud, including working envelope, handling capacity, working range and energy history (Fig. 10.3a). When path planning and optimisation is needed, the cloud is able to pull the data regarding positioning, kinematics/joint status from shop floor and profile specifications in cloud database. Then the cloud is able to take heavy computing task and output optimised path to the robotic cell.

For CNC machines (Fig. 10.3b), the limitations and physical characters can be maintained in the profile database, e.g. feed force, maximum acceleration, position tolerance and energy parameters, etc. In this way, localised machining process plans can be generated which is suitable for the specific machine tools.

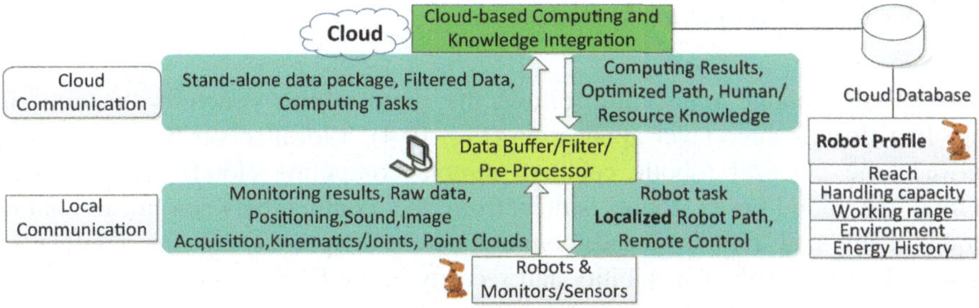


(a) Robot and robot profile

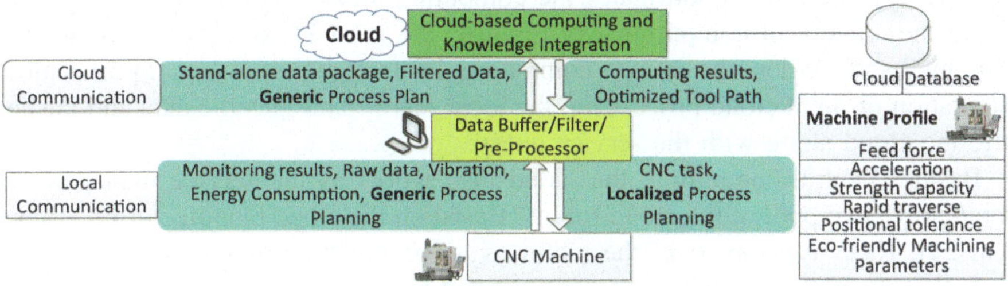


(b) Machine and machine profile

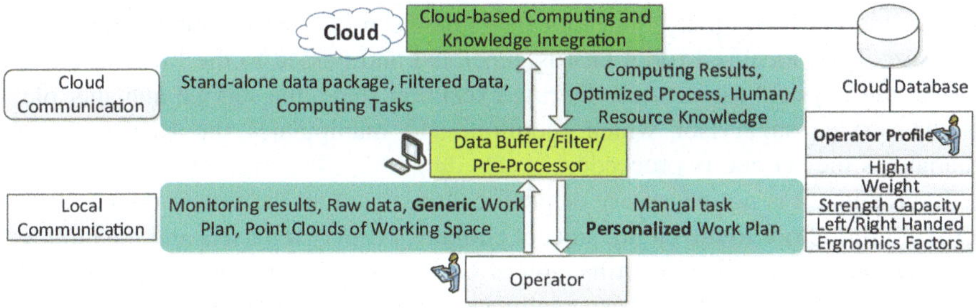


(c) Human resource and worker profile

**Fig. 10.3** Localised manufacturing plan based on cloud

As a specific type of resource, human resource can be integrated with ICMS as well, considering physical and ergonomic factors of individual operators specifically (Fig. 10.3c). Exclusive human factors differ among individual operators, e.g. height, weight, strength, handedness (left or right), and other ergonomic factors. Before an operator starts to work, his or her staff barcode can be quickly scanned and the personalised ergonomic specification can be identified in cloud database. Based on these profiles, a personalised work plan is generated. It is especially helpful during the implementation of human-robot collaborations, since robot movement strategies can be adjusted based on aforementioned human factors.

### 10.3.2 Cloud Robotic Application

In ICMS, the robotic application follows the cloud robotic structure illustrated in Fig. 10.1. In this three-layer application (Fig. 10.4), a local server bridges the gap between the physical robotic cell and remote computing cloud. As mentioned before, the cloud-based robots suffer from the conflict between local control requirements and remote cloud communication. Especially in the context of industrial robotic cells, the stability and security of the ROS are critically essential. Hence in ICMS, the ROS unit and hardware are locally connected, while a local server works as the data buffer and filter between the physical layer and the Cloud layer. The local server coordinates the computing loads between cloud and itself. When the amount of computing work is low, the local PC or server executes the task by itself. When the heavy computing load is required, e.g. optimisation, simulation or point cloud processing, the local server packages the data and pushes it to the cloud along with the service query.

Besides central computing, the cloud also works as the service manager and data/knowledge pool of the cloud robots. Based on the cloud manufacturing system [28, 29], RaaS is adopted in the cloud as a specific type of production service. Virtualised robotic applications are maintained in the service database in the cloud in terms of cloud services, e.g. assembly service, moving service, planning service, etc. When a user requests specific robot service via standardised API, the service broker interprets the query and allocates suitable robot service maintained in the database. Then localised robot task is generated and passed to the local server and then to the physical robotic cell. During execution stage, the service manager plays as the service supervisor which monitors and manipulates the RaaS unit and guarantees the service is proceeded as scheduled.

In practice, it is not necessary or practical to stream all the sensor data collected form the shop floor to the cloud, since it contains primary data sets and meaningless information. To optimise the utilisation of bandwidth, storage space and computing capability in the cloud, it is especially necessary to separate the low-level data

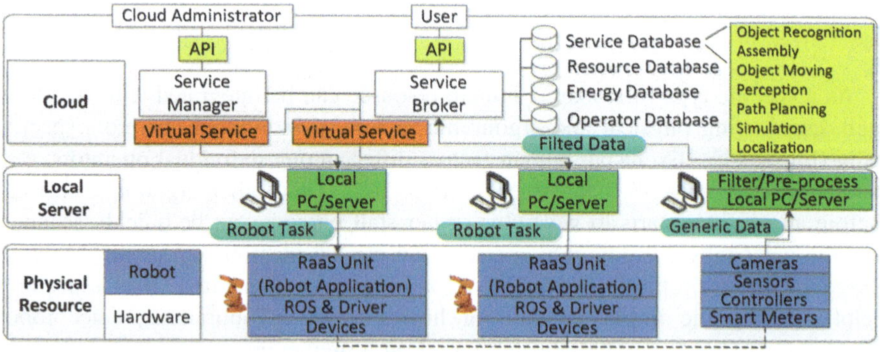


**Fig. 10.4** Cloud-based robotic application layout

selectively. If the data is valuable of being maintained in the cloud repository, the local server works as the data filter and buffer before the generic data is pre-processed and streamed to the cloud database.

With the help of sensor and monitor devices, e.g. cameras, sensors, controllers and smart meters, the trustworthy status of the robotic cell can be collected by the local server in real-time. The monitoring history and data can be stored and processed in the local machines if necessary. When high-performance computing power is needed, for example for path optimisation or energy analysis, the local machine works as the data buffer and pre-processes the data into standalone data package, and then sends the package to the cloud over the network. Thus the cloud computing capability can be utilised to deliver the query in faster speed with higher quality. Additionally, the valuable information and knowledge are maintained and shared in the cloud database, which supports the other cloud robot/manufacturing services in the future. After the computing task is completed, the cloud sends back the results to the local server, which utilises the results as inputs and references for the physical robot control. Thus it forms a feasible solution which guarantees the hardware stability and remote computing capability at the same time.

In the cloud computing paradigm, there are different deployment models, i.e. private, community, public and hybrid cloud. These models can be introduced into the robot clouds. During the implementation of cloud robots, different connection methods are deployed and evaluated. Local Area Network (LAN) is conventionally well appreciated due to its stability and high speed. In recent years Wi-Fi and Bluetooth are popular for connecting mobile and humanoid robots to the cloud thanks to their high speed and easy configuration. However, these connections are limited by stability, cost, distance and security issues. Especially in a factory environment, the data transmission scenarios are critical and a flexible and reliable deployment model is required. In the private robot cloud, the robots, server and cloud are connected inside the boundary of a manufacturing enterprise or company (Fig. 10.5). These devices can be configured based on the local specifications and environments. In a community robot cloud, multiple enterprises share the data and knowledge collaboratively while sensitive information needs to be restrained within the enterprise scope. Thus the local server plays an important role as the data filter. Filtered and pre-processed data is transmitted to the community cloud over the Internet. In the public cloud model, a big amount of data and information is

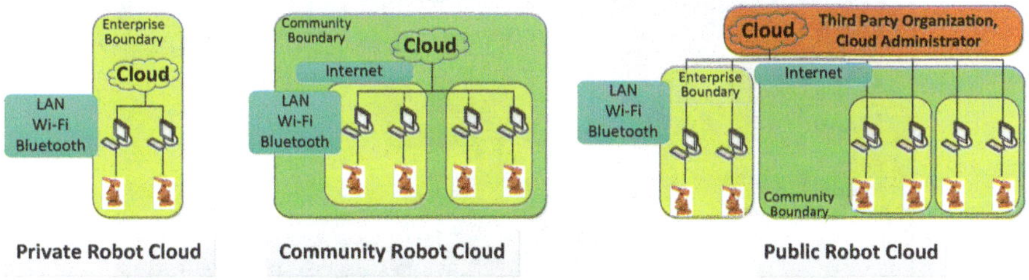


**Fig. 10.5** Private, community and public robot cloud

exchanged over the network and cloud. Hence it is necessary to identify and protect the data at different levels, especially when it is passed through the enterprise or community boundary.

## 10.4 Implementation and Case Studies

To validate and evaluate the aforementioned production system based on cloud, ICMS is implemented based on previous research works [30–32]. In the cloud that hosts virtual environments and service modules, 32 cores and 132 gigabyte memories are deployed to provide the computing power for the proposed system (Fig. 10.6). In this work, Java applet is utilised to develop the user interface since it offers light weight environment of ICMS and good mobility among different systems/environments. MySQL databases are established to maintain production specifications mentioned above. To secure the safety of the cloud system and privacy of users, Secure Sockets Layer Virtual Private Network (SSL VPN) is utilised to provide protected remote access to the cloud.

### *10.4.1 Cloud-Based Manufacturing Chain*

ICMS' cloud production service flow is illustrated in Fig. 10.7. A user firstly accesses to the cloud environment through VPN over the Internet. Command

**Fig. 10.6** Cloud servers in the black cabinet and an industrial robot

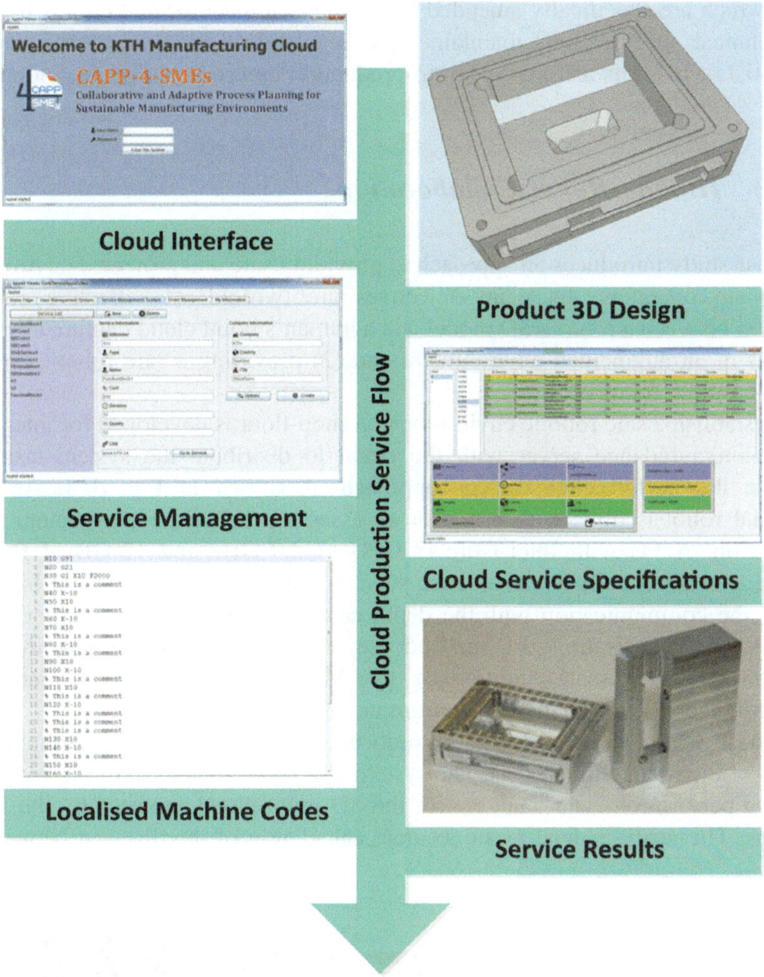


**Fig. 10.7** Cloud production service flow

dashboards are developed for cloud administrators and users, respectively. A cloud administrator is able to manage broadcasted services, customer orders and user profiles remotely. After the product 3D design is uploaded to the cloud, the user's requirements of machining service are interpreted by the smart manager mechanism. Multiple candidate solutions are identified in the cloud database. Among multiple machining providers, the user is able to filter the candidate pool based on different preference criteria, e.g. price, duration and quality priority.

Being part of the cloud service, process planning is generated based on generic feature information from the product 3D design. After the machining service provider is determined, generic production document is converted to localised NC codes which are specifically amended for the chosen machine and cutters based on the technical specifications maintained in the physical resource profile database (Fig. 10.3a). It forms a from-design-to-production environment on the cloud.

## 10.4.2 Human-Robot Collaboration

This case study introduces an approach to providing safe and protected environment for human operators. Its main objectives are twofold: (1) effective collision detection between a robot's 3D model and a human's point cloud captured by depth cameras in an augmented environment; and (2) providing a safe robotic environment by online robot control.

To establish a safe robotic environment, a shop-floor is developed for integrating a collision avoidance server with the cloud to distribute the system tasks and improve the performance of this approach as shown in Fig. 10.8. An ABB industrial robot is utilised to establish an assembly cell for experimentation and verification. A PC with Intel Core i7 CPU of 2.9 GHz and 12 GB of RAM is utilised as a local collision avoidance server. The server is responsible for maintaining the communication with the depth cameras and the robot. Furthermore, it prepares the visualisations for the end user and controls the robot to avoid the collision with the human. Additionally, two Microsoft Kinect depth cameras have been installed to capture the depth information of the robotic cell.

A local database server is also introduced to the system to define the specifications of the robot and its tools. The system also takes advantage of the computing cloud to perform efficient analyses for the 3D point cloud captured by the Kinect cameras. The analyses lead to an accurate calculation of the distance between the

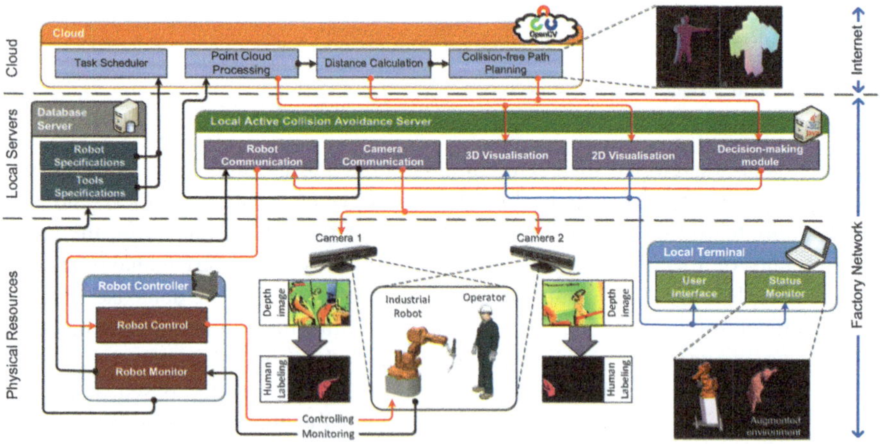


Fig. 10.8 Human-robot collaboration system

robot and the human. This allows the system to avoid any collision in the robotic cell and plan the robot's path accordingly. The cloud is utilised together with the database server to schedule the tasks (in this case assembly tasks) for the robotic cell based on the existence of the human and the availability of the robot and its tools. Sharing the availability information with other robotic cells improves the production performance in terms of speed and quality, and leads to a better machine utilisation. In the experimental results, the capability of collision-free path planning is validated for the robotic system, without the need for programming. The robot is also controlled online by the system to follow the operators hand and perform collaborative assembly tasks.

### 10.4.3 Minimisation of Robot Energy Consumption

This case study presented an approach to minimising the energy consumption of an industrial robot's movements. An optimisation module is introduced to choose the most energy-efficient robot joint configuration and fulfil the time-critical trajectory requirements defined by an operator. The capability of this system was measured by comparing the results of the optimisation module with those of commercial simulation software of the robot. On the other hand, the performance of the developed module was evaluated using three scenarios; with and without a payload at the robot's end-effector. Detailed analyses for the robot working envelope are performed to identify the regions with the lowest level of energy consumption.

A local server in the middle layer is installed and configured to prepare the robot parameters needed for the optimisation (Fig. 10.9). With the help of the well-known Denavit-Hartenberg notation, the server is able to describe mathematically the robot's joints and the relationships between them. The local server is also

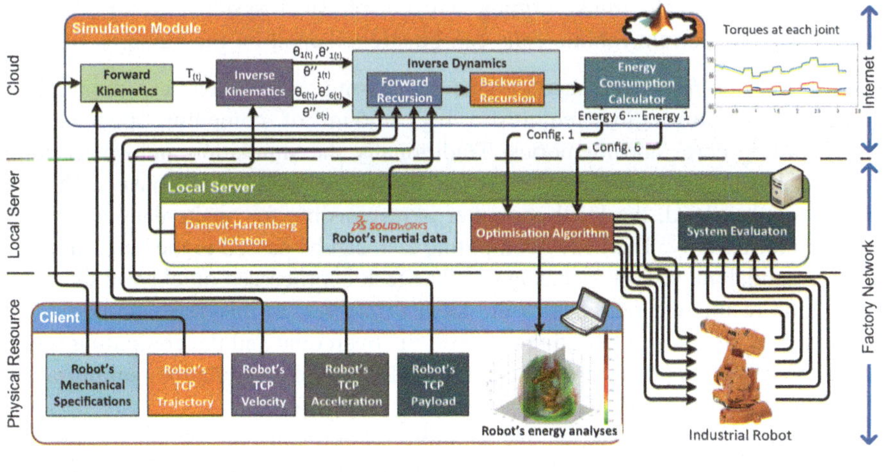


Fig. 10.9 Minimising robot energy consumption

responsible for calculating the inertial tensors of the robot's joints, which are important for solving the inverse dynamics of the robot.

The developed system benefits from the cloud capabilities to construct a MATLAB®-based simulation module. It consists of four parts: the forward kinematics is solved in the first part; the second part is dedicated to solve the inverse kinematics of the trajectory. The third part is responsible for determining forward and backward recursions to solve the inverse dynamics of the robot. The fourth part calculates the energy consumption of each robot's joints configuration and sends them back to the local server. The local server optimises the robot's calculated energy consumption and sends the suitable configuration to the robot. The developed approach allows the local operator to specify the task requirements (trajectory, velocity, acceleration and payload). The energy consumption of the robot's working envelop is analysed based on the proposed approach. Therefore, it is an effective tool to design the robotic layout and decide the locations of the equipment in the robotic cell to achieve optimised energy consumption.

## 10.5 Conclusions

Morden production industry calls for a new generation of manufacturing systems. Nowadays, fast-changing ICT technologies have dramatically altered the way people think and do business. However, most of current production systems still function as 20 years ago. As a disruptive technology, cloud offers an environment with remote access, resource pooling and customisation.

Cloud technologies provide a shared environment of manufacturing capability, computing power, knowledge and resource. It can contribute with innovative robotic technologies to factories of the future. It offers an environment to connect the computing and service resources in the cyber world to the machines and robots in the physical world, thus forming a cyber-physical system. As an enabling technology, the current robotic cells can be strengthened by the cloud as follows.

- Fast response/process speed: image processing, simulation and point cloud generation/optimisation requires a big amount of computing powers, e.g. computing cores and memories. Traditionally the simulation and optimisation task takes long time on local computers that stand beside robots onsite. With the help of the cloud, the local computer is responsible for onsite data acquisition and filtering only. The heavy computing task can be passed to the cloud in terms of a standalone task or metadata file. The task is processed quickly on the cloud and the results are fed back to the local PC/server. In this way, the bottleneck caused by the local computing capability is overcome and the computing power of the robotic cell is strengthened via cloud-based simulation, optimisation, image processing, point cloud generation, and so forth.
- Flexibility: the workload among multiple robotic cells can be balanced at two levels, i.e. computing level and task level. At the computing level, multiple

CPUs and RAMs can be shared by different robotic stations. The computing resource can be dynamically balanced between different tasks. Thus the total computing performance and reaction speed are improved, and initial investment on cloud are shared and distributed by multiple stations. At the task level, different jobs can be analysed by the cloud and dispatched to different robots based on the robot capability, task nature, availability, etc. It forms an optimised robotic collaboration network at global level.
- Human-robot collaboration and adaptiveness: with the help of centralised Cloud database, the detailed specifications of human operators can be maintained dynamically, for example the physical human factors (height, weight, capability), work habits (left-/right-handed), ergonomic statistics and records. This knowledge is especially helpful for decision making of human-robot collaborative tasks. Additionally, when the job is switched from one robotic cell to another, e.g. in case that the original robot is down, new TCP path can be quickly generated which is suitable for the new robotic environment on the cloud. It improves the adaptability and flexibility of the robots as well.
- Energy consumption: with the help of the cloud database, the energy consumption profile/record can be integrated on the cloud. The energy consumption maps can be shared and updated on the cloud dynamically. Thus, eco-friendly and energy-saving strategies can be made at higher level.

In this chapter, a cloud-based system is introduced especially for ubiquitous manufacturing. Integration mechanisms of physical resources are outlined, and customised production planning methods are presented. The introduced system is evaluated through three case studies. A local server-driven architecture is adopted to combat the conflicts between local connections and Internet communications. In practice, safety and security challenges for cloud robotics include resource constraints, information and learning constraints and communication constraints [33]. In the future, the cloud-based manufacturing systems can benefit from the related technologies utilised by computing and manufacturing cloud. Firewalls and access control keeps an ICMS system from unwilling access and effects. Meanwhile network encryption and private keys are helpful to keep sensitive data in specific working domains. In the past years, private cloud models were welcomed by the production enterprises since the company is enabled to protect the cloud infrastructure within their own fences in both cyber and physical worlds. In the future, the cloud manufacturing systems can be further supported by other successful methods, e.g. secure gateways, coding, antivirus software, etc.

## References

1. P. Mell, T. Grance, The NIST definition of cloud computing. NIST Spec. Publ. **800**, 7 (2011)
2. X. Xu, From Cloud Computing to Cloud Manufacturing. Rob. Comput. Integr. Manuf. **28**, 75–86 (2012)

3. S. Jordán et al., The rising prospects of cloud robotic applications, in *IEEE 9th International Conference on Computational Cybernetics (ICCC)* (2013), pp. 327–332
4. K. Goldberg, B. Kehoe, Cloud Robotics and Automation: A Survey of Related Work (EECS Department, University of California, Berkeley, Tech. Rep. UCB/EECS-2013-5, 2013)
5. C. Bohus et al., *Running Control Engineering Experiments Over the Internet* (Oregon State University, Dept. of Computer Science, Corvallis, OR, 1995)
6. J.J. Kuffner, Cloud-enabled robots, in *IEEE-RAS International Conference on Humanoid Robotics, Nashville, TN*, 2010
7. Y. Chen et al., Robot as a service in cloud computing, in *Fifth IEEE International Symposium on Service Oriented System Engineering (SOSE)* (2010), pp. 151–158
8. Z. Du et al., Design of a robot cloud center, in *10th International Symposium on Autonomous Decentralized Systems (ISADS)* (2011), pp. 269–275
9. Y.-L. Chu et al., An integrated java platform for telematic services, in *Fourth International Conference on Genetic and Evolutionary Computing (ICGEC)*, pp. 590–593 (2010)
10. C. Mouradian et al., An infrastructure for robotic applications as cloud computing services, in *IEEE World Forum on Internet of Things (WF-IoT)*, pp. 377–382 (2014)
11. L. Gherardi et al., A software product line approach for configuring cloud robotics applications, in *IEEE 7th International Conference on Cloud Computing (CLOUD)* (2014), pp. 745–752
12. E. Guizzo, T. Deyle, Robotics trends for 2012. IEEE Robot. Autom. Mag. **19**, 119–123 (2012)
13. G. Mohanarajah et al., *Rapyuta: A Cloud Robotics Platform* (2014)
14. D. Hunziker et al., Rapyuta: the roboearth cloud engine, in *IEEE International Conference on Robotics and Automation (ICRA)* (2013), pp. 438–444
15. F.Q. Ansari et al., A cloud based robot localization technique, in *Contemporary Computing*, ed. by M. Parashar et al. (Springer, 2012), pp. 347–357
16. Y. Chen, H. Hu, Internet of intelligent things and robot as a service. Simul. Model. Pract. Theory **34**, 159–171 (2013)
17. K. Bekris et al., Reducing roadmap size for network transmission in support of cloud automation. IEEE Robot. Autom. Mag. (2015)
18. Y. Kato et al., RSi-cloud for integrating robot services with internet services, in *IECON 2011-37th Annual Conference on IEEE Industrial Electronics Society* (2011), pp. 2158–2163
19. K. Kamei et al., Cloud networked robotics. Netw. IEEE **26**, 28–34 (2012)
20. N. Mavridis et al., The human-robot cloud: situated collective intelligence on demand, in *IEEE International Conference on Cyber Technology in Automation, Control, and Intelligent Systems (CYBER)* (2012), pp. 360–365
21. F. Ren, Robotics cloud and robotics school, in *7th International Conference on Natural Language Processing and Knowledge Engineering (NLP-KE)* (2011), pp. 1–8
22. O. Morariu et al., vMES: virtualization aware manufacturing execution system. Comput. Ind. **67**, 27–37 (2015)
23. B. Liu et al., Information fusion in a cloud computing era: a systems-level perspective. IEEE Aerosp. Electron. Syst. Mag. **29** (2014)
24. S. Jeong et al., Cloud computing based livestock monitoring and disease forecasting system. Int. J. Smart Home **7** (2013)
25. S. Nakagawa et al., An implementation of a distributed service framework for cloud-based robot services, in *IECON 2012-38th Annual Conference on IEEE Industrial Electronics Society* (2012), pp. 4148–4153
26. C. Agüero, M. Veloso, Transparent multi-robot communication exchange for executing robot behaviors, in *Highlights on Practical Applications of Agents and Multi-Agent Systems*, ed. by J. B. Perez et al. (Springer, 2012), pp. 215–222
27. W.-S. Jang, R.Y.C. Kim, *Template Design of Automatic Source Code Generation Based on Script Language used in Cloud Robot Compiling Environment* (2013)
28. X.V. Wang, X. Xu, An interoperable solution for cloud manufacturing. Robot. Comput. Integr. Manuf. **29**, 232–247 (2013)

29. X.V. Wang, X. Xu, ICMS: a cloud-based manufacturing system, in *Cloud Manufacturing: Distributed Computing Technologies for Global and Sustainable Manufacturing*, ed. by W. Li, J. Mehnen (Springer, London, 2013), pp. 1–22
30. M. Givehchi et al., Generic Machining process sequencing through a revised enriched machining feature concept. J. Manuf. Syst. (2015 in press)
31. M. Givehchi et al., Adaptive distributed process planning and execution for multi-tasking machining centers with special functionalities, Presented at the Flexible Automation and Intelligent Manufacturing (FAIM), 2015
32. M. Givehchi et al., Latest advances in function block enabled adaptive process planning, Presented at the 4th International Conference on Virtual Machining Process Technology (VMPT-2015), 2015
33. G. Hu et al., Cloud robotics: architecture, challenges and applications. Netw. IEEE **26**, 21–28 (2012)

# Chapter 11
# Context-Aware Human-Robot Collaborative Assembly

## 11.1 Introduction

Robotic systems are the key assets in various industrial sectors. How to utilise them efficiently and effectively is a practical challenge. Recently, the concept of Human-Robot Collaboration (HRC) has generated many interests. The existing literature suggests that human workers have incomparable problem-solving skills and sensory-motor capabilities, but are restricted in force and precision [1, 2]. Robotic systems however provide better fatigue, higher speed, higher repeatability and better productivity, but are restricted in flexibility. When integrated, HRC can release human workers from heavy tasks and establish communication channels between human workers and robots for better overall performance. By combining the advantages of both human workers and industrial robots, an HRC system has the potential to achieve higher productivity and better sustainability in modern factories.

Ideally, an HRC team should work similarly as a human-human collaborative team. However, traditionally, time-separation or space-separation approaches have been applied in HRC systems, which reduced productivity for both humans and robots [1]. In order to build an efficient HRC team, human-human collaboration teams can be analysed as examples. In human teamwork and collaboration, there are two theories: joint intention theory and situated learning theory [3–6]. To apply the theories to an HRC team, humans and robots should collaborate symbiotically:

- All team members in an HRC team should share the same plan of execution;
- An HRC team should have structured ways of communication; and
- All team members in an HRC team should be aware of the context of the collaborative environment.

Recent researches revealed that the acquisition and processing of 3D data is already available for industrial applications [7]. In HRC environment, the safety of human workers can be protected by depth sensors and compatible algorithms [8].



L. Wang and X. V. Wang, *Cloud-Based Cyber-Physical Systems in Manufacturing*,
https://doi.org/10.1007/978-3-319-67693-7_11

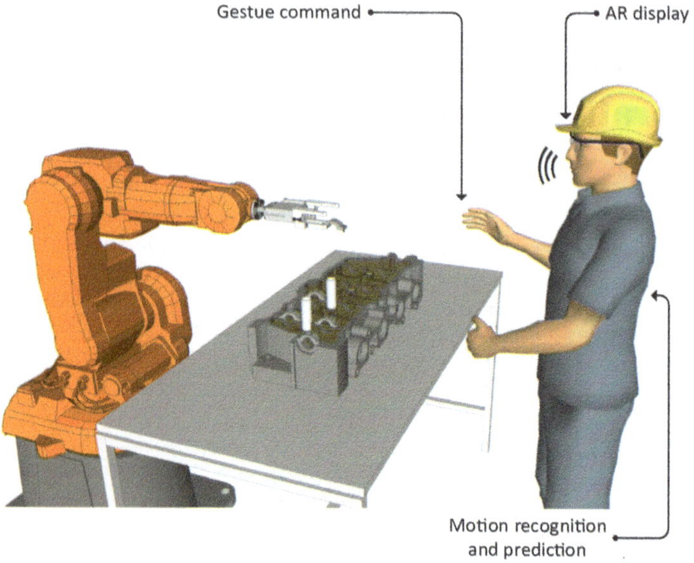


**Fig. 11.1** Symbiotic human-robot collaborative assembly based on context awareness

Recognised human motions can be used as input for industrial robot control [9]. However, current industrial robots still cannot establish the context awareness as we introduced previously. Human workers cannot control industrial robots intuitively, either. The information feedback channel from industrial robots to human workers is still limited.

In modern HRC manufacturing environment, products are highly customised and flexible. It requires that the human worker be the leader of the HRC team in an HRC manufacturing system. A human worker should be able to flexibly re-assign assembly tasks to industrial robots based on availability and capability. Thus, the assembly task should be available for re-assignment in the HRC system once the human worker's task re-assignment intent or command is detected. The information feedback from industrial robots to human workers should also be established.

As shown in Fig. 11.1, in this chapter, the context awareness based symbiotic HRC is achieved by three key technical components: gesture recognition, human motion prediction, and AR-based worker support system.

## 11.2 Gesture Recognition

Gesture is a communication method. Head nodding, hand gestures and body postures are effective communication channels in human-human collaboration [2, 10]. Gestures can be categorised into three types [11]:

- Body gestures: full body actions or motions,
- Hand and arm gestures: arm poses, hand gestures, and
- Head and facial gestures: nodding or shaking head, winking lips.

Gesture recognition refers to the mathematical interpretation of human motions by a computing device. In order to collaborate with human workers, robots need to understand human gestures correctly and act based on the gestures efficiently. In HRC environment, a natural way of gesture communication between robots and humans should be available.

### 11.2.1 Gesture Recognition for Human-Robot Collaboration

To recognise gestures in the HRC context, it is beneficial to investigate into a generic and simplified human information processing model. As shown in Fig. 11.2, Parasuraman et al. [12] generalised human information processing into a four-stage model. Based on the generic model in Fig. 11.2, a specific model for gesture recognition in HRC is introduced here. As shown in Fig. 11.3, there are five essential parts related to gesture recognition for HRC, i.e.:

- Sensor data collection: gesture raw data is captured by sensors.
- Gesture identification: in each frame, a gesture is located from the raw data.
- Gesture tracking: the located gesture is tracked during the gesture movement. For static gestures, gesture tracking is unnecessary.
- Gesture classification: tracked gesture movement is classified according to pre-defined gesture types.
- Gesture mapping: gesture recognition result is translated into robot commands and sent back to workers.

### 11.2.2 Sensor Technologies

Before gesture recognition process started, gesture raw data need to be collected by sensors. In this section, different sensors in the literature are analysed based on different sensing technologies. As shown in Fig. 11.4, there are two basic categories for data acquisition: image based and non-image based approaches.

**Fig. 11.2** A four-stage model of human information processing [12]

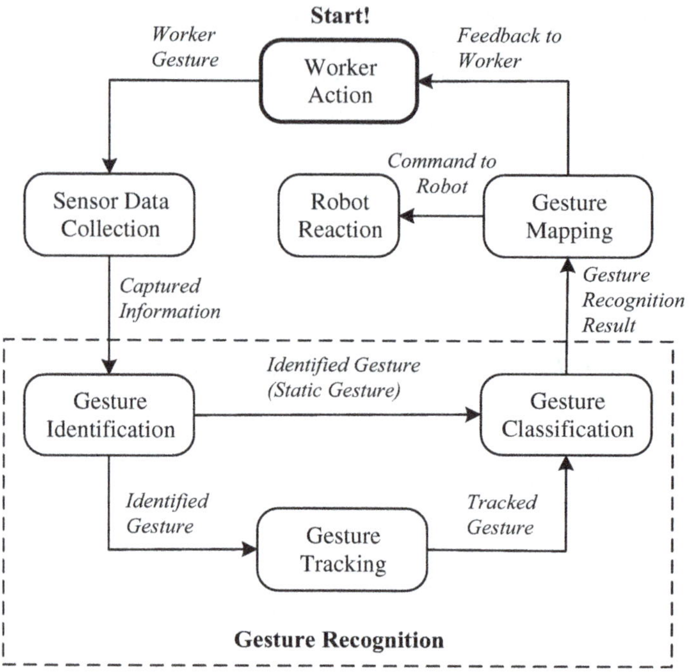


**Fig. 11.3** A process model of gesture recognition for human-robot collaboration

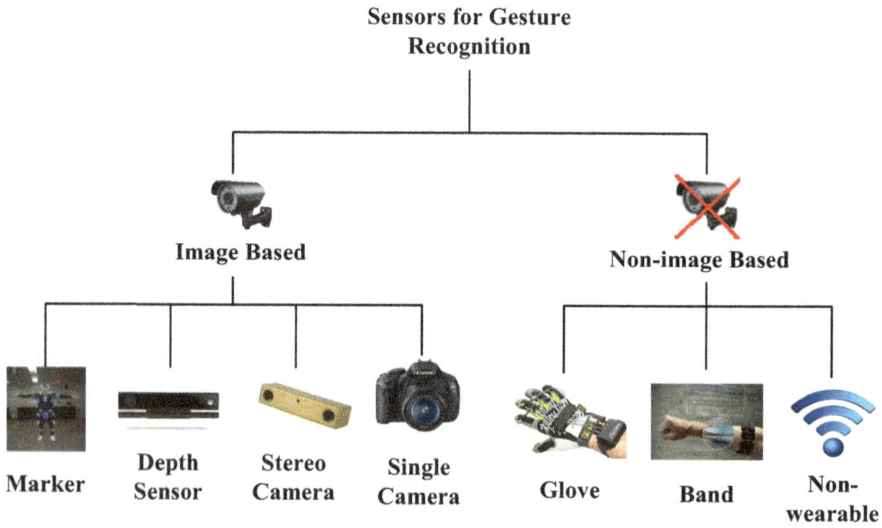


**Fig. 11.4** Different types of sensors for gesture recognition

### Image based approaches

Technologies are often inspired by nature. As human being, we use our eyes to recognise gestures. Therefore, for robots, it is reasonable to use cameras to "see" gestures. The image-based approaches are further divided into four categories.

- **Marker**: In marker-based approaches, the sensor is a normal optical camera. In most marker-based solutions, users need to wear obvious markers [11]. Today, we enjoy much faster graphical processing speed as compared with twenty years ago. There are more gesture recognition sensors available.
- **Single camera**: In the early 90th, researchers started to analyse gestures using a single camera [13, 14]. A drawback of single-camera approaches is the restriction of view angles, which affects a system's robustness [15]. Recent research, however, applied single-camera approaches in high-speed gesture recognition [16]. The system utilises a speed image sensor and a specially designed visual computing processor to achieve fast gesture recognition.
- **Stereo camera**: In order to achieve robust gesture recognition, researchers suggested stereo camera based approaches to construct 3D vision. Here, we define stereo camera based approaches as applications that use two cameras (optical stereo camera) to construct 3D depth information. Many stereo camera based approaches followed a similar workflow [17, 18]. Although stereo camera systems have improved robustness in outdoor environment, they still suffered from problems such as computational complexity and calibration difficulties [19].
- **Depth sensor**: Recently, depth sensing technologies have emerged rapidly. We define a depth sensor as a non-stereo depth sensing device. Non-stereo depth sensors enjoy several advantages compared to the traditional stereo cameras. For example, the problems of setup calibration and illumination conditions can be prevented [20]. Moreover, the output of a depth sensor is 3D depth information. Compared with colour information, the 3D depth information simplifies the problem of gesture identification [11]. A comparison of gesture identification accuracy by using colour and depth information can be found in [21]. There are two types of common non-stereo depth sensors: Time-of-Flight (ToF) camera, and Microsoft Kinect (or similar IR sensors).

### Non-image based approaches

Gesture recognition has been dominated by image-based sensors for a long time. Recent developments in MEMS and sensor technology have greatly boosted non-image based gesture recognition technologies.

- **Glove**: Glove-based gestural interfaces are commonly used for gesture recognition. Usually, glove-based approaches require wire connections, accelerometers and gyroscopes. However, a cumbersome glove with a load of cables can potentially cause problems in HRC [11, 22]. Glove-based approaches also introduced complex calibration and setup procedures [23].

- **Band**: Another contactless technology uses band-based sensors. Band-based sensors rely on a wristband or similar wearable devices. Band-based sensors use wireless technology and electromyogram, which avoids connecting cables. The sensor only needs to contact with wrist; user's hand and fingers can be released. An example of band-based sensor is Myo gesture control armband [24]. Recently, several band-based sensor gesture control systems were reported [25–27].
- **Non-wearable**: The third type of non-image based technologies adopts non-wearable sensors. Non-wearable sensors can detect gestures without contacting human body. Google introduced Project Soli, a radio frequency spectrum (radar) based hand gesture tracking and recognition system [28]. The device is capable of recognising different hand gestures within a short distance. MIT has been leading non-wearable gesture recognition technology for years. Electric Field Sensing technology was pioneered by MIT [29]. A recent discovery by Adib et al. [30–32] from MIT introduced WiTrack and RF-Capture system1 that capture user motion by radio frequency signals reflected from human body. The systems are able to capture human gestures even from another room through a wall with a precision of 20 cm. In summary, non-wearable sensor based technologies are promising and fast growing for gesture recognition.

**Comparison of sensor technologies**

A comparison of different sensor technologies is provided in Table 11.1. The advantages and disadvantages of different approaches are indicated. It is clear that there is no sensor fits all applications. Two observations of sensor technologies are provided based on the above analyses.

**Table 11.1** Advantages and disadvantages of different sensor technologies

| | Advantage | Disadvantage |
|---|---|---|
| Markers | Low computational workload | Markers on user body |
| Single camera | Easy setup | Low robustness |
| Stereo camera | Robust | Computational complexity, calibration difficulties |
| ToF camera | High frame rate | Resolution depends on light power and reflection |
| Microsoft Kinect | Fast emerging, software support for body gesture recognition | Cannot be used for hand gesture recognition over 2 m |
| Glove | Fast response, precise tracking | Cumbersome device with load of cables |
| Band sensor | Fast response, large sensing area | Band needs to contact with human body |
| Non-wearable | Avoid contact with human body | Low resolution, technology not mature enough |

- With indoor applications, depth sensor approaches are the most promising image-based technologies. Depth sensors possess advantages of easy setup calibration and easy data processing. A large application development community exists, which provides ready solutions.
- Non-wearable approaches are the most promising technology among non-image based approaches. They avoid direct contact with users. Non-wearable sensing is also a fast-growing field.

### *11.2.3 Gesture Identification*

Gesture identification is the first step in the gesture recognition process after raw data are captured from sensors. Gesture identification refers to the detection of gestural information and segmentation of the corresponding gestural information from the raw data. Popular technologies to solve the gesture identification problem are based on visual features, learning algorithms, and human models.

**Visual features**

Human hands and body have unique visual features. In image-based gesture recognition, gestures consist of human hands or body. Therefore, it is straightforward to utilise such visual features in gesture identification.

- **Colour**: Colour is a simple visual feature to identify a gesture from background information. However, colour-based gesture recognition systems are easily influenced by illumination and shadows in a complex HRC environment [33]. Another common problem in skin colour detection is that human skin colour actually varies among human races. Due to the problems above, in recent approaches, skin colour is only considered to be one of many cues in gesture identification.
- **Local features**: In image-based gesture recognition, illumination conditions greatly influence gesture identification quality. Therefore, many researchers utilise local features method that is not sensitive to illumination conditions. Local features approach is a detailed texture-based approach. It decomposes an image into smaller regions that are not corresponding to body parts [34]. As shown in Fig. 11.5, one of the most important local features is Scale Invariant Feature Transform (SIFT) [35]. The SIFT method is rotational, translational, scaling and partly illumination invariant. Several similar local feature approaches, for example, SURF and ORB are proposed in later years [36, 37]. Normally, local features approaches are also only considered as one of many cues in gesture identification. Several identification methods such as shape and contour methods, motion methods, and learning methods are based on local features.

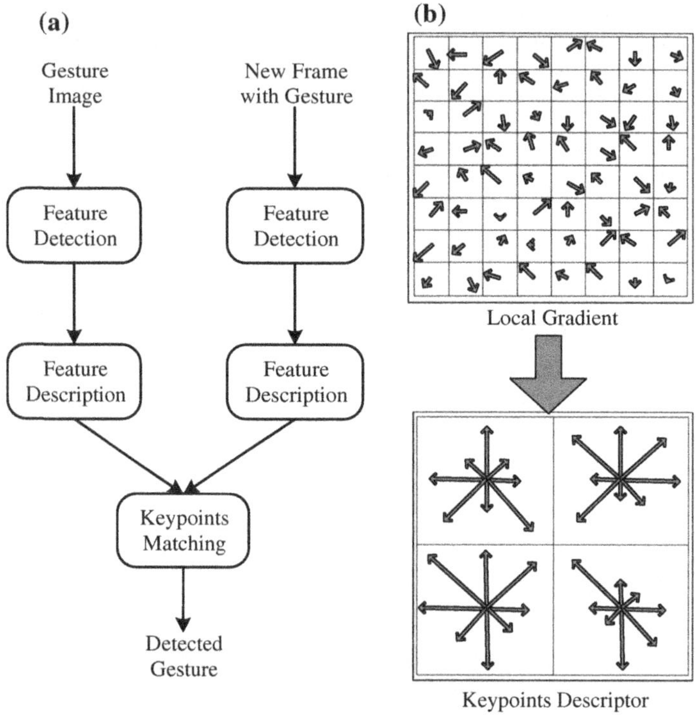


**Fig. 11.5** SIFT algorithm: **a** SIFT algorithm for gesture identification; **b** SIFT feature description example [35]

- **Shape and contour**: Another intuitive and simple way to identify gestures is to utilise the unique shape and contour of a human body in HRC environment. Shape model based approach matches a pre-constructed shape model and shape features from observation. A milestone for shape detection and matching was reported by Belongie et al. [38]. They introduced a shape context descriptor method. Shape context descriptor is used for detection of similar shapes in different images. The development of depth sensor provides opportunities to measure surface shapes. The 3D models generated from the technologies enable highly detailed representation of human body shape [39, 40].
- **Motion**: In certain HRC manufacturing environment, a human worker is the only moving object in the raw data. Therefore, the human motion is a useful feature to detect human gestures. Optical flow is a key technology for motion-based gesture identification. Optical flow does not need background subtraction, which is an advantage compared to shape and contour based approaches. Several gesture recognition applications were implemented based on optical flow method [41, 42]. Dalal and Thurau [43] introduced the famous Histograms of Oriented Gradients (HOG) method. The HOG descriptors divide image frames into blocks. For each block, a histogram is computed. Among

non-image based sensors, motion-based gesture identification is also a popular method [31, 44]. Usually, thresholding and filtering are applied to raw sensor data to identify human gestures.

### Learning algorithms

A recent trend of gesture identification is to use learning algorithms, especially for static gesture detection that can be represented in a single frame. The visual feature methods are based on various visual features, while learning algorithms utilise machine learning algorithms to identify gestures from raw sensor data. Although some algorithms are based on the visual feature methods, image background removal is not always necessary for learning algorithms. Learning algorithms such as Support Vector Machine (SVM), Artificial Neural Networks (ANN) and Random Decision Forests (RDF) are widely applied in gesture recognition systems [45–47].

### Skeleton model

To identify body gestures, a detailed model of the human body is useless. Different from the aforementioned approaches, skeleton model approach uses a human skeleton to recover human body poses [48]. Skeleton model is a simplified human body model that preserves only the most valuable information from a human body. Skeleton model approach also provides advantages for simplifying the gesture classification process [49]. With benefits mentioned above, the skeleton model approach has become an attractive solution for depth sensors [49, 50].

### Summary of gesture identification approaches

A gesture identification quality comparison case study was presented by Han [49]. It can be summarised that depth-based approach outperforms RGB-based approach. Skeleton model belongs to depth-based approach. Most of the visual features approaches belong to RGB-based approach. In Table 11.2, both advantages and disadvantages of different gesture identification methods are summarised. Moreover, according to different sensors, different gesture identification method should be applied. Due to the nature of HRC in the manufacturing environment, human workers are the most important members of an HRC team. Despite understanding human body gestures, the skeleton model approach will also monitor human movements, which provides a secure environment for the human-robot team. As mentioned earlier, skeleton model simplifies human body, while valuable

**Table 11.2** Advantages and disadvantages for different gesture identification methods

| | Advantage | Disadvantage |
|---|---|---|
| Visual features | Low computational workload | Low quality |
| Learning algorithms | Background removal can be avoided | Higher computational expenses |
| Skeleton model | Only the most important information is abstracted from a human body | Only possible to use with depth sensor based systems |

information is well preserved. Subsequent gesture classification process can be simplified by the skeleton models. Therefore, currently, skeleton model approach is an appropriate solution for gesture recognition in HRC manufacturing systems.

### *11.2.4 Gesture Tracking*

In gesture recognition, the notion of tracking is used differently in different literatures. We define the notion of tracking as the process of finding temporal correspondences between image frames. Specifically, we focus on the continuous gesture tracking problem that associates the identified gesture in the previous frames with the current frame. As for static gestures that can be represented by a single frame, gesture tracking is unnecessary. An example of gesture tracking is shown in Fig. 11.6.

**Single hypothesis tracking**

Single hypothesis tracking refers to a best-fit estimation with minimum-error matching. Therefore, in single hypothesis tracking, a gesture is represented by only one hypothesis. Most of the advanced tracking algorithms are based on the single hypothesis tracking technologies.

- **Mean shift**: Mean shift tracker is a basic tracking technology. Mean shift tracker performs matching with RGB-colour histograms [51]. For each new frame, mean shift tracker compares the Bhattacharyya distance between the target window histograms of the new frame with those of the old frame. A complete mathematical explanation can be found in [51].
- **Kalman filter**: Kalman filter (KF) is a real-time recursive algorithm used to optimally estimate the underlying states of a series of noisy and inaccurate measurement results observed over time. A complete KF mathematical derivation can be found in [52, 53]. Nowadays, KF has evolved and been applied in different fields such as aerospace, robotics, and economics.

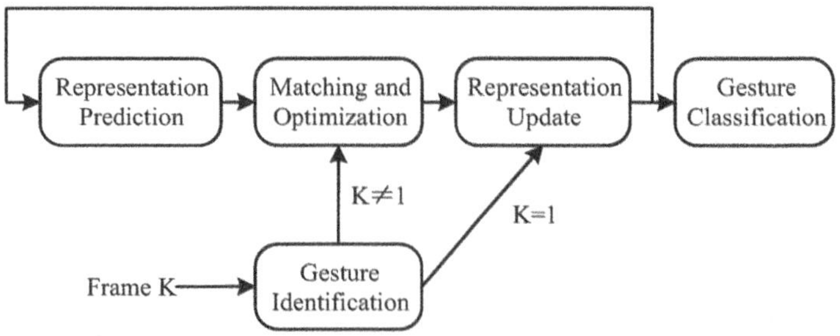


**Fig. 11.6** A gesture tracking example

- **Kalman filter extensions**: KF is given a prerequisite that the state vector is a linear model. Extend Kalman Filter (EKF) is a functional tracking algorithm even if the model is nonlinear [54]. Another algorithm that solves the same problem from a different angle is Unscented Kalman Filter (UKF) [55]. UKF solves the problem by applying a deterministic weighted sampling approach. The state distribution is represented using a minimal set of chosen sample points.

**Multiple hypotheses tracking**

In HRC manufacturing scenarios, many human workers are working in the same station at the same time [1]. To track multiple workers' gesture simultaneously, multiple hypotheses tracking technologies should be applied.

- **Particle filter**: Particle filter (PF) is a popular technology in solving robotic problems. Different from KF, PF does not make assumption on posterior model. The PF representation is a nonparametric approximation which can represent a broader space of distribution. Therefore, PF satisfies multiple hypotheses tracking requirement [56]. An example of PF is shown in Fig. 11.7. Several advanced tracking algorithms also apply PF to scan probability density function [57–59].
- **Particle filter extensions**: Many researchers attempted to combine PF with other algorithms, for example with mean shift tracker, genetic algorithm, PSO, ant colony optimisation, and other machine learning algorithms to solve the sample degeneracy and impoverishment problem [61]. Other researchers also improved PF resampling strategy [62, 63].

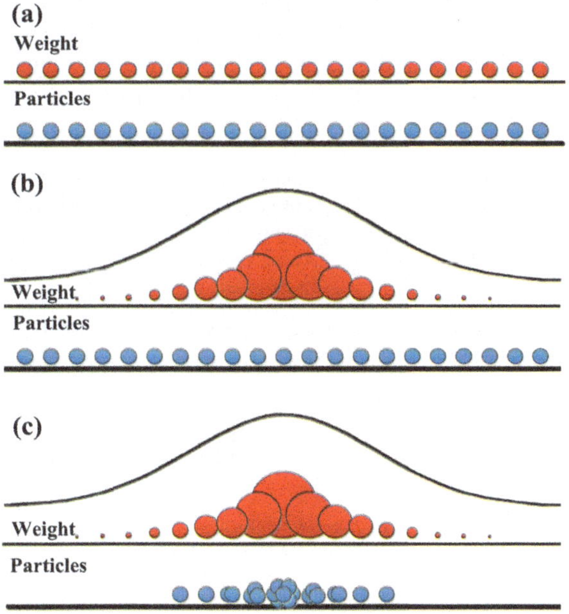


**Fig. 11.7** Particles and weight factors: **a** after particles initialisation; **b** after weight factor calculation; **c** after resampling [60]

### Advanced tracking methods

Recently, there are many advanced tracking methods introduced. Some of these advanced methods utilised part of the tracking algorithms mentioned above. Other methods improved tracking performance by detection or learning algorithms.

- **Extended model tracking**: For long-term tracking problems, many tracking algorithms would fail because target maintains fixed models. Extended model tracking saves target behaviour or appearance from the past few image frames. Therefore, more target information is reserved for target estimation. Incremental Visual Tracker uses extended model to preserve more details for tracking process [64]. Kwon et al. [58] presented Tracking by Sampling Tracker. The extended model is preserved by a sampling process. The tracker samples many trackers and accordingly the appropriate tracker is selected.
- **Tracking by detection**: Another type of tracking algorithms is built together with the gesture identification learning algorithms introduced in the earlier sections. For these tracking algorithms, a classifier or detector is applied in image frames to identify gesture from the background information [59]. One representative approach is Tracking, Learning and Detection Tracker [65]. The approach integrates the result of an object detector with an optical flow tracker. Another typical tracking-by-detection technology is to apply Multiple Instance Learning [66]. The learning algorithm can increase tracker robustness and decrease parameter tweaks.

### Comparison of different gesture tracking approaches

Smeulders et al. [67] presented a test result of different gesture tracking algorithms. The resulting score is normalised F-score. F-score provides us an insight of the average coverage of the tracked object bounding box and the ground truth bounding box. Therefore, the tracking algorithms with higher F-scores have better tracking quality. In Fig. 11.8, the test results in different video conditions are presented.

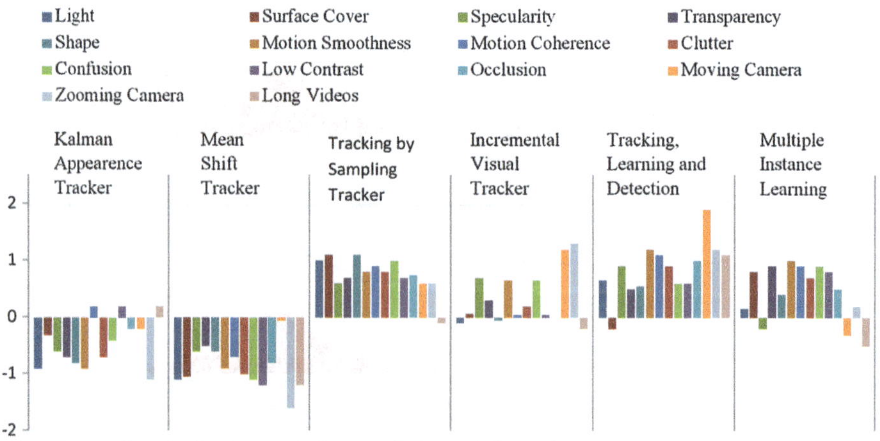


**Fig. 11.8** Test results of tracking algorithms in different video conditions [67]

Kalman Appearance Tracker and Mean Shift Tracker belong to the single hypothesis tracker. Tracking by Sampling Tacker and Incremental Visual Tracker belong to the extended model tracker. Multiple Instance Learning Tracker, and Tracking, Learning and Detection Tracker belong to the tracking-by-detection method. It is easy to observe that the single hypothesis trackers perform lower than the others. However, simple gesture tracking algorithms also provide a lighter computational load. Depending on computation power and tracking quality requirement, an appropriate gesture tracking algorithm can be selected for HRC manufacturing system.

### *11.2.5 Gesture Classification*

Gesture classification is the last and the most important step in gesture recognition. As a typical machine learning problem, gesture classification can be solved by many popular machine learning algorithms.

**K-nearest neighbours**

K-nearest neighbours (KNN) algorithm is a fundamental gesture classification algorithm that classifies input data according to the closest training examples [68].

**Hidden Markov model**

Hidden Markov model (HMM) is a popular gesture classification algorithm. HMM is a combination of an unobservable Markov chain and a stochastic process. An example of HMM is shown in Fig. 11.9. To solve the problem, Expectation-Maximisation (EM) algorithm is applied [69]. Many papers discussed HMM

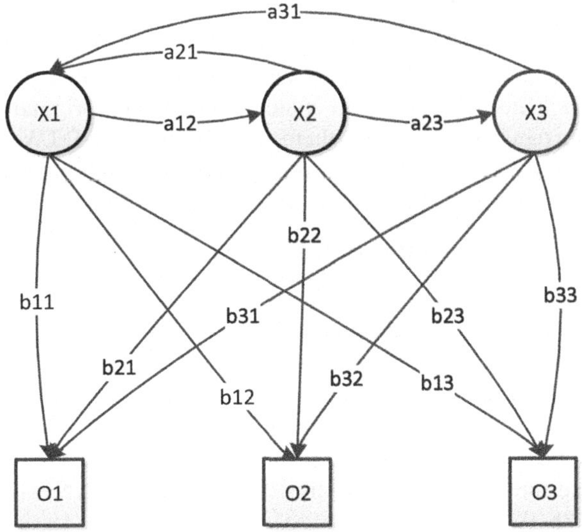


**Fig. 11.9** Example of hidden Markov model [100]

gesture recognition applications [70–72]. Some articles combined HMM with other classification approaches [71]. Others extended HMM algorithm into wider range of applications [72].

**Support vector machine**

Support Vector Machine (SVM) is a discriminative classifier defined by a separating hyperplane [73]. Classification decision boundaries are identified by maximising a margin distance. The optimal separation hyperplane maximises the margin of training data. The training examples closest to the optimal hyperplane are called support vectors. A common problem for SVM is that the number of support vectors grows linearly with the size of the training set. Some researchers proposed Relevance Vector Machine (RVM) to solve the problem [74]. SVM kernel trick was introduced by Scholkopf [75]. SVM kernel trick enables linear SVM in nonlinear problems. SVM kernel transforms low-dimensional training data into high-dimensional feature space with nonlinear method [76]. There are also research efforts that combined SVM with other classification methods to improve gesture classification performance [77–79].

**Ensemble method**

Ensemble method is another type of widely-used gesture classification algorithm. The primary assumption of ensemble method is that ensembles are more accurate than individual weak classifiers. One of the well-known ensemble methods is Boosting by Schapire et al. [80, 81]. The boosting algorithm starts with several weak classifiers. The weak classifiers are repeatedly applied. In a training iteration, part of training samples is used as input data. After the training iteration, a new classification boundary is generated. After all iterations, the boosting algorithm combines these boundaries and merges into one final prediction boundary.

**Dynamic time warping**

Dynamic time warping (DTW) is an optimal alignment algorithm for two sequences. DTW generates a cumulative distance matrix that warps the sequences in a nonlinear way to match each other. Originally, DTW is used for speech recognition. Recently, there have been many DTW applications in gesture recognition as well [82]. Some papers also introduced Derivative Dynamic Time Warping (DDTW) as an extension of DTW [83].

**Artificial neural network**

Artificial neural network (ANN) is a family of information processing models inspired by biological neural networks [84]. ANN consists of many interconnected processing unions (neurons) that work in parallel. Each union (neuron) receives input data, processes input data and gives output data. ANN can be used to estimate functions that depend on a large number of input data. Recently, many researchers have utilised ANN for gesture recognition [85–87]. Several papers also presented gesture recognition systems that combined ANN with other classification methods [88–90].

### Deep learning

Deep learning is an emerging and fast-growing branch of machine learning. Deep learning enables data modelling with high-level abstractions by using multiple processing layer neural networks. Moreover, different from traditional learning algorithms, deep learning needs little engineering by hands, which enables the possibility to take advantages of exponentially increasing available data and computational power [91]. Today, deep learning is applied in image recognition, speech recognition, particle accelerator data analysis, etc. [92]. Especially, deep learning is employed for solving the problem of human action recognition in real-time video monitoring, which contains a large number of data [93, 94]. Convolutional Neural Networks (CNN) and Recurrent Neural Networks (RNN) are two popular deep learning algorithms [91]. Several gesture recognition systems have applied above deep learning algorithms, recently [95, 96].

### Comparison of gesture classification approaches

Table 11.3 lists the advantages and disadvantages of varying gesture classification approaches. One of the potentials for HRC manufacturing systems is deep learning approach. The primary constraint of deep learning is the limited computational power. However, the exponentially increasing computation power can solve the problem quickly. The number of deep learning based gesture classification

**Table 11.3** Advantages and disadvantages of gesture classification approaches

| Approach | Advantages | Disadvantages |
|---|---|---|
| K-nearest neighbours | Simple | K needs to be chosen carefully |
| Hidden Markov model | Flexibility of training and verification, model transparency | Many free parameters need to be adjusted |
| Support vector machine | Different kernel function can be applied | Number of support vectors grows linearly with the size of training set |
| Ensemble method | Do not need large number of training data | Overfitting easily, sensitive to noise and outliers |
| Dynamic time warping | Reliable nonlinear alignment between patterns | Time and space complexity |
| Artificial neural network | Can detect complex nonlinear relationships between variables | "Black box" nature and cannot be used for small training data set |
| Deep learning | Do not need good design of features, outperform other machine learning methods | Need large number of training data and computationally expensive |

applications is growing rapidly. Another trend is to combine different classification algorithms. Every classification algorithm has own advantages and disadvantages. To utilise that, different classifiers can be combined to achieve better performance in a manufacturing environment. It is also important to coordinate gesture classification algorithms with gesture identification and gesture tracking algorithms.

### 11.2.6 Future Trends of Gesture Recognition

Although the above sections provided a general picture of gesture recognition for HRC, it is never easy to summarise such an interdisciplinary and fast-developing field in any capacity. Sensor related technologies usually started from hardware. Software technologies and algorithms are designed to utilise the performance of hardware. Therefore, we would predict future trends starting with sensor technologies.

- Depth sensor and skeleton model based gesture recognition: due to the nature of HRC, human workers are the most important members of any HRC team. Despite understanding human body gestures, depth sensor together with skeleton model approach will monitor human movements, which provides a safer environment for HRC. Moreover, skeleton model will simplify gesture classification process. Therefore, simpler gesture tracking and classification method can be applied.
- Non-wearable sensor and deep learning based gesture recognition: although non-wearable sensor technologies are not ready, it is still the most promising non-image based sensor. In HRC manufacturing systems, human workers should be able to communicate with robots naturally. For this very purpose, nothing should be attached to workers' body. Non-wearable sensors still suffer from low gesture identification and classification quality. The problem can potentially be solved using deep learning methods.
- Hard real-time gesture recognition system: one of the most important requirements in manufacturing is the real-time requirement. Especially, in HRC manufacturing systems, the safety of human workers is always the priority. Therefore, real-time gesture recognition system is another future direction. Currently, band and glove sensors provide the fastest response. Moreover, high-speed single-camera gesture recognition system is also emerging recently. In gesture identification, tracking and classification, quick and effective methods can be applied.
- Multi-sensor gesture recognition system: all the sensors have advantages and disadvantages. For instance, band sensor has large sensing area; Kinect has good performance in body gesture recognition. To best utilise the system performance, different gesture recognition sensors can be used in the same system.
- Algorithms combination approach: similar with sensors, different gesture classification algorithms also have their advantages and disadvantages. As we mentioned in gesture classification section, the combination of algorithms improves efficiency.

## 11.3 Human Motion Prediction

As discussed in the previous sections, an HRC manufacturing system is more customised and flexible than conventional manufacturing systems. An efficient HRC system should be able to understand a human worker's intention and assist the human when performing an assembly task.

Since a worker's (work-related) motions are limited and repetitive, a sequence of human motions can be modelled to represent an assembly task. Existing human motion recognition techniques can be applied to recognise the human motions associated with the assembly task. The recognised human motions are modelled by Hidden Markov model (HMM). A motion transition probability matrix is then generated after solving the HMM. Based on the result, human motion prediction becomes possible. The human intent is analysed with the input of predicted human motion. The predicted human intent can be used as input for robot motion planning. The robot can thus be controlled to support and collaborate with the human worker. The workflow of human motion prediction in HRC is shown in Fig. 11.10.

### *11.3.1 Assembly Tasks Sequence*

We intend to model an assembly task as a sequence of human motions. In this section, the problem is formulated. Task-level assembly and motion recognition is introduced. Based on formulated problem and analysis, statistic model solution is selected.

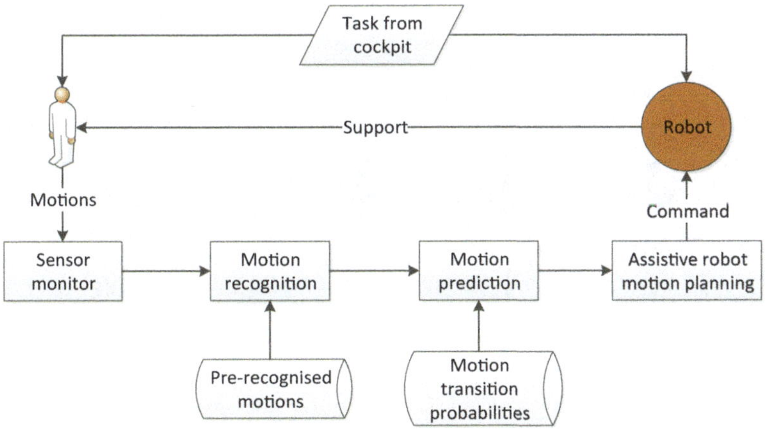


**Fig. 11.10** Workflow of human motion prediction in HRC

**Task-level assembly**

In modern manufacturing, operator instruction sheets (OIS) are widely used. An OIS provides detailed instructions and explanations of the assembly tasks in an assembly station. Generally, an OIS provides task-level instructions and references for the human worker who is working in the assembly station. According to the OIS, the task sequence of the assembly station is pre-defined and fixed. However, in an HRC manufacturing system, the human worker's motion can be different and flexible when doing a task. Different workers may prefer to perform the same task in a variety of ways. With current motion capturing sensors [7], human worker's motions can be obtained. Therefore, it is possible to generalise task-level human motions as a discrete model. An example of a task-level representation is shown in Fig. 11.11. To apply this approach to HRC, the human worker's motions need to be further recognised and described by a mathematical model. In the next section, the motion recognition process is introduced.

**Motion recognition**

As shown in Fig. 11.10, motion recognition is a pre-process of human motion prediction for HRC. The output information from motion recognition is the input of human motion prediction. Although motion recognition technologies are not the focus of this section, the motion recognition result still needs to be analysed and cleaned for human motion prediction.

As introduced in [9], there exist different motion observation and detection technologies. Strong and well-developed motion recognition technologies possess higher observation reliability but are limited in application feasibility, whereas

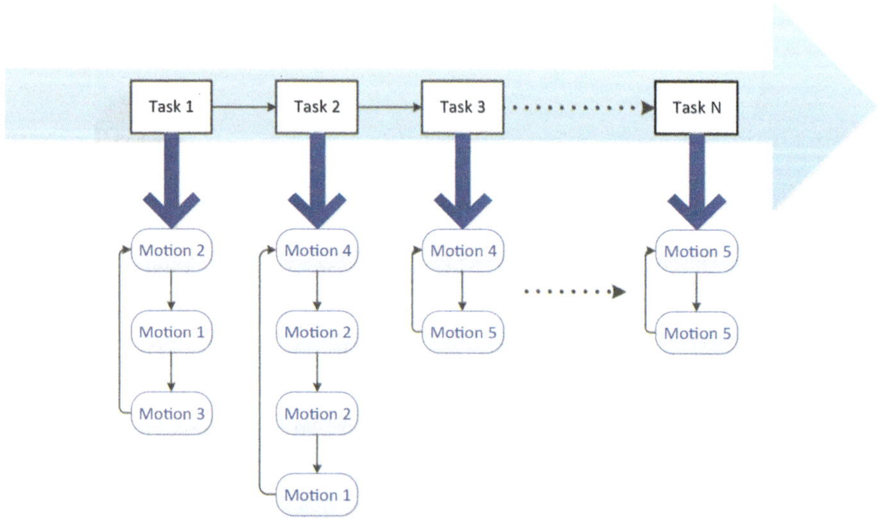


**Fig. 11.11** Example of task-level representation in an assembly station

weak and evolving technologies possess lower observation reliability but can be applied in more practical situations. For the strong motion recognition technologies, RFID tags can be used as an example. RFID tags are widely used in the current assembly line. RFID tag detection relies on the distance between the human body and the detector. In a part-taking motion, for example, an RFID tag is fixed on the worker's clothes, and an RFID reader is placed near the parts storage. The observation probability can be generalised as a step function:

$$P_h = \begin{cases} 1 & \textit{if detected} \\ 0 & \textit{otherwise} \end{cases} \tag{11.1}$$

For the weak motion recognition technologies, vision-based motion sensors can serve as an example. Vision-based motion sensors rely on the captured visual data. In the part-taking motion, the start of the motion can be defined when the arm starts to approach the parts storage. The end of the motion can be defined when the arm takes assembly part in hand. The closer to the end of the motion, the higher probability the motion is detected. The observation likelihood of a vision-based motion detector can be generalised as a continuous detection distribution:

$$P_l = \begin{cases} P(o_{s:e}) & \textit{if detected} \\ 0 & \textit{otherwise} \end{cases} \tag{11.2}$$

where $o_{s:e}$ represents the observation of a gesture from the start to the end.

**Statistical model selection**

As shown in Fig. 11.10, the input to human motion prediction is the result from human motion recognition. The output of human motion prediction is a prediction probability of a human worker's subsequent motion that can be used in the assistive robot motion planning. The human motion prediction problem can be regarded as a machine learning problem. The results of human motion recognition can easily be discretised. Therefore, several standard machine learning classification solutions can be applied, such as HMM, SVM, KNN, HSOM and dynamic bag-of-words, to the human motion recognition and prediction problem [97–99]. Among these algorithms, HMM is a well-developed discrete sequence based algorithm. Markov chain is a well fit to the applications of human motion prediction for HRC. The hidden state transition can also be used in scenarios that highly uncertain results are generated by some weak motion recognition technologies. Therefore, it is reasonable to utilise HMM for human motion prediction. The HMM algorithm for human motion prediction will be introduced in the next section.

### *11.3.2 HMM Human Motion Prediction*

In this section, a brief introduction to HMM is presented. The assembly task representation in the HMM context is analysed. Finally, an HMM solution for human motion prediction is illustrated.

**Hidden Markov model**

HMM is a statistical Markov model with hidden states. The states in HMM is not observable. The hidden states have different transition probabilities. The output generated from the states is observable. Each state has a probability distribution for generating different outputs. As the example shown in Fig. 11.9, an HMM can be defined from the following elements [100]:

- The states are denoted as $S = \{s_1, s_2, \ldots, s_N\}$. $N$ is the number of states in the model. The state sequence is $Q = \{q_1, q_2, \ldots, q_t\}$. The state at time $t$ is $q_t$.
- The observation symbols are denoted as $V = \{v_1, v_2, \ldots, v_M\}$. $M$ is the number of distinct observation symbols per state. The observation sequence is $O = \{o_1, o_2, \ldots, o_t\}$. $o_t$ is the observation at $t$.
- The state transition probability distribution is $A = \{a_{ij}\}$, where

$$a_{ij} = P(q_{t+1} = s_j | q_t = s_i), 1 \leq i, j \leq N. \tag{11.3}$$

- The observation symbol probability distribution is $B = \{b_j(k)\}$, where

$$b_j(k) = P(o_t = v_k | q_t = s_j), 1 \leq j \leq N, 1 \leq k \leq M. \tag{11.4}$$

- The initial state distribution $\pi = \{\pi_i\}$, where

$$\pi_i = P(q_1 = s_i), 1 \leq i \leq N. \tag{11.5}$$

It is possible to summarise from the above that one complete HMM requires the specification of parameters $N$ and $M$, observation symbols, and probability measures $A$, $B$, and $\pi$. A compact notation is introduced to indicate the complete model parameters:

$$\lambda = (A, B, \pi) \tag{11.6}$$

**Task representation**

As shown in Fig. 11.12, the representation of a human worker's motions is a linear sequence. For each task, different motion sequences can be observed from various

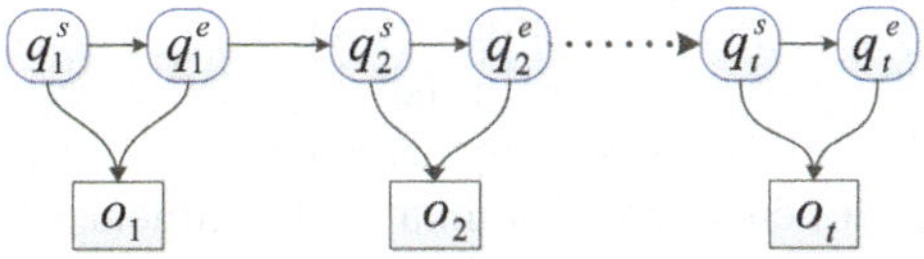


**Fig. 11.12** An HMM model representation of a human worker's motions

human workers. In this model, the human worker's motions are modelled as a Markov process that each motion starts after the end of the previous motion. $q_t^s$ represents the start of a motion $t$. $q_t^e$ represents the end of the motion $t$. The motion is presented between $q_t^s$ and $q_t^e$. On a shop floor, the industrial robot needs to respond in a continuous time domain. However, the system models an HRC task as a discrete HMM model. Therefore, during each motion, only one observation $o_t$ is generated. The time between two motions is ignored.

The observation of motion $q_t$ is $o_t$. Therefore, the observation probability given the start and end of the motion can be described as:

$$P\left(o_{s:e}^t | q_t^s, q_t^e\right) \tag{11.7}$$

Equation (11.1) can be explained in an HMM model:

$$P_h\left(o_{s:e}^t | q_{s:e}^t\right) = \begin{cases} 1 & ifdetected \\ 0 & otherwise \end{cases} \tag{11.8}$$

where $P_h\left(o_{s:e}^t | q_{s:e}^t\right)$ represents strong motion recognition technologies. As explained earlier, the motion recognition technologies represented by Eq. (11.8) possess higher observation reliability.

The observation of HMM model requires discretised probability input. Therefore, Eq. (11.2) can be discretised as:

$$P_l\left(o_{s:e}^t | \mathrm{q}_{s:e}^t\right) = \begin{cases} L_h & if\ L_h \leq P(o_{s:e}) \leq 1 \\ L_l & if\ L_l < P(o_{s:e}) < L_h \\ 0 & if\ 0 \leq P(o_{s:e}) \leq L_l \end{cases} \tag{11.9}$$

where $P_l\left(o_{s:e}^t | q_{s:e}^t\right)$ is the probability of motion recognition results. $P(o_{s:e})$ represents the observation probability of a motion from the start to the end. $L_h$ and $L_l$ are parameters that represent the limits of high and low detection probabilities. The parameters can be adjusted according to different motion detectors. As explained in earlier sections, the weak motion recognition technologies represented by Eq. (11.9) possess lower observation reliability.

**HMM solution**

As introduced by Rabiner [100], the following three fundamental problems can be solved by HMM in real applications:

- Given observation sequence $O = \{o_1, o_2 \ldots o_t\}$ and a model $\lambda = (A, B, \pi)$, how to compute the probability of the observation sequence $P(O|\lambda)$;
- Given observation sequence $O = \{o_1, o_2 \ldots o_t\}$ and a model $\lambda = (A, B, \pi)$, how to choose the optimal state sequence $Q = \{q_1, q_2 \ldots q_t\}$; and
- How to adjust model parameters $\lambda = (A, B, \pi)$ to maximise $P(O|\lambda)$.

Here, the observation sequence $O = \{o_1, o_2 \ldots o_t\}$ is known, whereas $A$ and $B$ need to be learned. The prediction of a human worker's motion mainly relies on $A$. Therefore, the third problem needs to be solved.

### *11.3.3 Experiment*

Car engine assembly is a complicated process. In this section, a car engine assembly case is utilised to demonstrate the potential of human motion prediction as an HRC application. The parts before assembly are shown in Fig. 11.13a, the right corner of which shows four electric control plugs. Each plug needs to be plugged in the engine and fastened with one screw. Figure 11.13a also shows a plastic cover to be placed on top of the engine and fastened with screws. Figure 11.13b shows the car engine after the assembly task.

To represent the case study above, five different worker motions are defined:

1. Take a screwdriver
2. Take the plastic part (take a big part)
3. Take an electric control plug (take a small part)
4. Take a screw
5. Assembly the screw with the screwdriver (assembly).

Four of the motions (take a screwdriver, take a plastic part, take an electric control plug, take a screw) can be detected by RFID tags. One of the motions (assembly) can be detected by vision-based motion observer. Therefore, the states

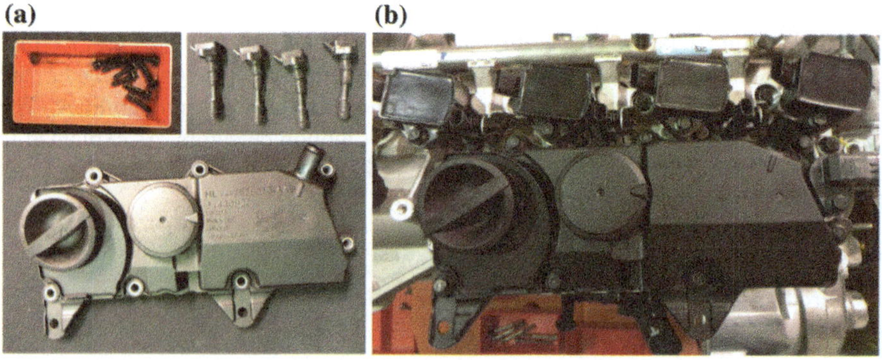


**Fig. 11.13** Example assembly task, **a** parts before assembly; **b** engine after assembly

**Table 11.4** States and observation symbols defined for the assembly task

| State | State meaning | Observation symbol | Symbol meaning |
|---|---|---|---|
| $s_1$ | Take screwdriver | $v_1$ | Take screwdriver observed |
| $s_2$ | Take big part | $v_2$ | Take big part observed |
| $s_3$ | Take small part | $v_3$ | Take small part observed |
| $s_4$ | Take screw | $v_4$ | Take screw observed |
| $s_5$ | Assembly | $v_5$ | Assembly observed (with low probability) |
| | | $v_6$ | Assembly observed (with high probability) |

and the observation symbols are defined in Table 11.4. The five different states are: $S = \{s_1, s_2, s_3, s_4, s_5\}$. According to the previous sections and Eqs. (11.8) and (11.9), the six different observation symbols are: $V = \{v_1, v_2, v_3, v_4, v_5, v_6\}$.

In this case study, a human worker is invited to perform the same assembly task for ten times. The initial motion is defined as $s_1$: take a screwdriver. The record of the assembly task is used for HMM training. The trained state transition probability distribution matrix graph is illustrated in Fig. 11.14a. The state observation probability graph is shown in Fig. 11.14b. The differences between weak and strong motion recognition technologies are well illustrated by the observation probability graph. Also, it can be reflected from the state transition probability matrix that the worker explored many different assembly motion sequences. Compared to $s_1$ and $s_5$, $s_2$, $s_3$, $s_4$ have less uncertainty with regard to the next state. $s_5$ has many different next states. However, $s_5$ is the end of a 'sub-sequence'. Therefore, it is reasonable to have different possibilities after assembly. It is also noticed from the observation symbol probability matrix that sometimes $v_6$ is observed after $v_5$.

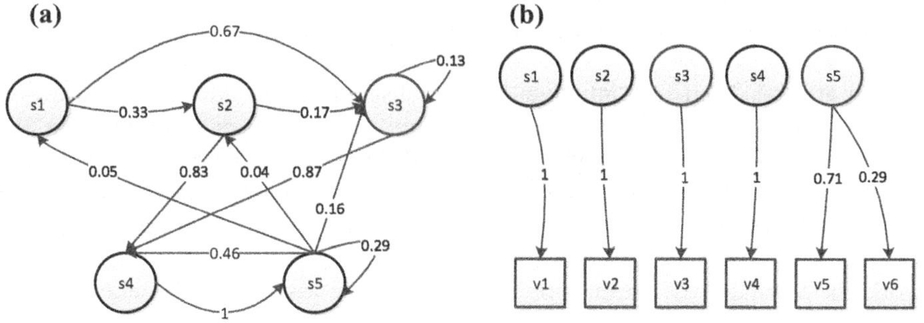


**Fig. 11.14** **a** HMM states transition probability matrix graph of the assembly case; **b** HMM states observation probability graph of the assembly case

### 11.3.4 Discussions

The case study showcased one example of human motion prediction for HRC. Although the worker explored different assembly motion sequences, there are still patterns that can be used to predict the worker's motion for HRC. As mentioned earlier, $s_2$, $s_3$, $s_4$ have comparatively certain next states. Therefore, it is possible to control an industrial robot to prepare or help accordingly. $s_5$ has many state transition probabilities. However, the transition probabilities to $s_1$ and $s_2$ are rather small. In this HRC application, these two possibilities can be ignored. In this case study, the source of the components is not considered yet. In the HRC application, the number of the components is fixed for each assembly task. The available assembly components can be used to improve motion prediction result. It is worth to mention that the unstableness of the current vision-based motion recognition technology also affects the prediction result. To solve the problem, $v_6$ and $v_5$ are defined as observations with high probability and low probability, respectively. By changing detection probability limit, it can be utilised to eliminate a false alarm. It is also possible to combine different motion detection and recognition technologies to increase the system robustness. A false-correction system can be designed to further improve the robustness of the HRC system.

## 11.4 AR-Based Worker Support System

A worker support system should be able to provide information feedback and support to human worker instantly and intuitively in the HRC context. This section introduces the potential of adopting augmented reality (AR) technologies in worker support system for HRC manufacturing. As shown in Fig. 11.15, in an HRC manufacturing system, the human worker is placed at the central control position. Based on the assembly tasks from a cockpit, the task sequence planning system generates robot control commands and human worker assembly instructions. With commands from task sequence planner, an industrial robot is controlled to assist human worker at task level. The assembly instructions and robot control information are translated into intuitive AR visualisation by an AR-based instruction system. The human worker is supported by AR visualisation to perform the assembly tasks with the support of the industrial robot. During the assembly operation, human worker is monitored by a worker monitoring system. The worker monitoring system ensures worker's safety [8], recognises worker's commands [9] and predicts worker's intentions. The task re-planner can be activated once human worker's command is detected or worker's task re-assignment intention is predicted. According to the assembly context, modified robot control commands and assembly instructions can be generated by the task re-planner. Modified assembly instructions are translated again by AR-based instruction system. Therefore, after the re-assignment of assembly task, the human worker is instantly supported by the

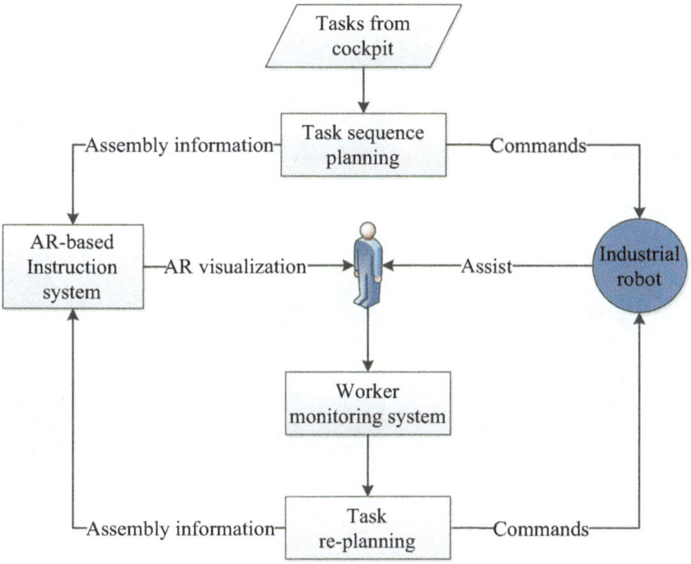


**Fig. 11.15** Information flow of AR-based HRC worker support system

updated AR visualisation and the industrial robot. Utilising the intuitive visualisation, the highly customised and flexible HRC manufacturing is achievable with the help from AR-based worker support system.

## *11.4.1 System Architecture*

As mentioned in the previous section, an HRC worker support system should be able to provide instant and intuitive information feedback and support to a human worker. The worker support system requires a real-time display for the human worker to facilitate assembly. In this section, the AR-based HRC worker support system is introduced. The system architecture of the AR-based HRC worker support system is shown in Fig. 11.16. The AR-based HRC worker support system mainly consists of four sub-systems: AR-based instruction system, task sequence planning and re-planning system, worker monitoring system, and industrial robot control system.

*AR-based instruction system*: the AR-based instruction system mainly consists of assembly information registrar and AR device. The assembly information registrar receives human worker assembly instruction and robot information generated from task sequence planner. In Fig. 11.16, the assembly information consists of human worker assembly instructions and robot information. The assembly information registrar places the received assembly information at the correct location once the corresponding assembly parts are detected in the world coordinate system. The

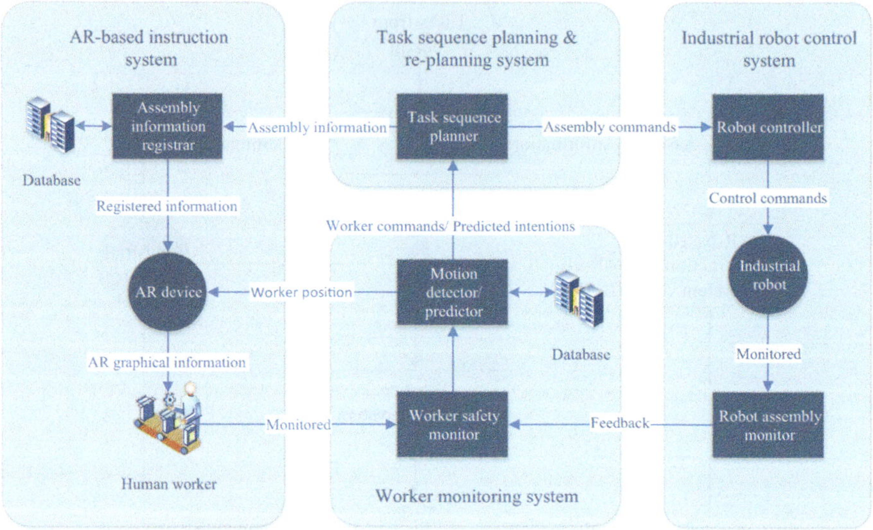


**Fig. 11.16** System architecture of AR-based HRC worker support system

detection is achieved by comparing the assembly parts captured by the sensor and the assembly parts saved in a database. The detailed assembly information registrar technologies are explained in next section. The AR device receives the registered graphical information and displays the information according to the world coordinate system. Finally, the human worker performs the assembly task with the support of the AR graphical information and the industrial robot.

*Task sequence planning and re-planning system*: the task sequence planning and re-planning system receives the predicted human worker intention and outputs human worker assembly information and industrial robot assembly commands. The task sequence planner decides the updated assembly plan according to the initial assembly task and the recognised worker commands or predicted worker intentions. The assembly plan is further translated into assembly instructions for the human worker and assembly commands for the robot. The assembly information for the robot is also sent to the AR-based instruction system to keep the human worker in the loop. In Fig. 11.16, the assembly information sent from the task sequence planner to the assembly information registrar includes both worker assembly instructions and robot information.

*Worker monitoring system*: the worker monitoring system consists of motion recogniser, predictor and worker safety monitor. A depth sensor is utilised as the data input of the worker monitoring system. The worker safety monitor calculates the distance between a worker and a robot in real-time to ensure a safe distance between the two in the HRC environment [8]. The robot can be slowed down or stopped if the distance between the human and the robot is smaller than the safe distance. The processed worker motions are sent to motion recogniser and predictor for further analysis. The motion recogniser recognises human gesture commands and assembly motions. Worker intentions are predicted based on the analysis of recognised worker's assembly motions and the worker's assembly motions history. Recognised worker commands and predicted worker intentions are sent to task sequence planner for further processing. Estimated human worker's position is sent to an AR device for AR visual registration.

*Industrial robot control system*: the industrial robot control system consists of a robot controller, a robot, and a robot assembly monitor. The robot controller receives assembly commands from the task sequence planner. The output of the robot controller is translated control commands for the robot. The robot assists the human worker during assembly. The assembly process is observed by the robot assembly monitor in real-time. The observed information is sent to the worker monitoring system.

With the above mentioned four sub-systems, the AR-based HRC worker support system is fully integrated.

### 11.4.2 AR Assembly Information Registrar

The AR assembly information registrar empowers a human worker to access the assembly instructions and robot information from an AR device. It is an essential component in the AR-based instruction system. As shown in Fig. 11.17, the AR

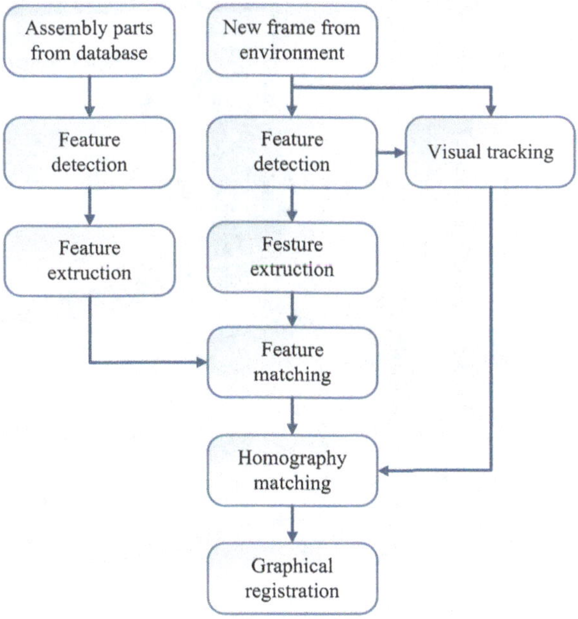


**Fig. 11.17** Information flowchart of AR-based instruction registrar

assembly information registrar enables the registration of graphical instructions and robot information with the real-world assembly parts. Since the AR assembly instructions need to be registered around the real assembly parts, the virtual assembly parts need to be stored in the database for recognition. Feature detection and extraction should be applied on the virtual assembly parts. For each new video frame of the assembly environment, the same feature detection and extraction process are applied. The processed video frames are matched with all the virtual assembly parts. If one or more assembly parts are recognised, the homography is computed. The visual tracker provides a smooth homography matching between different video frames. Based on the matched homography, the registration of graphical instruction and robot information becomes possible.

### *11.4.3 Case Study*

As shown in Fig. 11.18, the case study setup consists of a camera, a robot, a screen and part of a car engine to be assembled. The assembly information registrar is developed by computer vision library OpenCV and game engine Unity. The object and tool models are saved in a database. As explained in the previous section, the feature detection and extraction is done once the database receives object and tool models. For every new image frame, the feature detection and extraction is performed. If there is an object matched by the feature matching algorithm, the

**Fig. 11.18** Setup of case study

Table 11.5 Planned assembly sequence

| Number | Task description | Task type |
|---|---|---|
| 1 | Take tool 1 | Human worker |
| 2 | Take object 1, place it on the engine | Industrial robot |
| 3 | Assembly object 1 on engine | Human worker |
| 4 | Take object 2, place it on the engine | Industrial robot |
| 5 | Assembly object 2 on engine | Human worker |
| 6 | Put back tool 1 | Human worker |

homography will be computed. The 3D text or object will be registered at the location according to the homography. The screen will display the registered 3D text or object. A human worker will be instructed and informed by the 3D text or object.

The case study is designed to demonstrate the usability of the AR-based instruction system. A pre-designed assembly sequence plan is shown in Table 11.5. The goal of the designed case study is to assemble objects 1 and 2 on the car engine with help of an industrial robot. Two assembly parts and a tool are presented in the setup. Above each assembly part and tool, a 3D text is registered. The text is used for intuitive assembly instructions. A cylinder shaped red object is registered above the text label of object 1, which indicates the next assembly part planned for the robot. The first step of the planned assembly sequence is to take tool 1. After tool 1 is taken by the worker, the robot will be activated and pick object 1 and place it on the engine. The worker will be instructed to assemble object 1 on the engine. Upon completion, the robot will be activated and pick object 2 and place it on the engine. The worker will be instructed to assemble object 2 on the engine. Finally, the worker will be instructed to put back tool 1, which indicates the end of the assembly sequence.

## 11.5 Conclusions

This chapter introduces both gesture recognition and AR-based worker support. A generic model of gesture recognition for human-robot collaboration is also reported. There are four essential technical components in the model of gesture recognition: sensor technologies, gesture identification, gesture tracking and gesture classification. Reviewed approaches are classified according to the four essential technical components. In the section part, assembly tasks are modelled as a sequence of human motions. Existing human motion recognition techniques are applied to recognise the human motions. Hidden Markov model is used in the motion sequence to generate a motion transition probability matrix. Based on the result, human motion prediction becomes possible. The predicted human motions are evaluated and applied in task-level human-robot collaborative assembly.

Finally, the potential of adopting AR technologies in a worker support system is explored. The robot commands and worker instructions can be virtually augmented for human workers intuitively and instantly. The designed AR-based worker support system is demonstrated in a case study.

As the future work of gesture recognition and human motion prediction, the assembly resources can also be considered into human worker's motion prediction. Assembly resources give a major constraint on human worker's motion. The human motion prediction could be more accurate. The combination of different motion recognition technologies can increase the system's robustness. Therefore, the fusion of motion recognition technologies can also be considered in the future. Besides, the human motion prediction system needs to be tested in a real production environment with more workers and assembly tasks involved. By applying to different scenarios, the system's reliability can be further improved.

## References

1. J. Krüger, T.K. Lien, A. Verl, Cooperation of human and machines in assembly lines. CIRP Ann. Technol. **58**, 628–646 (2009)
2. S.A. Green, M. Billinghurst, X. Chen, G.J. Chase, Human-robot collaboration: a literature review and augmented reality approach in design. Int. J. Adv. Robot. Syst. 1–18 (2008)
3. P.R. Cohen, H.J. Levesque, Teamwork. Nous 487–512 (1991)
4. L.S. Vygotsky, *Mind in society: the development of higher psychological processes* (Harvard University Press, 1980)
5. P.R. Cohen, H.J. Levesque, persistence, intention, and commitment. Reason. About Actions Plans 297–340 (1990)
6. C. Breazeal et al., Humanoid robots as cooperative partners for people. Int. J. Humanoid Robot. **1**, 1–34 (2004)
7. Z.M. Bi, L. Wang, Advances in 3D data acquisition and processing for industrial applications. Robot. Comput. Integr. Manuf. **26**, 403–413 (2010)
8. B. Schmidt, L. Wang, Depth camera based collision avoidance via active robot control. J. Manuf. Syst. **33**, 711–718 (2014)
9. H. Liu, L. Wang, *Gesture recognition for human-robot collaboration: a review* (J. Ind. Ergon, Int, 2017). doi:10.1016/j.ergon.2017.02.004
10. A. Bauer, D. Wollherr, M. Buss, Human–robot collaboration: a survey. Int. J. Humanoid Robot. **5**, 47–66 (2008)
11. S. Mitra, T. Acharya, Gesture recognition: a survey. IEEE Trans. Syst. Man, Cybern. Part C Appl. Rev. **37**, 311–324 (2007)
12. R. Parasuraman, T.B. Sheridan, C.D. Wickens, A model for types and levels of human interaction with automation. IEEE Trans. Syst. Man Cybern. Part A Syst. Humans. **30**, 286–297 (2000)
13. T.E. Starner, *Visual Recognition of American Sign Language Using Hidden Markov Models* (1995)
14. T. Starner, J. Weaver, A. Pentland, Real-time american sign language recognition using desk and wearable computer based video, in *IEEE Trans. on Pattern Analysis and Machine Intelligence*, vol. 20 (1998), pp. 1371–1375
15. N.R. Howe, M.E. Leventon, W.T. Freeman, Bayesian reconstruction of 3D human motion from single-camera video. NIPS **99**, 820–826 (1999)

16. Y. Katsuki, Y. Yamakawa, M. Ishikawa, High-speed human/robot hand interaction system, in *Proceedings of the Tenth Annual ACM/IEEE International Conference on Human-Robot Interaction Extended Abstracts* (2015), pp. 117–118
17. M. Elmezain, A. Al-Hamadi, J. Appenrodt, B. Michaelis, A hidden markov model-based continuous gesture recognition system for hand motion trajectory, in *19th International Conference on Pattern Recognition* (2008), pp. 1–4
18. Y. Matsumoto, A. Zelinsky, An algorithm for real-time stereo vision implementation of head pose and gaze direction measurement, in *Proceedings Fourth IEEE International Conference on Automatic Face and Gesture Recognition* (2000), pp. 499–504
19. J.P. Wachs, M. Kölsch, H. Stern, Y. Edan, Vision-based hand-gesture applications. Commun. ACM **54**, 60–71 (2011)
20. J. Suarez, R.R. Murphy, Hand gesture recognition with depth images: a review. *IEEE RO-MAN* 411–417 (2012)
21. P. Doliotis, A. Stefan, C. McMurrough, D. Eckhard, V. Athitsos, Comparing gesture recognition accuracy using color and depth information, in *Proceedings of the 4th International Conference on Pervasive Technologies Related to Assistive Environments* (2011), p. 20
22. T. Sharp et al., Accurate, robust, and flexible real-time hand tracking, in *Proceeding CHI* (2015), p. 8
23. A. Erol, G. Bebis, M. Nicolescu, R.D. Boyle, X. Twombly, Vision-based hand pose estimation: a review. Comput. Vis. Image Underst. **108**, 52–73 (2007)
24. T. Labs, Myo (2015) https://www.myo.com/
25. Y. Zhang, C. Harrison, Tomo: wearable, low-cost electrical impedance tomography for hand gesture recognition, in *Proceedings of the 28th Annual ACM Symposium on User Interface Software & Technology* (2015), pp. 167–173
26. N. Haroon, A.N. Malik, Multiple hand gesture recognition using surface EMG signals. J. Biomed. Eng. Med. Imaging **3**, 1 (2016)
27. S. Roy, S. Ghosh, A. Barat, M. Chattopadhyay, D. Chowdhury, Artif. Intell. Evol. Comput. Engin. Syst. 357–364 (2016)
28. Google, Project soli (2015) https://www.google.com/atap/project-soli/
29. J. Smith et al., Electric field sensing for graphical interfaces. Comput. Graph. Appl. IEEE **18**, 54–60 (1998)
30. F. Adib, C.-Y. Hsu, H. Mao, D. Katabi, F. Durand, Capturing the human figure through a wall. ACM Trans. Graph. **34**, 219 (2015)
31. F. Adib, D. Katabi, *See through walls with WiFi!* ACM. **43** (2013)
32. F. Adib, Z. Kabelac, D. Katabi, R.C. Miller, 3D tracking via body radio reflections. Usenix NSDI **14** (2014)
33. J. Letessier, F. Bérard, Visual tracking of bare fingers for interactive surfaces, in *Proceedings of the 17th annual ACM symposium on User interface software and technology* (2004), pp. 119–122
34. D. Weinland, R. Ronfard, E. Boyer, A survey of vision-based methods for action representation, segmentation and recognition. Comput. Vis. Image Underst. **115**, 224–241 (2011)
35. D.G. Lowe, Object recognition from local scale-invariant features, in *Proceedings of 7th IEEE International Conference on Computer Vision*, vol. 2 (1999), pp. 1150–1157
36. H. Bay, T. Tuytelaars, L. Van Gool, *Computer vision—ECCV* (2006), pp. 404–417
37. E. Rublee, V. Rabaud, K. Konolige, G. Bradski, ORB: an efficient alternative to SIFT or SURF, in *IEEE International Conference on Computer Vision (ICCV)* (2011), pp. 2564–2571
38. S. Belongie, J. Malik, J. Puzicha, Shape matching and object recognition using shape contexts. Pattern Anal. Mach. Intell. IEEE Trans. **24**, 509–522 (2002)
39. B. Allen, B. Curless, Z. Popović, Articulated body deformation from range scan data. ACM Trans. Graph. **21**, 612–619 (2002)

40. I. Oikonomidis, N. Kyriazis, A.A. Argyros, Efficient model-based 3D tracking of hand articulations using Kinect. BMVC **1**, 3 (2011)
41. R. Cutler, M. Turk, *View-Based Interpretation of Real-Time Optical Flow for Gesture Recognition* (1998), p. 416
42. J.L. Barron, D.J. Fleet, S.S. Beauchemin, Performance of optical flow techniques. Int. J. Comput. Vis. **12**, 43–77 (1994)
43. C. Thurau, V. Hlaváč, Pose primitive based human action recognition in videos or still images, in *IEEE Conference on Computer Vision and Pattern Recognition* (2008), pp. 1–8
44. Q. Pu, S. Gupta, S. Gollakota, S. Patel, Whole-home gesture recognition using wireless signals, in *Proceedings of the 19th Annual International Conference on Mobile Computing & Networking* (2013), pp. 27–38
45. R. Ronfard, C. Schmid, B. Triggs, *Computer Vision* (2002), pp. 700–714
46. S.-J. Lee, C.-S. Ouyang, S.-H. Du, A neuro-fuzzy approach for segmentation of human objects in image sequences. Syst. Man Cybern. Part B Cybern. IEEE Trans. **33**, 420–437 (2003)
47. D. Tang, H.J. Chang, A. Tejani, T.-K. Kim, Latent regression forest: structured estimation of 3D articulated hand posture, in *IEEE Conference on Computer Vision and Pattern Recognition* (2014), pp. 3786–3793
48. J. Taylor, J. Shotton, T. Sharp, A. Fitzgibbon, The vitruvian manifold: inferring dense correspondences for one-shot human pose estimation, in *IEEE Conference on Computer Vision and Pattern Recognition* (2012), pp. 103–110
49. J. Han, L. Shao, D. Xu, J. Shotton, Enhanced computer vision with microsoft kinect sensor: a review. Cybern. IEEE Trans. **43**, 1318–1334 (2013)
50. Y. Li, Hand gesture recognition using Kinect, in *IEEE 3rd International Conference on Software Engineering and Service Science* (2012), pp. 196–199
51. D. Comaniciu, V. Ramesh, P. Meer, Real-time tracking of non-rigid objects using mean shift. IEEE Conf. Comput. Vis. Pattern Recognit. **2**, 142–149 (2000)
52. S. Thrun, W. Burgard, D. Fox, *Probabilistic Robotics* (MIT Press, 2005)
53. R.E. Kalman, A new approach to linear filtering and prediction problems. J. Fluids Eng. **82**, 35–45 (1960)
54. S. Haykin, *Kalman Filtering and Neural Networks*, vol. 47 (Wiley, 2004)
55. E. Wan, R. Van Der Merwe, The unscented Kalman filter for nonlinear estimation, in *IEEE Adaptive Systems for Signal Processing, Communications, and Control Symposium* (2000), pp. 153–158
56. K. Okuma, A. Taleghani, N. De Freitas, J.J. Little, D.G. Lowe, *Computer Vision* (Springer, 2004), pp. 28–39
57. S. Oron, A. Bar-Hillel, D. Levi, S. Avidan, Locally orderless tracking, in *IEEE Conference on Computer Vision and Pattern Recognition* 1940–1947 (2012)
58. J. Kwon, K.M. Lee, Tracking by sampling trackers, in *IEEE International Conference on Computer Vision* (2011), pp. 1195–1202
59. J. Kwon, K.M. Lee, F.C. Park, Visual tracking via geometric particle filtering on the affine group with optimal importance functions, in *IEEE Conference on Computer Vision and Pattern Recognition* (2009), pp. 991–998
60. R. Gao, L. Wang, R. Teti, D. Dornfeld, S. Kumara, M. Mori, M. Helu, Cloud-enabled prognosis for manufacturing. CIRP Ann. Technol. **64**(2), 749–772 (2015)
61. T. Li, S. Sun, T.P. Sattar, J.M. Corchado, Fight sample degeneracy and impoverishment in particle filters: a review of intelligent approaches. Expert Syst. Appl. **41**, 3944–3954 (2014)
62. T. Li, T.P. Sattar, S. Sun, Deterministic resampling: unbiased sampling to avoid sample impoverishment in particle filters. Sig. Process. **92**, 1637–1645 (2012)
63. Rincón J.M. Del, D. Makris, C.O. Uruñuela, J.-C. Nebel, Tracking human position and lower body parts using Kalman and particle filters constrained by human biomechanics. Syst. Man. Cybern. Part B Cybern. IEEE Trans. **41**, 26–37 (2011)
64. D.A. Ross, J. Lim, R.-S. Lin, M.-H. Yang, Incremental learning for robust visual tracking. Int. J. Comput. Vis. **77**, 125–141 (2008)

65. Z. Kalal, J. Matas, K. Mikolajczyk, Pn learning: bootstrapping binary classifiers by structural constraints, in *IEEE Conference on Computer Vision and Pattern Recognition* (2010), pp. 49–56
66. B. Babenko, M.-H. Yang, S. Belongie, Visual tracking with online multiple instance learning, in *IEEE Conference on Computer Vision and Pattern Recognition* (2009), pp. 983–990
67. A.W.M. Smeulders et al., Visual tracking: an experimental survey. Pattern Anal. Mach. Intell. IEEE Trans. **36**, 1442–1468 (2014)
68. L.E. Peterson, K-nearest neighbor. Scholarpedia **4**, 1883 (2009)
69. A.D. Wilson, A.F. Bobick, Parametric hidden markov models for gesture recognition. Pattern Anal. Mach. Intell. IEEE Trans. **21**, 884–900 (1999)
70. S. Lu, J. Picone, S. Kong, Fingerspelling Alphabet Recognition Using A Two-level Hidden Markov Modeli in *Proceedings of the International Conference on Image Processing, Computer Vision, and Pattern Recognition* (2013), p. 1
71. J. McCormick, K. Vincs, S. Nahavandi, D. Creighton, S. Hutchison, Teaching a digital performing agent: artificial neural network and hidden Markov model for recognising and performing dance movement, in *Proceedings of the 2014 International Workshop on Movement and Computing* (2014), p. 70
72. S.-Z. Yu, Hidden semi-Markov models. Artif. Intell. **174**, 215–243 (2010)
73. M.A. Hearst, S.T. Dumais, E. Osman, J. Platt, B. Scholkopf, Support vector machines. IEEE Intell. Syst. their Appl. **13**, 18–28 (1998)
74. M.E. Tipping, Sparse Bayesian learning and the relevance vector machine. J. Mach. Learn. Res. **1**, 211–244 (2001)
75. B. Schiilkopf, The kernel trick for distances, in *Proceedings of the 2000 Conference* on *Advances in Neural Information Processing Systems*, vol. 13 (2001), p. 301
76. A. Cenedese, G.A. Susto, G. Belgioioso, G.I. Cirillo, F. Fraccaroli, Home automation oriented gesture classification from inertial measurements. Autom. Sci. Eng. IEEE Trans. **12**, 1200–1210 (2015)
77. K. Feng, F. Yuan, Static hand gesture recognition based on HOG characters and support vector machines, in *2nd International Symposium on Instrumentation and Measurement, Sensor Network and Automation* (2013), pp. 936–938
78. D. Ghimire, J. Lee, Geometric feature-based facial expression recognition in image sequences using multi-class adaboost and support vector machines. Sensors **13**, 7714–7734 (2013)
79. O. Patsadu, C. Nukoolkit, B. Watanapa, Human gesture recognition using Kinect camera, in *International Joint Conference on Computer Science and Software Engineering* (2012), pp. 28–32
80. R.E. Schapire, *Nonlinear estimation and classification* (Springer, 2003), pp. 149–171
81. Y. Freund, R.E. Schapire, A decision-theoretic generalization of on-line learning and an application to boosting. J. Comput. Syst. Sci. **55**, 119–139 (1997)
82. S. Celebi, A.S. Aydin, T.T. Temiz, T. Arici, Gesture recognition using skeleton data with weighted dynamic time warping. VISAPP **1**, 620–625 (2013)
83. E.J. Keogh, M.J. Pazzani, Derivative dynamic time warping. SDM **1**, 5–7 (2001)
84. S.S. Haykin, *Neural Networks and Learning Machines*, vol. 3 (Pearson Education Upper Saddle River, 2009)
85. T.H.H. Maung, Real-time hand tracking and gesture recognition system using neural networks. World Acad. Sci. Eng. Technol. **50**, 466–470 (2009)
86. H. Hasan, S. Abdul-Kareem, Static hand gesture recognition using neural networks. Artif. Intell. Rev. **41**, 147–181 (2014)
87. T. D'Orazio, G. Attolico, G. Cicirelli, C. Guaragnella, A neural network approach for human gesture recognition with a kinect sensor. ICPRAM 741–746 (2014)
88. A.H. El-Baz, A.S. Tolba, An efficient algorithm for 3D hand gesture recognition using combined neural classifiers. Neural Comput. Appl. **22**, 1477–1484 (2013)

89. K. Subramanian, S. Suresh, Human action recognition using meta-cognitive neuro-fuzzy inference system. Int. J. Neural Syst. **22**, 1250028 (2012)
90. Z.-H. Zhou, J. Wu, W. Tang, Ensembling neural networks: many could be better than all. Artif. Intell. **137**, 239–263 (2002)
91. Y. LeCun, Y. Bengio, G. Hinton, Deep learning. Nature **521**, 436–444 (2015)
92. J. Schmidhuber, Deep learning in neural networks: an overview. Neural Netw **61**, 85–117 (2015)
93. J. Tompson, M. Stein, Y. Lecun, K. Perlin, Real-time continuous pose recovery of human hands using convolutional networks. ACM Trans. Graph **33**, 169 (2014)
94. K. Simonyan, A. Zisserman, Two-stream convolutional networks for action recognition in videos. *Adv. Neural Inf. Process. Syst.* 568–576 (2014)
95. J. Nagi et al., Max-pooling convolutional neural networks for vision-based hand gesture recognition, in *IEEE International Conference on Signal and Image Processing Applications* (2011), pp. 342–347
96. A. Jain, J. Tompson, Y. LeCun, C. Bregler, *Computer Vision* (2014), pp. 302–315
97. K. Li, Y. Fu, Prediction of human activity by discovering temporal sequence patterns. IEEE Trans. Pattern Anal. Mach. Intell. **36**, 1644–1657 (2014)
98. M.S. Ryoo, Human activity prediction: early recognition of ongoing activities from streaming videos, in *IEEE International Conference on Computer Vision* (2011), pp. 1036–1043
99. W. Ding, K. Liu, F. Cheng, J. Zhang, Learning hierarchical spatio-temporal pattern for human activity prediction. J. Vis. Commun. Image Represent. **35**, 103–111 (2016)
100. L.R. Rabiner, A tutorial on hidden Markov models and selected applications in speech recognition. IEEE Proc. **77**, 257–286 (1989)

# Part IV
# CPS Systems Design and Lifecycle Analysis

# Chapter 12
# Architecture Design of Cloud CPS in Manufacturing

## 12.1 Introduction

Cloud Computing (CComputing) is a model for enabling ubiquitous, convenient and on-demand network access to a shared pool of configurable computing resources (e.g. networks, servers, storage, applications, and services) that can be rapidly provisioned and released with minimal management effort or service provider interactions [1, 2]. It provides resources to a user on the "pay-as-you-go" basis. There are three common types of CComputing structure, i.e. Infrastructure as a Service (IaaS), Platform as a Service (PaaS), and Software as a Service (SaaS). IaaS provides a bunch of physical and virtual machines, based on which users are able to install and deploy their own operation systems and working environments. A PaaS model packages a computing platform including operating system, programming language execution environment, database, and web server. A PaaS client is able to develop and run its applications at the software layer. Finally, SaaS simplifies the utilisation of a large amount of software applications remotely, elastically and seamlessly.

During the past years, many successful CComputing business cases are found worldwide [3–7]. Among various types of models, the key characteristic of CComputing is that of pay-as-you-go. In the increasingly globalised manufacturing context, customer-oriented manufacturing is a promising approach to improving the service quality and competitiveness, in particular for the Small and Medium-sized Enterprises (SMEs). Thus, a new concept of advanced manufacturing model is proposed worldwide, namely Cloud Manufacturing (CManufacturing). In the first half of this paper, recent CManufacturing research is reviewed, followed by discussions on related works that support building and maintaining a CManufacturing system. In the second half, a novel manufacturing platform is introduced to archive manufacturing interoperability in the cloud paradigm.



L. Wang and X. V. Wang, *Cloud-Based Cyber-Physical Systems in Manufacturing*,
https://doi.org/10.1007/978-3-319-67693-7_12

CManufacturing is a model for enabling ubiquitous, convenient and on-demand network access to a shared pool of configurable manufacturing resources (e.g. manufacturing software tools, manufacturing equipment, and manufacturing capabilities) that can be rapidly provisioned and released with minimal management effort or service provider interactions [8]. Like the CComputing concept, manufacturing infrastructure, platform and software application in CManufacturing can be offered as a service to a CUser. By extending the concept to a broader scope, all the production objects and features can be treated as services, hence Everything-as-a-Service (XaaS). The rest of this section discusses the CManufacturing structure and related technologies.

### 12.1.1 State-of-the-Art Cloud Manufacturing Approaches

Cloud concept presents a promising future for computing business and the same can be said for manufacturing business. Tao et al. [9] proposed a framework of CManufacturing with discussions of key advantages and challenges for future CManufacturing systems. CManufacturing is described as a computing and service-oriented manufacturing model developed from existing advanced manufacturing models (e.g. Application Service Provider, Agile Manufacturing, Networked Manufacturing, and Manufacturing Grid), enterprise information technologies under the support of cloud computing, Internet of Things, virtualisation and service-oriented technologies, and advanced computing technologies. It is predicted that a CManufacturing system would reduce the cost and increase the utilisation rate of resources. Li et al. [10] proposed a service-oriented networked manufacturing model. The paper also discussed a number of methods to support the model. Intelligent agent, Product Lifecycle Management (PLM), resource modelling and evaluating technologies are considered as the supporting technologies for cloud architecture.

A cloud-based manufacturing research project [11] was launched in Europe in 2010, sponsored by the European Commission. The goal of this project is to provide users with the ability to utilise the manufacturing capabilities of configurable and virtualised production networks. CManufacturing is described as a service-oriented IT environment as a basis for the next level of manufacturing networks by enabling production-related interenterprise integration down to shop floor level [12]. A set of software-as-a-service applications have been developed. In the proposed system, customised production of technologically complex products is enabled by dynamically configuring a manufacturing supply chain [13, 14]. It is believed that the development of a front-end system with a next level of integration to a cloud-based manufacturing infrastructure is able to better support on-demand manufacturing of customised products.

To facilitate a CManufacturing environment, existing resources need to be scaled, modelled and adapted into the cloud. Wu and Yang [15] proposed a method to describe and scale manufacturing resources in a cloud. Hu et al. [16] analysed the factors that affect the classification of virtual resources in CManufacturing. Examples are introduced to validate the effect of these factors to task assignment. Luo et al. [17] discussed a CManufacturing system from the viewpoint of network, function and running. A multi-dimensional information model was proposed to describe manufacturing abilities [18]. This knowledge-based data model helps provide a user with manufacturing services via network.

To control and manage the flexibility of the resource service composition in CManufacturing, Zhang et al. [19] proposed architecture considering major uncertain dynamic changing factors in the lifecycle of a resource service. Multi-agent is proved to be an effective tool in solving problems through sharing knowledge during the implementation of CManufacturing [20]. An Agent-based mechanism provides flexible and effective sharing and utilising of elastic resources.

After resource modelling, the next challenge is resource integration. Fan and Xiao [21] proposed an integrated architecture of CManufacturing based on a federation principle. Federation integration rules are applied before resources are connected to the system. Thus, joining or exiting of a resource would not affect operation of the whole cloud environment. To maintain the CManufacturing resources, an Optimal Allocation of Computing Resources (OACR) system was proposed [22]. In OACR, improved Niche immune algorithm is introduced to solve the resource scheduling problem in a grid system or CComputing system, associated with the Niche strategy.

### *12.1.2 Supporting Technologies for Cloud Manufacturing*

Although CManufacturing is a relatively new concept, it draws upon technologies such as virtual enterprise, distributed and collaborative manufacturing systems. Xu [8] reviewed the systematic requirements of CManufacturing systems. Advanced manufacturing technologies are discussed to fulfil these specifications and support a CManufacturing environment. Research contributions are reviewed regarding collaborative manufacturing and interoperable systems [23]. Manufacturing systems are re-evaluated from the cloud perspective, e.g. IaaS and SaaS. In addition, ERP (Enterprise Resource Planning), SOA (Service-Oriented Architecture), and modelling systems are also relevant to the concept of CManufacturing.

After the manufacturing activities are properly modelled, the next step is to integrate their operational processes. To represent a business in a Manufacturing Cloud (MCloud), the first step is to understand and model an enterprise. ERP systems have been studied extensively [24–28], including the inter-organisation performance [29] and the behaviour throughout the supply chain with multiple stakeholders [30]. With the help of ERPs, inter-organisation behaviours/reactions can be modelled and mapped in a standardised manner as neutral APIs (Application Protocol Interfaces). Based on these APIs, MCloud can be established via integrating these reactions in standardised semantics, without changing the organisational structure of an enterprise.

Papazoglou and Van den Heuvel [31] proposed a framework named Enterprise Service Bus. It is an integration platform that utilises web services standards to support SOA applications within an enterprise. The extended SOA system can be further adopted by CManufacturing to enable capabilities e.g. service orchestration, ‘intelligent’ routing, provisioning, integrity and security of message as well as service management. In an SOA system, business procedures can be modelled and componentised to support seamless business integration [32]. Schmidt et al. [33] proposed architecture declaring clear definitions of service capability and requirements in a service-oriented context. Models have been proposed to evaluate the quality/feedback in the business-to-business context [34].

It has been suggested that a commonly used data model/schema should be utilised for a wide range of products [35]. Data management should be encapsulated by schema and manipulation rules in a data model. In the perspective of CManufacturing, international standards, e.g. STEP/STEP-NC have a role to play in ensuring product data interoperability. STEP (the Standard for Exchange of Product data [36]) is one of such standards, providing mechanisms for describing product information throughout the lifecycle. Different Application Protocols have been developed for different applications/domains. As an extension of STEP, STEP-NC [37] is developed to support CNC (Computer Numerical Control) manufacturing. Compared with previous standards, these data models offer a set of effective tools for interoperability solutions in the computer-aided manufacturing context [38].

### *12.1.3 Recap*

CManufacturing is not just an implementation of CComputing in manufacturing. Manufacturing enterprises and related resource/capability need to be described, componentised, virtualised and integrated in an MCloud. Some existing research work provides some enabling tools to the CManufacturing concept. This said, there is still a lack of Cloud solution for the entire manufacturing supply chain. It is necessary to implement a supervision mechanism to organise and control Cloud Services (CService) at upper level. Moreover, an interoperable environment is also needed to integrate current and future manufacturing resources. The next section describes a proposed service-oriented CManufacturing system named Interoperable Cloud-based Manufacturing System (ICMS).

## 12.2 Cloud Manufacturing Framework

Nowadays, a manufacturing enterprise would not survive without Computer Aided applications (CAx) technologies. Deploying CAx software on the Cloud improves the performance in terms of flexibility, extendibility, integrity and easy/unlimited data storage. With a cloud structure, software is easily maintained and utilised on a cloud server. Version updating, maintaining and integrating is remotely done by the cloud provider, which replaces periodic services by onsite maintenance specialists. Thus, the cost of IT infrastructure is cut down via reduced management and maintenance effort. Additionally, thanks to the pay-as-you-go basis of a CService, the cost of expensive applications is spread over multiple CUsers. Costly but rarely used software can be priced by the amount of usage.

CManufacturing faces a tougher challenge than implementing manufacturing-related software in CComputing. Unlike software programme and IT infrastructure, physical machines, monitors, and facilities cannot be readily deployed on the cloud. There is also a need to understand the intermediate processes from raw material to finished products.

## *12.2.1 Manufacturing Capability and Manufacturing Resource*

Zhang et al. [39] identified manufacturing ability as a kind of resource. In practice, the main reason for acquiring a manufacturing facility is the functionality of the equipment but not the equipment itself. It is therefore necessary to recognise a Cloud resource, its capability and services at different levels. In the cloud background, the definitions of a resource, capability and service are given below.

- Manufacturing Resource (MResource): material and nonmaterial manufacturing supplies including equipment, machine, device and intelligent properties.
- Manufacturing Capability (MCapability): ability of transforming one form into another in manufacturing domain. It is realised via related MResources.
- Cloud Manufacturing Service (CMService): self-contained, configurable and on-demand manufacturing service package to fulfil user's original needs. A CMService can be random, short-term, long-term, or strategic.

The containment relationships of MResource, MCapability and CMService can be summarised as shown in Fig. 12.1. MResources are contained within MCapability as one of the essential requirements, since MCapability is realised and implemented via MResource. MCapabilities are re-packaged and deployed in the MCloud as CMService as a convenient feature that can be rapidly provisioned and released by a CUser.

A CManufacturing system encapsulates and implements MCapability in the cloud as CMService packages. Manufacturing Capability is composed of Design, Production, Experimentation, Management, and Communication Capability.

- Design Capability (DC) refers to domain-specific design knowledge, expertise of the organisation and past experience from previous design activities.
- Production Capability (PC) relies on the speed and quality of creating an output, i.e. product or service, to fulfil a production order.

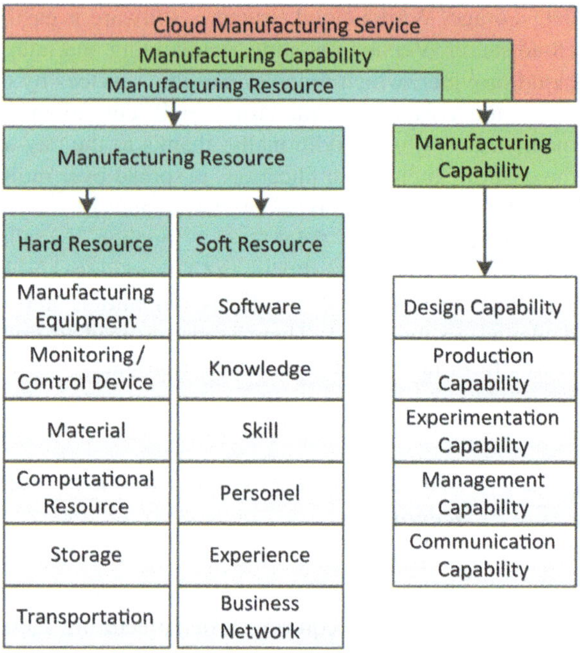


**Fig. 12.1** MCapability and MResource

- Experimentation Capability (EC) entails the experimentation knowledge and specialists.
- Management Capability (MC) includes planning, organising, staffing, leading and controlling of an organisation. It relies on the ability of the operational business and organisational activities.
- Communication Capability (CC) refers to the data exchangeability between applications/devices. It includes data transportation, speed, storage, conversion and QoS.

From the resource's perspective, each kind of manufacturing capability requires support from the related MResource(s). For each type of MCapability, its related MResource(s) comes in two forms, soft resources and hard resources. The soft resources include:

- Software: software applications throughout the product lifecycle including design, analysis, simulation, process planning, etc.
- Knowledge: experience and know-how needed to complete a production task, i.e. engineering knowledge, product models, standards, evaluation procedures and results, customer feedback, etc.
- Skill: expertise in performing a specific manufacturing task.
- Personnel: human resource engaged in manufacturing process, i.e. designers, operators, managers, technicians, project teams, customer service, etc.
- Experience: performance, quality, client evaluation, etc.

- Business Network: business relationships and business opportunity networks that exist in an enterprise.

The hard resources contain:

- Manufacturing Equipment: facilities needed for completing a manufacturing task, e.g. machine tools, cutters, test and monitoring equipment and other fabrication tools.
- Monitoring/Control Resource: devices used to identify and control other manufacturing resource, for instance, RFID (Radio-Frequency IDentification), WSN (Wireless Sensor Network), virtual managers and remote controllers.
- Computational Resource: computing devices to support production process, e.g. servers, computers, storage media, control devices, etc.
- Materials: inputs and outputs in a production system, e.g. raw material, product-in-progress, finished product, power, water, lubricants, etc.
- Storage: automated storage and retrieval systems, logic controllers, location of warehouses, volume capacity and schedule/optimisation methods.
- Transportation: movement of manufacturing inputs/outputs from one location to another. It includes the modes of transportation, e.g. air, rail, road, water, cable, pipeline and space, and the related price, and time taken.

To formulate MCapability, a MCapability Description Model (MCDM) as a 5-tuple is adopted,

$$\begin{aligned} MCapability = \{&DC(R_{SoftDC}, R_{HardDC}), \\ &EC(R_{SoftEC}, R_{HardEC}), PC(R_{SoftPC}, R_{HardPC}), \\ &MC(R_{SoftMC}, R_{HardMC}), CC(R_{SoftCC}, R_{HardCC})\} \end{aligned} \tag{12.1}$$

where R is MResource for all the resources required to carry out a task, including hard resource $R_{Hard}$ and soft resource $R_{Soft}$.

High-performance service needs sufficient resource and suitable methodology to exploit it. Hence, an effective MCapability is contributed by the domain-specific ability and its related resource. MCDM includes the capability of both an individual enterprise and an alliance made up of multiple participants. This means an MCapability meeting a CUser's need could be provided by a single Service Provider (SProvider) or a union of them. A comprehensive cloud solution is required to take care of all the capabilities and resources mentioned above and provide an optimal solution. Eventually, identified MCapabilities are packaged as CMServices and deployed in the MCloud. During the conversion from current manufacturing status into CManufacturing, existing capabilities and resources should be integrated and utilised in the CManufacturing environment. Thus, an interoperable, service-oriented CManufacturing system can be realised.

## *12.2.2 Cloud Architecture*

As mentioned above, cloud technology provides an opportunity to reshape manufacturing business, in particular SMEs. Combined with SOA, it is capable of creating new economic growth for customised production or One-of-a-Kind Production (OKP) businesses. Specialised and customised demands can be better served due to the flexible and fast-reaction nature of a CManufacturing system. Compared with the Business-to-Business (B2B) and Business-to-Consumer (B2C) models, an X2C (Everything-to-Cloud) model is presented. The preliminary concept of ICMS has been reported in [23]. As public cloud infrastructure, ICMS consists of three layers, i.e. Smart Cloud Manager (SCM), User Cloud (UCloud), and MCloud (Fig. 12.2).

### 12.2.2.1 Customer and Enterprise User

At the UCloud layer, the CMService consumer is divided into two categories: Customer User (CU) and Enterprise User (EU). ICMS takes care of traditional manufacturing tasks for CUs as well as collaborative production requests from multiple organisations. By combining the Consumer-to-Cloud and Business-to-Cloud models, ICMS provides an X2C structure from the industrial context.

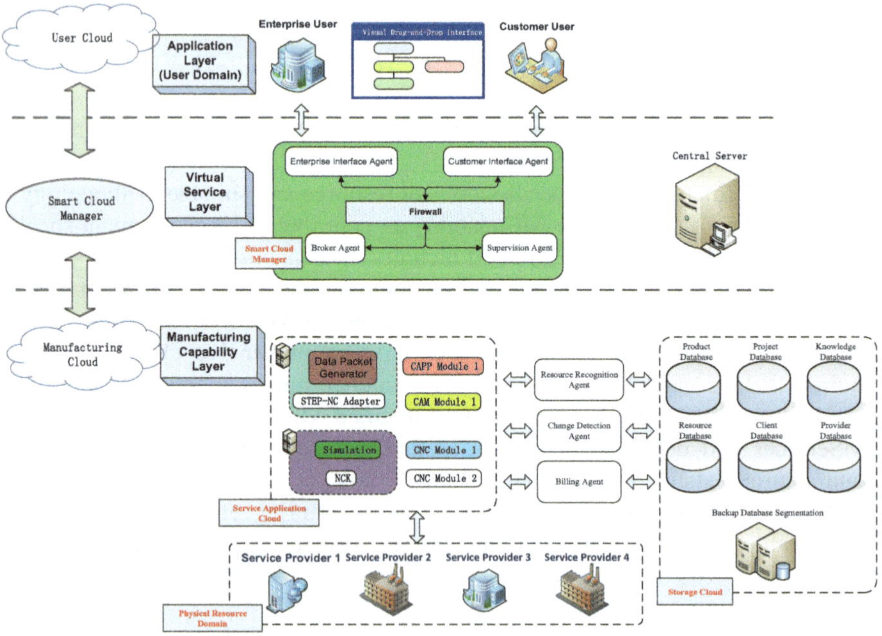


**Fig. 12.2** ICMS architecture

CU is defined as a customer or organisation with the request of a self-contained production task. Assisted by the Customer Interface Agent (CIA) of SCM, the manufacturing request of a CU is analysed and located by SCM, and provided by the MCloud. Thus, it forms a Request-Find-Provide service chain. Original user's requests are taken care of by SCM. SCM searches for potential solutions and feeds back the results to the user. The user is able to optimise the solution based on his/her original needs and finalise the service request. ICMS provides a user with a big range of flexible manufacturing capabilities. Customised and original requirements can be realised easily, compared with the traditional manufacturing practice. For industry, it offers new opportunities especially for OKP enterprises and SMEs. The enterprises are loosely integrated in MCloud as ICMS SProviders. MCapabilities and business opportunities are integrated and broadcasted in a larger resource pool, which enhances the competiveness of the entire team. Thus, more manufacturing objects can be achieved with minimum additional investment and effort.

Besides CUs, ICMS takes care of organisations/enterprises (EU) who are seeking additional MCapabilities and supports. In practice, customers occasionally come to a manufacturing enterprise requiring products or capability that the enterprise by itself cannot fulfil. With the help of the Enterprise Interface Agent (EIA), an EU can search for qualified SProviders who are able to "fill in the gap". The EU is able to recognise related MResources and allocate the temporary partner (s) for the task. In this case, the original EU plays a role of the "leading company" in the virtual organisation. The leading company is in charge of interacting with the customer, and collaborating with other participants as a coordinator. From the ICMS perspective, the leading company is considered as the EU, who will be assisted by the SCM module. This way, the CUsers are able to accomplish bigger and more demanding production tasks that are otherwise not possible by a single enterprise. As a matter of fact, the partner network of a company is made boundary-less (Fig. 12.3).

#### 12.2.2.2 Smart Cloud Manager

Intelligent agent technology is capable of supporting manufacturing procedures/decisions [40–42]. The SCM module is constructed by intelligent agents. In an ideal system, user should have full confidence of the system's intelligence. The interaction between a human being and the system intelligence should be minimised as long as the service request is well-defined by the cloud user. Intelligence kernel is capable of optimising and executing the task with the preference variables from the user. However, manufacturing decisions are difficult to make due to the complexity of manufacturing processes, variety of machines and devices, and the uncertainty of resources status. When multiple resources and variables are involved, it is even harder to predict a reliable and optimum service solution for the user. Thus, to fully utilise AI and human expertise/knowledge, a decision-making model is in need. SCM works in a neutral manner and consists of EIA, CIA, Broker Agent (BA), Supervision Agent (SA), and Firewall Module (see Fig. 12.4). EIA works with EUs,

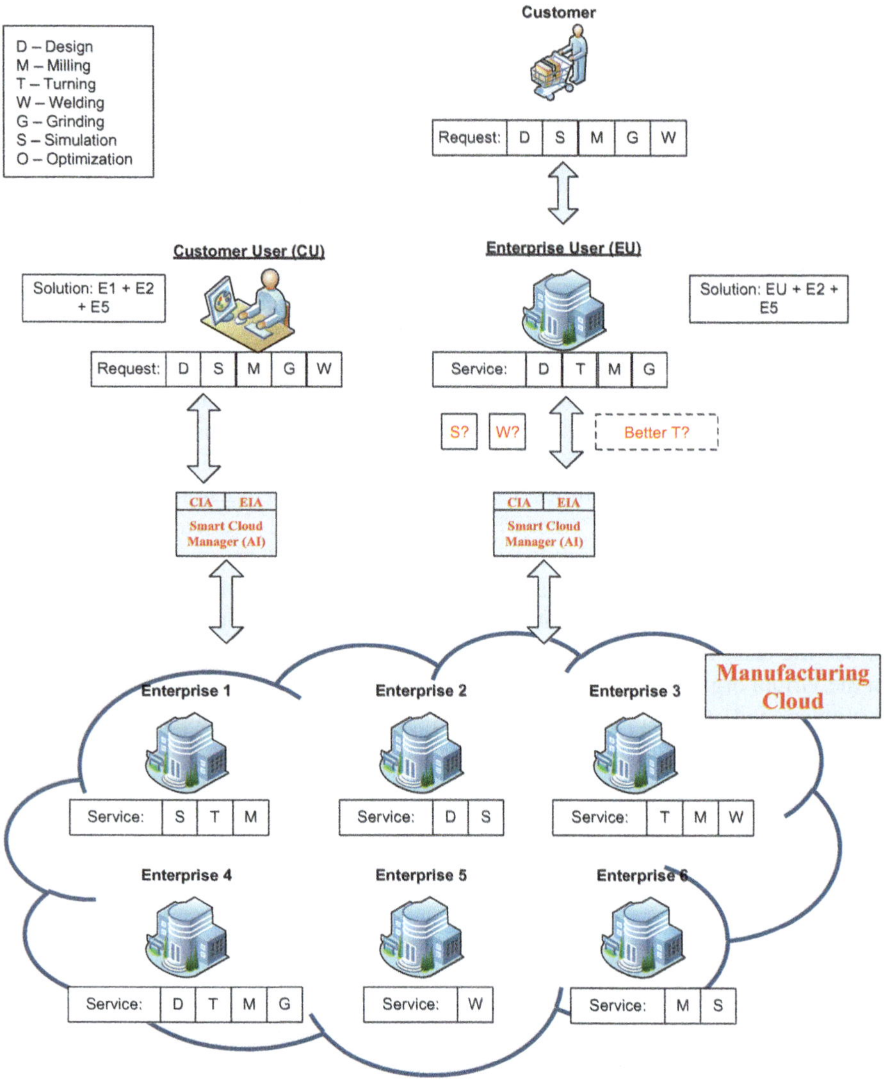


**Fig. 12.3** Customer user and enterprise user

and CIA handles requests from CUs. Although the GUIs (Graphical User Interface) and algorithms of EIA and CIA are different, the service procedures are almost the same (from the SCM perspective). After the user's request is collected by the Interface Agent (IA), BA communicates with the Provider Database and maps the requirement to the available CMServices. As long as the user modifies and confirms the service package, an ICMS Service Template (ST) is generated and delivered to the SCM. Based on ST, Supervision Agent starts up and works with the Service Application Cloud (SACloud). Specific CMServices are organised and launched to

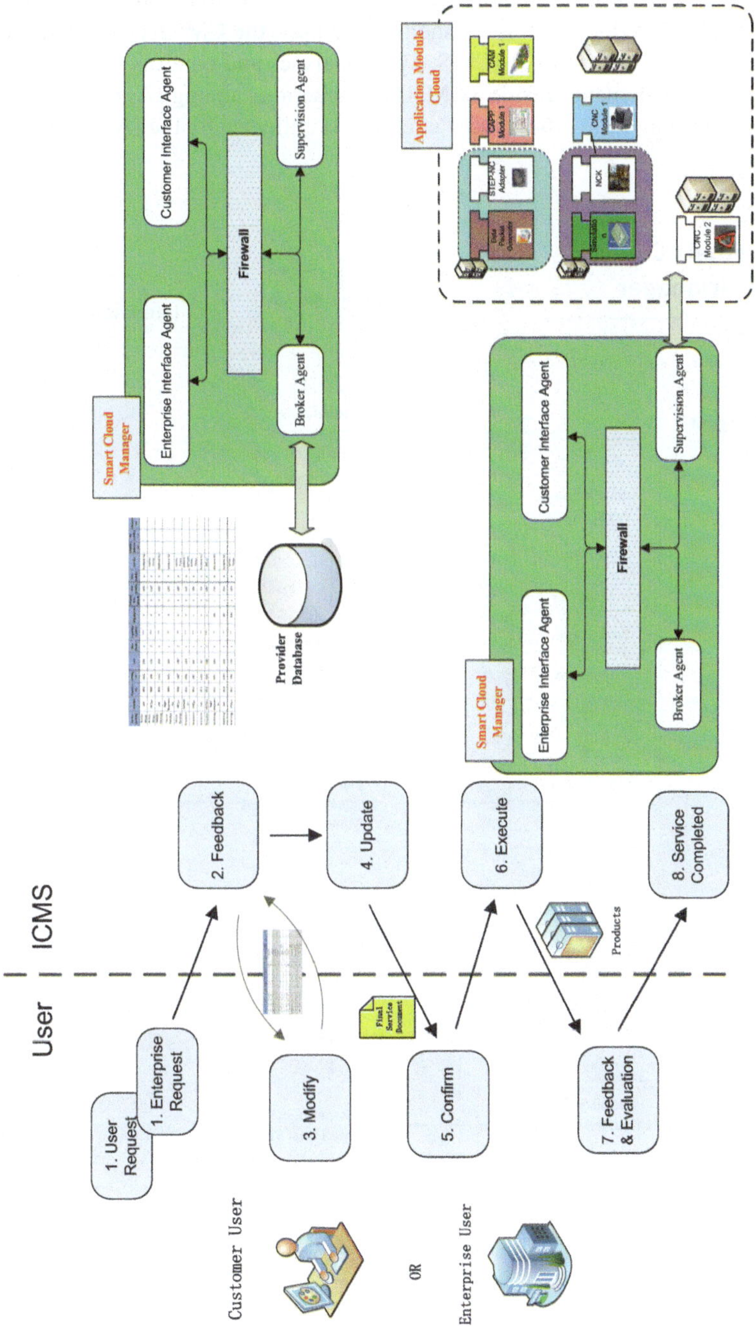


**Fig. 12.4** Cloud service procedure

meet the user's expectations. The final service output, which can be a product, computing data or a technical document, is then sent to the cloud user. After the feedback/evaluation document is finished by the user, the CMService is terminated.

As the supervisor or brain of ICMS, SCM analyses and controls the CMServices to fulfil the user's demand. Inside SCM, the interactions among IA, BA, and SA are summarised in Fig. 12.5. After a user's request is collected by the IA, the details are

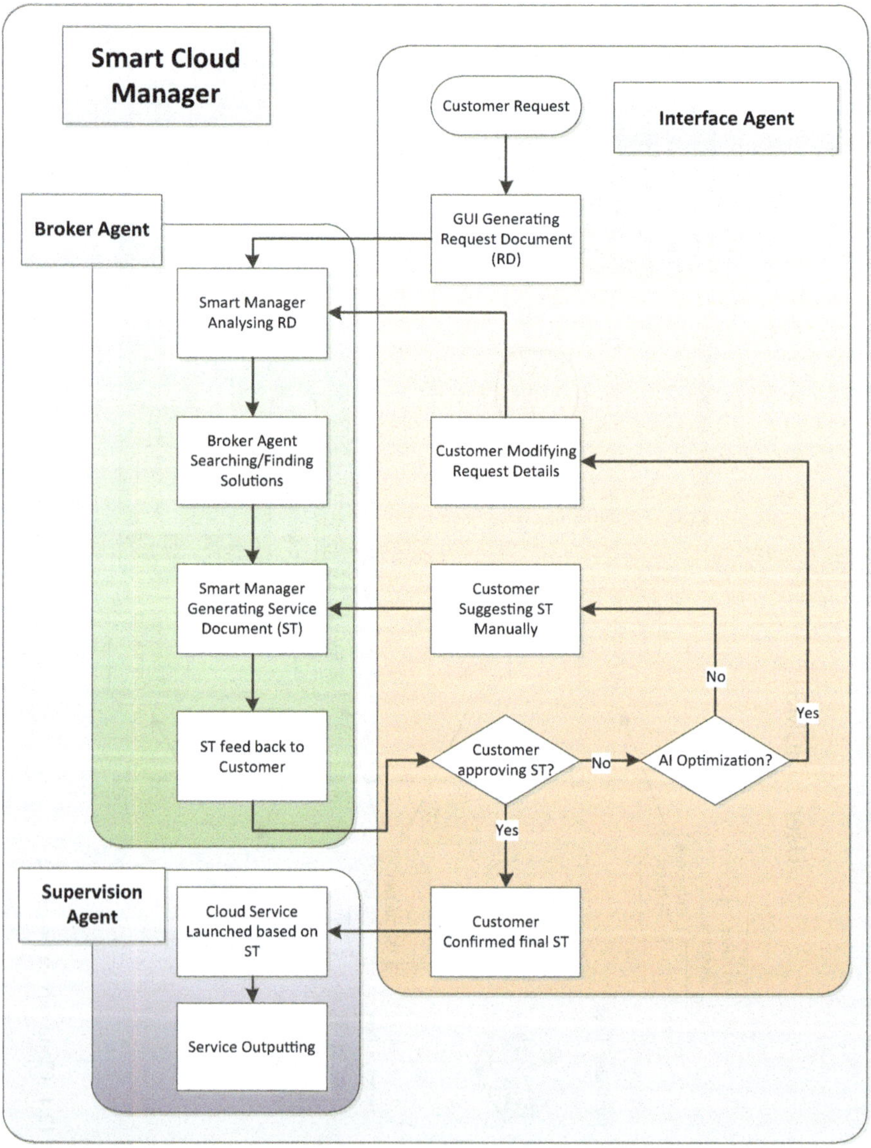


**Fig. 12.5** Logic flow within smart cloud manager

converted into a standard format. Based on these details, an internal request document is generated and sent to BA. According to the request document, BA searches in the MRsource database for potential solutions. Afterward, an initial ST file is created and sent back to the user. Cloud consumer is able to view all these solutions along with the suggestions from SCM. Based on the factors such as cost, quality, functionality etc., SCM recommendation is visible to the user in different levels of details. If the cloud customer is not satisfied with any of the suggestions provided, he/she is able to modify the ST.

At this stage, the cloud customer is able to either optimise the ST via BA intelligence or do it manually. If the customer prefers to utilise AI continuously, CUser is requested to modify his/her original searching request by providing more details or to modify technical variables. Then, the altered request condition is sent back to BA, who will process one more round of analysis and service detection. On the other hand, if the cloud customer chooses to improve the ST manually, he/she can work on it via GUI and allocate a preferred provider. This way, both of expert knowledge and optimisation are utilised in SCM.

As long as the user confirms ST, the specific Cloud Services are launched by the SA. SA is responsible for monitoring and controlling all the activities of the cloud service modules. By marking and manipulating the event and data flows of all the application modules, the ST is executed accurately as it is defined.

## 12.3 Interoperability and Other Issues

In recent years, research has been carried out worldwide in an attempt to develop an interoperable and collaborative environment with heterogeneous software applications. In the following part of this section, recent research works are reviewed and discussed.

### *12.3.1 Standardised File Formats*

System integration and interoperability is addressed as one of the key needs to be met [43]. A widely recognised information model is in need, especially for a collaborative and distributed environment.

To work on multiple versions and views of a shared model, [44] proposed a collaborative architecture to allow experts to share and exchange design information. In this architecture, product design is exchanged through a standardised, constraint-based model to maintain complex relationships in multidisciplinary collaborative design. Thanks to this data model, conflicts happening during synchronisation process can be resolved via the notification mechanism.

Besides the design applications, research has also been carried out to integrate the whole CAD/CAM/CNC chain. For facilitating a web-based design-manufacturing environment, [45] proposed a web-based system using a data structure similar to that of ISO14649 data. In this system, files in neutral formats are passed along a serial software chain composed by WebCADFeatures, WebCAPP and WebTuring applications. To integrate more applications seamlessly and efficiently, [46] proposed an Open Computer-Based Manufacturing system (OpenCBM in short). In this system, standardised file formats are chosen to reduce the cost of data transferring and exchange.

In a heterogeneous environment, data exchange is a challenging issue when proprietary software tools are integrated within the same architecture. [47] presented a method for semantically mapping different business documents to a conforming document format, given inevitable existence of multiple product representations. In this research, XML format is adopted to support web-based applications and an SOA (Service-Oriented Architecture) model through WWW (World Wide Web).

### *12.3.2 STEP/STEP-NC to Bridge the Gap*

Since standardised format is a potential solution to realising interoperability, the International Organisation for Standardisation (ISO) has been making its effort in the development of some international standards. STEP (the Standard for the Exchange of Product data [36]) is such an example. It has been established to describe the entire product data throughout the lifecycle of a product. STEP contains different Application Protocols (APs) which provide data models for targeted applications, activities or environments. Compared with previous standards, these data models offer a set of effective tools for computer-aided interoperability solutions [38].

Zhang et al. reviewed the fundamental structure of STEP data models [48]. Recently, a system named INFELT STEP was proposed to maintain the integration of CAD/CAM/CNC operations based on STEP data models [49]. In this three-layered system, different sections are defined in each layer to provide interfaces between different CAD, CAPP, CAM and CNC software packages. INFELT STEP has a distinct capability of enabling collaboration of different enterprise-wide CAD/CAPP/CAM/CNC systems in the design and production of a product using multiple APs of the STEP standard.

In the past few years many companies have implemented PDM (Product Data Management) systems, focusing on cost-cutting and shortening the product development cycle. To provide a solution via a common method of sharing standard product and design information, a STEP-compliant PDM system was developed to fulfil the demand for logically integrated product data which is stored physically in a distributed environment [50]. In this system, a STEP-based PDM schema was defined in XML format to support the Web service connecting PDM systems of several partners through an open network accessible via the Internet. As another implementation via XML, [51] developed an approach providing efficient data exchange in which the Web is utilised as a communication layer. Combining the STEP concept with XML, this work supports the integration of decentralised business partners and enables the information flow within the value added chain [52]. Additionally, the standard formats in STEP and XML is also utilised in the virtual reality (VR) platform to realise seamless design integration [53, 54].

As the extension of XML, Automation Markup Language (AutomationML) is utilised to support remote data management. A web based AutomationML server is developed to provide data space to support the entire production engineering process [54, 55].

Moreover, the data model for computerised numerical controllers, otherwise known as STEP-NC [37], was established as an international standard in 2003. As a data model to connect CAD/CAM systems with CNC machines, STEP-NC completes the integrated loop of CAD/CAM/CNC. It has been proven that STEP-NC provides contribution to both system interoperability and data traceability [56]. Hence, it becomes possible to implement interoperability in a STEP/STEP-NC complaint environment [57].

Nessehi et al. proposed a framework to combat the incompatibility problem among CAx systems [58]. In this framework, STEP-NC data model is utilised as the basis for representing manufacturing knowledge augmented with XML schema while a comprehensive data warehouse is utilised to store CNC information. The system consists of manufacturing data warehouse, manufacturing knowledgebase, intercommunication bus, and diverse CAx interfaces as main structures [59]. Mobile agent technology is used to support the intercommunication bus and CAx interfaces.

Recently, Mokhtar and Houshmand [60] studied a similar manufacturing platform utilising an axiomatic design theory to realise interoperability in the CAx chain. Two basic approaches are considered, utilising interfaces and utilising neutral format based on STEP. The methodology of axiomatic design is proposed to generate a systematic roadmap of an optimum combination of data exchange via direct (using the STEP neutral format) or indirect (using bidirectional interfaces) solution in the CAx environment.

In addition to the approaches mentioned above, more methods have been developed to strengthen the interoperability along STEP/STEP-NC based CAD/CAM/CNC chain. For instance, Vichare et al. [61] developed data models to describe all the elements of a CNC machine tool. In this approach called UMRM (Unified Manufacturing Resource Model), machine specific data is defined in the form of an STEP-compliant schema. This data model acts as a complementary part to the STEP-NC standard to represent various machine tools in a standardised form, which provide a universal representation of the manufacturing information at the tail of CAD/CAM/CNC chain.

### 12.3.3 Approaches to Achieving Product Information Sharing

To achieve an effective product data sharing environment, Do and Chae [62] developed a product data management architecture supporting collaborative product design. In this architecture, additional data model is proposed as an extension linked to the STEP standard. With the help of this system, different configurations or modifications made by various engenderers can be brought together. Hardware engineers and software programmers are able to share the same user environment, on a consistent database during the process of collaborative product development. In another piece of work regarding to EC (Engineering Change), Hwang et al. [63] proposed a data model representing and propagating EC information. In this collaborative product development environment, a neutral reference model is developed based on the STEP data structure. The EC conducted by collaborating companies can be applied and reflected in the product design. Within the reference model, a neutral skeleton model and an external reference model are developed to support the distributed collaborative design environment.

Choi et al. [64] defined a standard data format using XML for a neutral file containing product, process and resource (PPR) information, named PPRX (PPR eXchange). The information model mapped from ISO 10303-214 STEP models supports PPR information exchanges between commercial heterogeneous PLM (Product Lifecycle Management) systems and other systems. With the XML-based data exchange methodology, information exchange can be made without loss, which reduces unnecessary effort and supports effective integration and information sharing.

As an example of specific application protocol of STEP, Jardim-Goncalvas et al. [65] proposed a knowledge framework called funSTEP which provides enterprise and manufacturing systems with a semantically seamless communication with other stakeholders up and down the supply chain. Based on STEP AP236 standard [66], semantically enriched international product data standards, and knowledge representation elements are utilised as a basis for achieving seamless enterprise interoperability.

To speed up a specific task, a web service architecture called WSC (Web Service for CAD), was proposed by Kim et al. [67]. It can support collaborative product development of CAD assembly and part data. XML product models have been developed based on multiple STEP APs. Using these models, parallel processing is deployed to make an assigned task run faster because more than one processor can be used to run the tasks. The result of the experiment demonstrates that it is possible to retrieve product data partially, and improve the computing performance by processing these data subsets.

In order to display a specific range of data, ST-Developer [68] enables users to view allocated types of entities defined in a STEP Part 21 file. By using the functionality called "working set", which is embedded in the STEP file browser, user is able to exclude/include assigned type of entities, e.g. showing all the machining working steps in the document only, or hiding all the material information in a project. It demonstrates the capability of the object-oriented STEP data structure in the context of processing data subsets.

When the product file is shared and exchanged in the collaborative environment, the quality of the data itself needs to be considered as well. Kikuchi et al. [69] proposed Product Data Quality model as a resource model of the STEP international standards. In this way, quality of product data and in particular the shape data can be modelled and stored along with the product document.

Besides the aforementioned STEP-based achievements, a data structure called Linked Data was proposed by Graube et al. [70]. Through this generic data structure, distributed information spaces from different domains are condensed into an interlinked cloud, while there are two integration methods. The first is to merge them into a single Linked Data Cloud using appropriate adapters and converters, and the second is the complete migration of the databases to native Linked Data stores. In this approach, graph theory is utilised, which is possible to describe object-oriented data structure such as STEP.

Lee et al. [71] proposed a web-based neutral data format amongst heterogeneous simulation software. The data model is named NESIS (NEutral SImulation Schema). Defined and categorised product elements are in different levels, which clearly describes product, process, sim_list (multiple simulation versions) and configuration information. In the four-layer, NESIS acts as the central internal data structure of the system. At the Client Layer, interfaces are developed to enable collaboration of commercial simulation applications and NESIS, which act as central internal data structure of the system. Developed using Java programming language, these interfaces automatically generate simulation models using simulation information and related data that are received from NESIS, and conversely send simulation information and related data that are generated by commercial simulation applications. Thanks to the interfaces and natural data format, communication between various software applications is realised, and the reusability of simulation data model is achieved.

In the middleware of a CAE system developed by Song et al. [72], a structure of the proprietary file format was proposed to interface multiple CAE software tools. Using VRML, heterogeneous CAE data is translated into chunks described by entity-attribute data structures which are similar to STEP structures. It is proven to be effective deriving data in a chunk-type form. Even if new entities are subsequently added, the structure is enabled to be read in a supplementary form, in spite of the addition of new entities; these entities are described as new chunks. To integrate heterogeneous business organisations, a Collaboration Point (CP) concept is proposed by Li et al. [73]. CP is located on the boundary of different organisations, acting as the interface for processes to be interoperated across various organisations. The operational processes of the enterprise and cloud services could be described by business process models of CP. Then, the common activities of two kinds of processes could be identified. CPs are introduced and connected to the common activities as the modelling interpretation of interoperation. This interface can support data exchange, command transferring, monitoring and so forth.

To recap, utilising neutral product document is a propitious methodology to achieve system interoperability. Standardised data formats (e.g. STEP and STEP-NC) enable seamless data exchange environment along the CAD/CAM/CNC chain. However, there is a lack of solution integrating product information throughout the lifecycle at high levels, and providing user-specific data at the same time. In the following section, an advanced product data exchange mechanism is introduced to meet user's specific needs.

## 12.4 Standardisation for Cloud Manufacturing

At the Manufacturing Capability Layer, MCapabilities are integrated as self-contained service modules in the SACloud. The operational processes throughout the supply chain stay in form of CMService applications at this layer. By controlling the service input and output, CMServices are shared and published in the high-performance resource pool. The plug-and-play ability of service applications enables flexibility and adaptability to cope with uncertain and changing manufacturing market.

In Storage Cloud (SCloud), database maintains the product/project data as well as the information of the MResources meshed in SACloud. To model and recognise these application modules, three smart agents are developed, i.e. Resource Recognition Agent (RRA), Change Detection Agent (CDA) and Billing Agent (BiA). RRA is responsible for identifying newly-published capability and termination of existing ones. Since MCapabilities are loosely merged in the SACloud, the performance of the entire MCloud is not affected by binding or detaching an individual CMService. CDA is in charge of detecting and updating resource changes, such as its availability, price adjustment, facility maintaining, etc. Thus, the up-to-date data of resource can be supplied while SCM is searching for applications for the CUser. BiA works with Storage Cloud and SCM directly. When ST is generated, BA provides the service quote based on the predefined service description in SC. Thus, real-time information exchanging between MCapabilities and MCloud is enabled.

To clearly describe and present CMService, CProvider and CUser's requests, efficient data models are needed. A number of modelling languages can be used, e.g. Web Services Description Language (WSDL) for web-service description, Web Ontology Language (OWL) for knowledge representations, Web Services Business Process Execution Language (WS-BPEL) for executable business processes with web services, and EXPRESS. Among them, EXPRESS is chosen since it provides more robust modelling methods. EXPRESS is a standardised data modelling language for product data which is formalised in an ISO standard [74]. As the graphical notation of EXPRESS, Provider & Service models in EXPRESS-G provide the portability with standard data models such as STEP and STEP-NC.

As shown in Fig. 12.6, the enterprise which provides CMServices is defined as a SProvider. Therefore, a manufacturing enterprise can be described in cloud

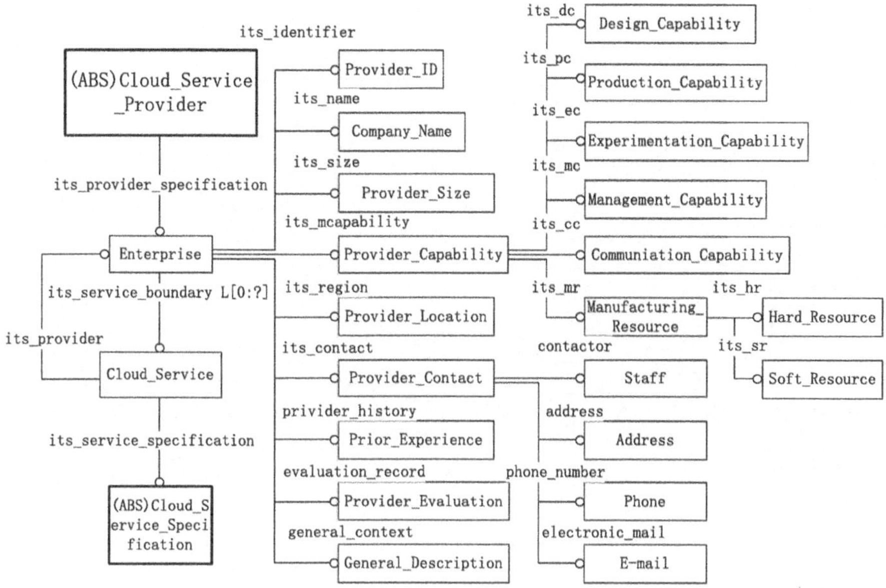


**Fig. 12.6** Cloud service provider model in EXPRESS-G

terminology, as provider profile and service properties. The provider specifications describe the information of the organisation, while service specifications present the MCapability in terms of service that it provides. Note that one company has a unique Enterprise Entity, while its service entities can be multiple. Hence, the organisation consistency and service variety are maintained concurrently. Entity Enterprise outlines the properties of a CManufacturing via the entities like Provider_ID, Company_Name, Provider_Size, Provider_Capability, Provider_Location, Provider_Contact, Prior_Experience, Provider_Evaluation and Provider_Description.

Entity Provider_ID provides a unique identifier in the MCloud for a SProvider. Based on its Provider_ID, all the CMServices from a provider and its related service history can easily be traced.

Entity Provider_Capabilities describes the MCapability of a SProvider via sub-entities Design_Capability, Experimentation_Capability, Production_Capability, Management_Capability, and Manufacturing_Resource, which are compliant with the aforementioned MCDM model. Entity Hard_Resource and Entity Soft_Resource describe the MResources that support a specific MCapability. These entities can be connected to a standardised data model directly, for example ISO14949-201 for machine tools, and ISO10303-45 for material and engineering properties. Hence, the MCapability of a SProvider is described in an explicit and scalable data model.

Entity Prior_Experience records the service history of one SProvider which is visible to the cloud administrator and provider itself, but not entirely to the users.

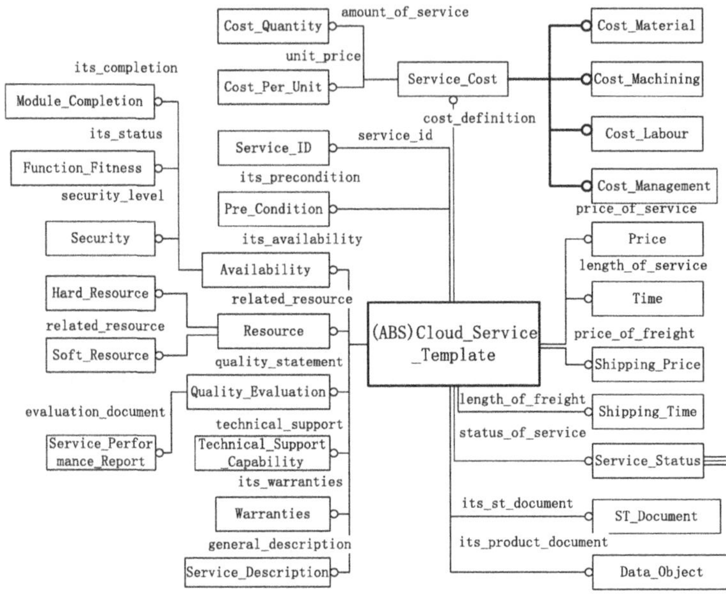


**Fig. 12.7** Cloud service model

Entity Provider_Evaluation documents the feedback from these service consumers. Based on these two entities, the performance of the service experiences are modelled explicitly.

As the second category of enterprise attributes, the recognition of the specific CMService is modelled via Entity Cloud_Service_Specification and its entities, i.e. Service_ID, Service_Cost, Price, Time, Shipping_Price, Shipping_Time, Service_Status, Service_Document, Data_Object, Pre_Condition, Availability, Resource, Quality_Evaluation, Technical_Support_Capability, Warranties and Service_Description (Fig. 12.7).

Entity Service_Cost documents the value of CMService in the monetary form. This entity provides an explicit model of the value that has been used to accomplish a service object. Service_Cost is only visible for SProviders to understand their MCapability internally, and to make reasonable price for external CUsers. Service_Cost is described with the help of entities Cost_Per_Unit and Cost_Quantity. The Service_Cost model structure is inherited by four sub-types, which divides the service cost into four categories, Cost_Material, Cost_Machining, Cost_Labour, and Cost_Management.

Entity Service_Status contains the information about the running stages after a CMService is launched. Stages of implementation are described via entities Initial, Ready, Running, Skipped, Completed, Overload, Paused, and Cancelled.

Entity ST_Document keeps the path and version of an ST file as abovementioned. When a CUser is working on an ST, all versions of the ST are recorded by this entity. Thus, the service/modification history is maintained.

Entity Data_Object records the technical document(s) related to the CMService. The optional attribute of this entity is Entity Project, which is compliant with the top level of a neutral data format defined in ISO10303 [75].

Entity Pre_Condition defines the requirements prior to the start of CMService. Limitations or preparations of the service input are recorded and published, e.g. limits of size, material preparation, heat treatment and so forth.

Entity Availability represents the availability and working condition of a CMService. This entity is dynamically updated by CDA. With the help of CDA, CUser is informed by the trustworthy situation of availability without major delays. CUsers are able to select the available CMService only, or queue in the list waiting for the preferred package till it is ready-to-be-used. The availability information is further described by its attributes, i.e. Module_Completion, Function_Fitness and Security.

Entity Resource defines the manufacturing resource that is required for a specific service. Its structural attributes (Hard_Resouce and Soft_Resource) are compliant with the resource representation of entity Enterprise. Thus, the resource specifications, from both service point of view and enterprise point of view, are shaped and integrated in SCloud.

To describe the user's query of a CMService, Cloud_Service_Request model is used (Fig. 12.8). Cloud_Service_Request is compliant with the SProvider and Cloud_Service_Template data structure. As a bridge between user's original

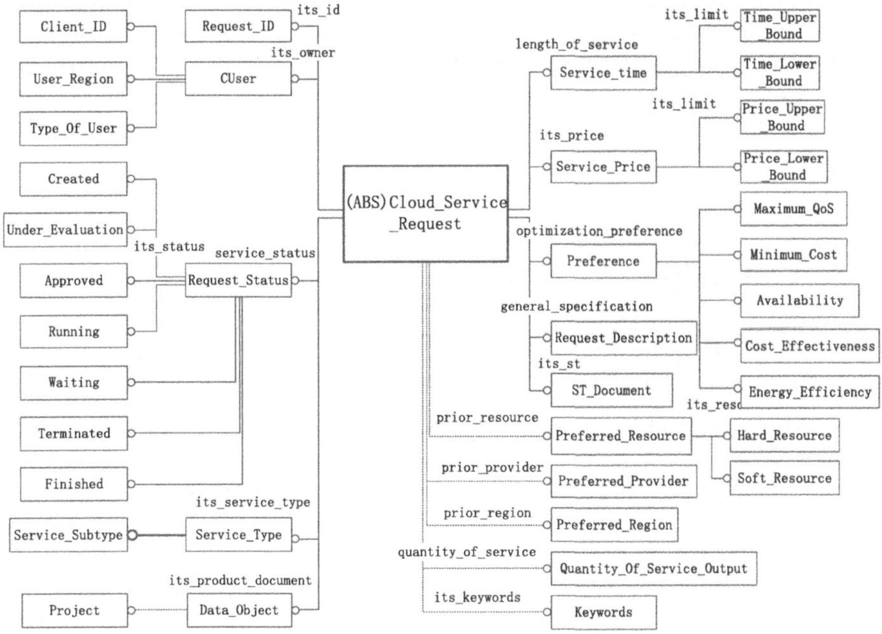


**Fig. 12.8** Service request model

demand and CMService in the MCloud, it provides a neutral and standardised methodology to document the query. Via GUI, the service description is arranged in the structured statement and transmitted to the SCM. Based on this piece of data, SCM is able to suggest the solution from the resource pool based on the terms and mapping preference. The request data is shaped via entities such as Request_ID, CUser, Request_Status, Service_Type, Data_Object, Service_Time, Service_Price, Preference, Request_Description, Service_Document, and optional entities Preferred_Resource, Preferred_Provider, Preferred_Region, Quantity_Of_Service_Output, and Keywords.

Entity Request_ID gives a unique serial number for a CUser's request. When a new query case is created, a permanent Request_ID is assigned. Users are able to resume, modify, and review the request case. Additionally, all the related cloud behaviour and history are traceable based on a Request_ID.

Entity Request_Status maintains the operational condition of a request. The variables of a process status are Created, Under_Evaluation, Approved, Running, Waiting, Terminated, and Finished.

Optional Entity Preferred_Resource keeps the user's predilection of resource and facility, for example specific machine tools, testing method or design software. The structure of this entity is compliant with the Hard_Resource and Soft_Resource entities of SProvider model aforementioned. Thus, the user's request can be directly connected to the MCapabilities in the cloud.

Thanks to the service request, CMService and SProvider models, the data is modelled from the initialisation to implementation stage in the SCloud. Information packages can be submitted, retrieved, and maintained over the Internet regardless the locations of the central database and server. For data storage queries, customer's private data is not maintained in the SCloud directly. In the background, data centres are hosted by third parties that are integrated as SProviders in the MCloud. Thus, the storage task is integrated as one of the CMServices in the virtualised service pools.

## 12.5 Conclusions

To recap, ICMS provides a flexible and distributed environment for shared MCapabilities. In particular, it offers a number of benefits as listed below.

- Data Interoperability: manufacturing business is commonly troubled by data interoperability issues. CAx applications are widely utilised throughout the production stages. However, these applications are provided by multiple providers using different programming languages and document formats, leading to a heterogeneous data environment. Software tools using different kernels are difficult to communicate with each other. Data loss and errors often occur during format conversions. By using ICMS, standardised (STEP-based) communication methodologies have been deployed to support collaborative interactions in the cloud environment. Moreover, ICMS offers explicit specifications of manufacturing resources in the MCloud. Detailed descriptions, e.g. input and output format requirements, are visible to all the CUsers. Interoperable problems can be easily identified and avoided. Users are able to choose the SPs that can smoothly communicate with each other, or alternately allocate reliable data conversion service beforehand as one of the CMServices. Therefore, interoperability is achieved even before a CMService is launched.

- Globalisation/Sub-Contracting: with the help of Internet of Things, manufacturing services/capabilities are virtualised in the MCloud. Compared with web-based manufacturing, ICMS provides a more distributed and flexible environment which knocks down the boundaries between organisations/enterprises. It is easier to find business partners/sub-contractors based on their performance of service, regardless of who and where they are.
- Customised Service and Specialised Demand: customisation is becoming more and more important in modern manufacturing, especially for SMEs. In a machine shop, specific cutter/machine tools are required for a particular job. With SCM, it is easy to locate required facilities in the resource pool. Therefore, specialised objects are achieved without additional investment on costly facilities and expertise.
- Facility Utilisation: resource can be shared in a cloud. Technical details and availability can be dynamically updated and published in the SCloud. Thus, manufacturing resources/capabilities can be better utilised. Production tasks can be easily balanced between high-usage facilities and the low-usage ones. From the user's perspective, CUsers are able to choose the available qualified providers for urgent jobs, or to wait for the preferred facility in the queue. Therefore, the facility utilisation is improved by widely shared environment and reasonable schedule.
- Global Optimisation: since services are broadcasted in the cloud, service solution can be improved and optimised based on the virtualised service modules implemented in the cloud. SCM predicts the service performance features beforehand, e.g. cost/time caused by preparing, machining, transporting and packing stages. So much so, the global solution is optimised based on particular factors or user's preferences.
- Cost-Saving: by adopting the CManufacturing concept, the manufacturing cost can be reduced. With the shared MCapabilities available in the cloud, optimised business solution is easily found according to optimised results. Since the features of SProviders are virtualised in the SC, it is more likely to find supplier with better performance, cheaper labour, higher productivity, and better geographical location. As a consequence of time-critical or cost-critical optimisation strategies, the performance of the service solution is predicted and improved from a higher level and in a bigger scope of cloud. Besides the cost of the service itself, the cost of strategic decision is reduced as well. With the technical specifications highly integrated in the SCloud, the cost of management, analysis, and comparison decreases, too.

- Better Enterprise Performance: when it comes to cost/time management, ICMS improves not only the experience of the CUsers but also the enterprise's performance as a CMService provider. The MCapabilities are accessible in the cloud, bringing more business opportunities. With the help of SCM, a SProvider is able to increase its production volume and react rapidly to market changes.

In this chapter, a cloud-based system is developed especially for ubiquitous manufacturing. Integration mechanisms of physical resources are proposed. A local server-driven architecture is developed to combat the conflicts between local connections and Internet communications. In practice, safety and security challenges for cloud includes Resource Constraints, Information and Learning Constraints and Communication Constraints [76]. In future works, the cloud-based manufacturing systems can benefit from the related technologies utilised by computing and manufacturing cloud. Firewalls and access control keeps an ICMS system from unwilling access and attack. Meanwhile network encryption and private keys are helpful to keep sensitive date in specific working domains. In the past years, private cloud models were welcome by the production enterprises since the company is enabled to protect the cloud infrastructure within their own fences in both cyber and physical worlds. In the future, the cloud manufacturing systems can be further supported by other successful methods, e.g. secure gateways, coding, antivirus software, etc.

## References


1. P. Mell, T. Grance, The NIST definition of cloud computing. NIST Spec. Publ. **800**, 7 (2011)
2. P. Mell, T. Grance, *Perspectives on Cloud Computing and Standards* (National Institute of Standards and Technology (NIST), Information Technology Laboratory, 2009)
3. Apple, *iCloud* (2012, April). Available https://www.icloud.com/
4. Amazon, *Amazon Elastic Compute Cloud (EC2)* (2012, April). Available http://aws.amazon.com/ec2/
5. Google, *Google App Engine—Google Code* (2012, Nov). Available http://code.google.com/intl/en/appengine/
6. Microsoft, *Windows Azure Platform_Microsoft Cloud Services* (2012, Nov). Available http://www.microsoft.com/windowsazure/
7. Oracle, *Sun Cloud Developer Homepage* (2012, Nov). Available http://developers.sun.com/cloud/
8. X. Xu, From *Cloud Computing to Cloud Manufac*turing. Robot. Comput. Integr. Manuf. **28**, 75–86 (2012)
9. F. Tao et al., Cloud Manufacturing: a Computing and Service-oriented Manufacturing Model. Proceedings of the Institution of Mechanical Engineers, Part B: Journal of Engineering Manufacture **225**, 1969–1976 (2011)
10. B. Li et al., Cloud manufacturing: a new service-oriented networked manufacturing model. Comput. Integr. Manuf. Syst. **16**, 1–7 (2010)
11. W. Terkaj et al., Virtual Factory Data Model in *Proceedings of the Second International Workshop on Searching and Integrating New Web Data Sources (VLDS 2012)*, Istanbul, Turkey, 2012

12. M. Meier et al., ManuCloud: the next-generation manufacturing as a service environment. ERCIM News **83**, 33–34 (2010)
13. O.E. Ruiz et al., EGCL: an extended g-code language with flow control, functions and mnemonic variables. World Acad. Sci. Eng. Technol. **67**, 455–462 (2012)
14. A.L.K. Yip et al. A front-end system to support cloud-based manfuacturing of cutstomized products in *Proceedings of the 9th International Conference on Manufacturing Research ICMR 2011*, Glasgow, UK, 193–198, 2011
15. L. Wu, C. Yang, A solution of manufacturing resources sharing in cloud computing environment. Coop. Des. Vis. Eng. Lect Notes Comput. Sci. **6240**, 247–252 (2010)
16. C.S. Hu et al., Study of classification and modeling of virtual resources in cloud manufacturing. Appl. Mech. Mater. **121–126**, 2274–2280 (2012)
17. Y.L. Luo et al., Study on multi-view model for cloud manufacturing. Adv. Mater. Res. **201**, 685–688 (2011)
18. Y.L. Luo et al., Research on the knowledge-based multi-dimensional information model of manufacturing capability in CMfg. Adv. Mater. Res. **472**, 2592–2595 (2012)
19. L. Zhang et al., Flexible Management of Resource Service Composition in Cloud Manufacturing, in *Proceedings of the 2010 IEEE IEEM*, 2278–2282, 2010
20. Q. Liu et al., Resource Management Based on Multi-agent Technology for Cloud Manufacturing, in *International Conference on Electronics, Communications and Control (ICECC)*, Zhejiang, China, 2821–2824, 2011
21. W.H. Fan, T.Y. Xiao, Integrated *Architecture of Cloud Manufacturing Based on Federation Mode*. Comput. Integr. Manuf. Syst. **17**, 469–476 (2011)
22. Y. Laili et al., A study of optimal allocation of computing resources in cloud manufacturing systems. Int. J. Adv. Manuf. Technol. **63**, 1–20 (2012)
23. X.V. Wang, X. Xu, ICMS: a cloud-based manufacturing system, in *Cloud Manufacturing: Distributed Computing Technologies for Global and Sustainable Manufacturing*, ed. by W. Li, J. Mehnen (Springer, 2012 in press)
24. G. Buonanno et al., Factors affecting ERP system adoption: a comparative analysis between SMEs and large companies. J. Enterp. Inf. Manag. **18**, 384–426 (2005)
25. T.F. Gattiker, D.L. Goodhue, What happens after ERP implementation: understanding the impact of interdependence and differentiation on plant-level outcomes. MIS Q. **29**, 559–585 (2005)
26. D.G. Ko et al., Antecedents of knowledge transfer from consultants to clients in enterprise system implementations. MIS Q. **29**, 59–85 (2005)
27. H. Liang et al., Assimilation of enterprise systems: the effect of institutional pressures and the mediating role of top management. MIS Q. **31**, 59–87 (2007)
28. C.C. Wei et al., An AHP-based approach to ERP system selection. Int. J. Prod. Econ. **96**, 47–62 (2005)
29. A. Paulraj et al., Inter-organizational communication as a relational competency: antecedents and performance outcomes in collaborative buyer-supplier relationships. J. Oper. Manag. **26**, 45–64 (2008)
30. S.B. Modi, V.A. Mabert, Supplier development: improving supplier performance through knowledge transfer. J. Oper. Manag. **25**, 42–64 (2007)
31. M.P. Papazoglou, W.J. Van Den Heuvel, Service oriented architectures: approaches, technologies and research issues. VLDB J. **16**, 389–415 (2007)
32. L. Cherbakov et al., Impact of service orientation at the business level. IBM Syst. J. **44**, 653–668 (2005)
33. M.T. Schmidt et al., The enterprise service bus: making service-oriented architecture real. IBM Syst. J. **44**, 781–797 (2005)
34. P. Rauyruen, K.E. Miller, Relationship quality as a predictor of B2B customer loyalty. J. Bus. Res. **60**, 21–31 (2007)
35. P.A. Bernstein, S. Melnik, Model Management 2.0: Manipulating Richer Mappings. Presented at the SIGMOD 07' international conference on management of Data, Beijing, China, 2007

36. ISO, ISO 10303 -1, *Industrial Automation Systems and Integration—Product Data Representation and Exchange—Part 1: Overview and Fundamental Principles*. (ISO, Geneva, 1994)
37. ISO 14649-1, *Industrial Automation Systems and Integration—Physical Device Control—Data Model for Computerized Numerical Controllers—Part 1: Overview and Fundamental Principles*, 2003
38. W. Gielingh, An assessment of the current state of product data technologies. CAD Comput. Aided Des. **40**, 750–759 (2008)
39. L. Zhang et al., Key technologies for the construction of manufacturing cloud. Comput. Integr. Manuf. Syst. **16**, 2510–2520 (2010)
40. R.D. Allen et al., The application of STEP-NC using agent-based process planning. Int. J. Prod. Res. **43**, 655–670 (2005)
41. L. Monostori et al., Agent-based systems for manufacturing. Ann. CIRP **55**, 697–720 (2006)
42. A. Nassehi et al., The application of multi-agent systems for STEP-NC computer aided process planning of prismatic components. Int. J. Mach. Tools Manuf. **46**, 559–574 (2006)
43. H. Panetto, A. Molina, Enterprise Integration and Interoperability in Manufacturing Systems: trends and Issues. Comput. Ind. **59**, 641–646 (2008)
44. M. Sadeghi et al., A collaborative platform architecture for coherence management in multi-view integrated product modelling. Int. J. Comput. Integr. Manuf. **23**, 270–282 (2010)
45. A.J. Álvares et al., An Integrated Web-based CAD/CAPP/CAM system for the remote design and manufacture of feature-based cylindrical parts. J. Intell. Manuf. **19**, 643–659 (2008)
46. C. Brecher et al., Module-based Platform for Seamless Interoperable CAD-CAM-CNC Planning, in *Advanced Design and Manufacturing Based on STEP*, ed. by X.W. Xu, A.Y.C. Nee (London: Springer, 439–462, 2009)
47. S.C. Oh, S.T. Yee, Manufacturing interoperability using a semantic mediation. Int. J. Adv. Manuf. Technol. **39**, 199–210 (2008)
48. Y. Zhang et al., Understanding the STEP-NC data model for computer numerical control, in *Advanced Computer Control* (*ICACC*), 2nd International Conference on, 2010, 300–304
49. O.F. Valilai, M. Houshmand, INFELT STEP: an integrated and interoperable platform for collaborative CAD/CAPP/CAM/CNC machining systems based on STEP standard. Int. J. Comput. Integr. Manuf. **23**, 1095 (2010)
50. J. Yang et al., OpenPDM-based product data exchange among heterogeneous pdm systems in a distributed environment. Int. J. Adv. Manuf. Technol. **40**, 1033–1043 (2009)
51. S. Makris et al., On the information modeling for the electronic operation of supply chains: a maritime case study. Robot. Comput. Integr. Manuf. **24**, 140–149 (2008)
52. G. Chryssolouris et al., Towards the Internet based supply chain management for the ship repair industry. Int. J. Comput. Integr. Manuf. **17**, 45–57 (2004)
53. D. Mavrikios et al., A New concept for collaborative product and process design within a human-oriented collaborative manufacturing environment. Future Prod. Dev. 301–310 (2007)
54. M. Pappas et al., A Collaboration Platform for Product Design Evaluation, Demonstration and Customization. Presented at the IFAC workshop on manufacturing modelling, management and control, Budapest, Hungary, 2007
55. S. Makris, K. Alexopoulos, AutomationML server-A prototype data management system for multi disciplinary production engineering. Procedia CIRP **2**, 22–27 (2012)
56. B. Asiabanpour et al., An overview on five approaches for translating CAD data into manufacturing information. J. Adv. Manuf. Syst. **8**, 89–114 (2009)
57. S.T. Newman et al., Strategic advantages of interoperability for global manufacturing using CNC technology. Robot. Comput. Integr. Manuf. **24**(2008), 699–708 (2008)
58. A. Nassehi et al., Toward interoperable CNC manufacturing. Comput. Integr. Manuf. **21**, 222–230 (2008)
59. S.T. Newman, A. Nassehi, Universal manufacturing platform for CNC machining. Ann. CIRP **56**, 459 (2007)
60. A. Mokhtar, M. Houshmand, Introducing a roadmap to implement the universal manufacturing platform using axiomatic design theory. Int. J. Manuf. Res. **5**, 252–269 (2010)

61. P. Vichare et al., A unified manufacturing resource model for representation of computerized numerically controlled machine tools. Proc. Inst. Mech. Eng. Part B J. Eng. Manuf. **223**, 463–483 (2009)
62. N. Do, G. Chae, A product data management architecture for integrating hardware and software development. Comput. Ind. **62**, 854–863 (2011)
63. J. Hwang et al., Representation and propagation of engineering change information in collaborative product development using a neutral reference model. Concur. Eng. **17**, 147–157 (2009)
64. S.S. Choi et al., XML-based neutral file and PLM integrator for PPR information exchange between heterogeneous PLM systems. Int. J. Comput. Integr. Manuf. **23**, 216–228 (2010)
65. R. Jardim-Goncalves et al., Knowledge framework for intelligent manufacturing systems. J. Intell. Manuf. **22**, 725–735 (2011)
66. ISO 10303-236, *Industrial Automation Systems and Integration—Product Data Representation and Exchange—Part 236: Application Protocol: Furniture Catalog and Interior Design*, (2003)
67. B.C. Kim et al., Web service with parallel processing capabilities for the retrieval of cad assembly data. Concur. Eng. Res. Appl. **19**, 5–18 (2011)
68. STEP Tools, *ST-Developer—STEP Tools, Inc* (2011, Nov). Available http://www.steptools.com/products/stdev/
69. Y. Kikuchi et al., PDQ (Product Data Quality): representation of data quality for product data and specifically for shape data. J. Comput. Inf. Sci. Eng. **10**, 1–8 (2010)
70. M. Graube et al., Linked Data as Integrating Technology for Industrial Data, in *2011 International Conference on Network-Based Information Systems* (*NBiS*), Tirana, Albania, 162–167, 2011
71. J.Y. Lee et al., NESIS: A neutral schema for a web-based simulation model exchange service across heterogeneous simulation software. Int. J. Comput. Integr. Manuf. **24**, 948–969 (2011)
72. I.-H. Song et al., Development of a lightweight CAE middleware for CAE data exchange. Int. J. Comput. Integr. Manuf. **22**, 823–835 (2009)
73. Q. Li et al., Towards the business-information technology alignment in cloud computing environment: an approach based on collaboration points and agents. Int. J. Comput. Integr. Manuf. **24**, 1038–1057
74. ISO, ISO 10303-11, *Industrial Automation Systems and Integration—Product Data Representation and Exchange—Part 11: Description Methods: The EXPRESS Language Reference Manual* (2004)
75. ISO 10303-21, *Industrial Automation Systems and Integration—Product Data Representation and Exchange—Part 21: Implementation Methods: Clear Text Encoding of the Exchange Structure* (2002)
76. G. Hu et al., Cloud robotics: architecture, challenges and applications. Netw. IEEE **26**, 21–28 (2012)

# Chapter 13
# Product Tracking and WEEE Management

## 13.1 Introduction

The amount of Waste Electrical and Electronic Equipment (WEEE) has grown significantly in recent years, due to increased Electrical and Electronic Equipment (EEE) and its shorter lifecycle. Different types of EEE are principally classified as shown in Table 13.1. The replacements of these devices (e.g. televisions, computers, cell phones, etc.) are more frequent than ever before because of the fast-changing market demand and planned obsolescence. New products offer attractive functionalities and convenience to the consumer, but also push the in-service products from Middle-of-Life (MOL) to End-of-Life (EOL) phase. From the manufacturers' perspectives, shorter lifecycle brings greater profits and keeps their positions on the competitive market. Compared with high-value products, most household appliances have shorter lifecycle and need litter maintenance service, e.g. cell phones, kettles, and lightening bulbs. When it stops service, the consumer intends to dispose them directly instead of repairing, since it is more cost-effective to purchase a new one in most cases. Even though selling new products brings profits to the manufacturers, it also contributes huge amount of WEEE out of the total tons of waste.

The huge volume of WEEE leads to global environmental issues on many scales. According to the statistics of the US Environmental Protection Agency [1], 438 million new electronic devices were sold in 2009 in America, which represented a doubling of sales from 1997. 2.37 million tons of them reached EOL in 2009, but only 25% of them were collected for recycling. Among different kinds of electrical and electronic products, the recycling rate of mobile devices (cell phones, smart phones, PDAs) was lowest, even less than 9%.

Thus it is important to manage and control WEEE with practical strategies. In the EU, handling WEEE is a high priority for all member states. Countries such as Switzerland, Denmark, Netherlands, Norway, Belgium, Sweden and Germany already have an established Extended Producer Responsibility (EPR) for WEEE.



L. Wang and X. V. Wang, *Cloud-Based Cyber-Physical Systems in Manufacturing*,
https://doi.org/10.1007/978-3-319-67693-7_13

**Table 13.1** Principal EEE categories

| Category | Examples |
|---|---|
| Information and communication | Computer, tablet, mobile phone |
| Large household appliances | Refrigerator, air conditioner, washing machine |
| Small household appliances | Iron, dryer, rice cooker |
| Lighting equipment | Household luminary, outdoor lighting, automotive lighting |
| Electrical and electronic tools | Volt-ohm-millimetre, soldering iron |
| Toys, leisure and sports equipment | Coin slot machines, car racing set |
| Automatic dispensers | Water dispenser, coffee machine |
| Medical equipment | Ultrasound machine, heart-lung machine |

In the case of WEEE facilities, many developed countries including the USA, Europe and Japan have mature technologies for the treatment of this waste stream [2]. However, in developing countries primitive activities predominate, as in the case of the largest e-waste recycling place in Guiyu, China where the practices include: manually classification and dismantling of e-waste, manual separation and solder recovery for mounted printed circuit board, precious metal extraction by acid, among others [3]. Also the informal sector has a predominant presence in these activities, as in the case of Nigeria, Ghana and Thailand [2]. Traditionally, the recycling of WEEE mainly stays at material level. The target of recycling is either separating hazardous elements from resources, e.g. mercury and brominated flame retardant or extracting valuable materials that can be utilised again, e.g. gold, silver, plastics, steel and aluminium. The risk in WEEE treatment is largely due to its toxicity. During WEEE recycling, three groups of substances may be released: the constituents of the EEE, the substances used in the recycling techniques, and the by-products formed during transformation of the original constituents [4, 5]. The toxicity of these substances is related to the presence of heavy metals and halogenated flame retardants. When treated by poorly controlled processes, it leads to damage and risk in multiple scales: soil and sediment pollution [6, 7], water [8], air [9], and human health [10, 11]. Additionally, the pollution may also infiltrate into the environment directly through municipal solid waste disposal [12].

The traditional path of WEEE is limited to recycling, for the sake of obtaining raw materials. In practice, it is possible to treat WEEE as used products, before it is considered as a discharged waste [13]. The EOL processes include the secondary market processing and component recovery (such as repair, reconditioning, and remanufacturing) or material recovery (recycling) [14]. According to the EU WEEE Directive, after electronics reach the end phase of their lifecycle, they should be filtered based on their status and their economic and functional potentials. Then the WEEE is processed via different paths after proper treatment, but principally WEEE is recycled. In practice, it is also important to consider other EOL processing routes, for example the BS 8887 standard serials give six EOL routes as follows, along with the likely change at warranty level compared with the original product [15], including reuse, remanufacture, recondition, repurpose, recycle, and dispose. In this roadmap, WEEE are handled not only as a waste, but also as a special category of product that can be reused through an extended lifecycle [13]. Although the term WEEE indicates the equipment as a waste, a huge proportion of the equipment can be defined as Used Electrical and Electronic Equipment (UEEE), which plays an important role for component recovery or extended usage. These activities are at a higher level than recycling in the environmental hierarchy of EOL strategies [14]. Considering this, such understanding can be included in the new perception of the EEE lifecycle. It is possible to put UEEE back to the market via proper recovery processes and treatments (Fig. 13.1).

Recovery activities aim to get usage of the components from UEEE, before they are disposed as waste, i.e. repair, reconditioning and remanufacturing. The assessment and utilisation can be extended to the functionality level [14]. After being disassembled, the parts from UEEE can be reutilised for different purposes based on their warranty conditions. Even though the product as a whole has reached its end of lifecycle, many parts inside it may still be functional [16]. One of the most successful business examples of profitable component recovery is the remanufacturing of parts in the automotive industry, and large mechanical and electromechanical products. Since the waste vehicles are able to provide profitable parts for repairing other vehicles still in service, these parts are not treated as waste any longer. They can be tagged as used products and be offered back to the market with reasonable warranties. It is similar in UEEE scenarios. Instead of treatment for valuable or dangerous materials, many parts within UEEE can be reutilised at functionality level. It is necessary to establish a platform that understands both the nature of WEEE/UEEE and integrates related processes. Then a collaborative environment can be established to maintain the data/knowledge and support component recovery and recycling processes.

In this chapter, a cloud based system is introduced to support not only the management of WEEE, considered as a waste, by recycling, but also the fraction of WEEE that is an UEEE by recovery and related processes, including remanufacturing, reconditioning and repairing. The architecture of the proposed system,

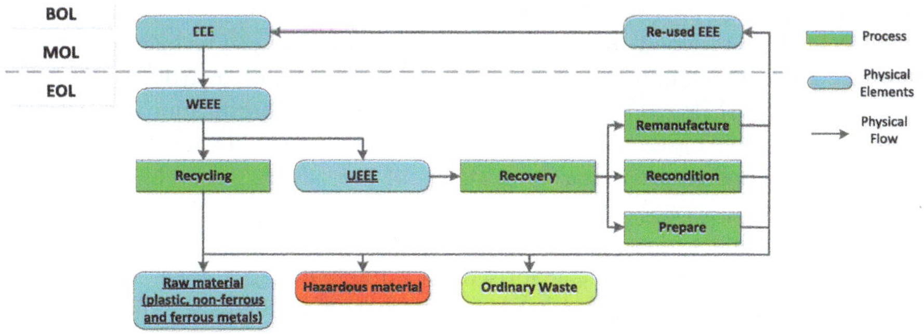


**Fig. 13.1** WEEE physical flow and UEEE

namely WEEE Recovery/Recycling Cloud (WR2Cloud), and developments are presented below.

## 13.2 System Framework

As mentioned above, an intelligent recovery and recycling system is required to support the management of WEEE at both material and functionality level. In recent years, cloud technology has been introduced in different scenarios since it provides the capability of scalable and flexible services in a customised manner. The cloud concept was initially proposed to describe the large number of computers that are connected via runtime communications over a network [17]. Then the cloud was extended to other areas supporting customised services, e.g. manufacturing [18–25]. Cloud manufacturing can be understood as the manufacturing model that enables scalable, on-demand access to manufacturing services, both digitally and physically [26]. As a specific category of manufacturing, the reproduction based on WEEE, especially UEEE can also be supported by the cloud via its integrated manufacturing solutions, high-level data management/control and flexible service models [27–31]. Thus, a cloud-based system is to fulfil the needs of WEEE.

### *13.2.1 System Requirements and Roles*

In the manufacturing paradigm, raw material is treated as the input for the start of a physical flow. The order, payment and transportation of raw materials can proceed within a mature supplier network. This is one of the major differences between traditional manufacturing and the WEEE component recovery/recycling business. In WEEE component recovery/recycling, used products are owned by random end users. These users' locations are usually unknown and their behaviour, e.g. in terms of where and when they would discard their used products are unpredictable. It is specifically difficult to maintain the knowledge of WEEE and organise related services due to the interrupted information flow.

With the help of cloud, all the data of individual WEEE can be maintained in an integrated and shared information pool. The centralised cloud repository also keeps the Beginning-of-Life (BOL) and Middle-of-Life (MOL) specifications of products. In this way, remote customers, who are connected to the Cloud via the network, can access and so update the status at all stages of a product's lifecycle. The system would also assist service scheduling after the product reaches EOL. The system requirement for data management can be summarised as shown in Fig. 13.2. The maintenance and update are categorised into different stages since an electronic device travels among different stakeholders throughout its lifecycle. After an EEE is manufactured, it is the manufacturer's responsibility to register the basic information about the EEE, e.g. important components, technical specifications and bill of

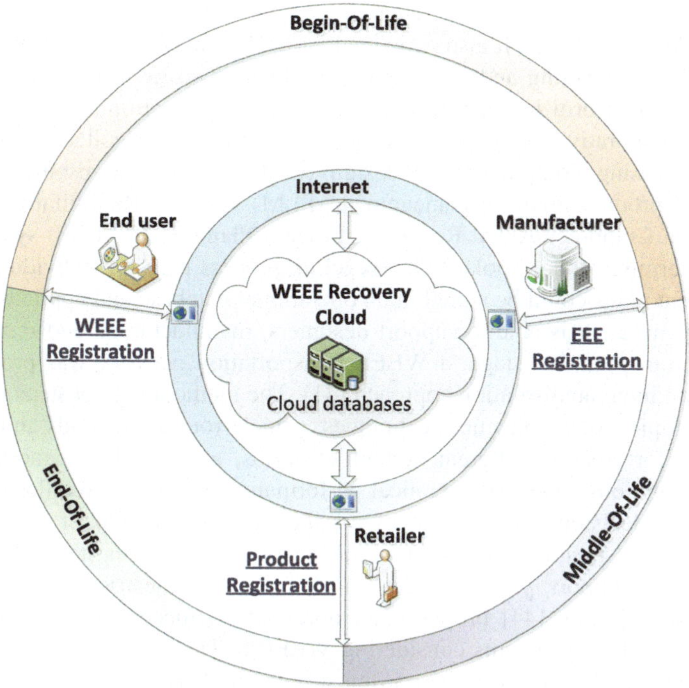


**Fig. 13.2** Cloud-based WEEE data maintenance

material. After the EEE is sold to a customer, the retailer needs to update the EEE as a final product and update the basic information about the customer, including name and contact information. Thus data is valuable for both maintenance and recovery services in the future. When the product meets the end of the lifecycle, the user is able to interact with the cloud, turn the registration from product to WEEE and organise related recycle or recovery service accordingly.

Even though the cloud-based solutions are implemented based on some web-based technologies, it needs to be pointed out that there are fundamental differences between traditional WEEE systems and the recovery/recycling cloud. From the stakeholder's perspective, in traditional web-based recovery and recycling systems [32, 33], the users of the system or platform are mainly recyclers or remanufacturers. The system helps them to communicate with remote resources over the network. Toyasaki et al. [34] highlighted the value of information systems for product recovery management. Conceptual models and methodological concepts were introduced to assist a green WEEE supply chain [35, 36].

At the BOL stage, Van Shaik and Reuter [37] developed dynamic models of e-waste recycling system performance based on product design. This model allowed for the design-driven modelling of material liberation in the shredding process. Yang et al. [38] proposed an intelligent product. It was based on a service enabling scheme which uses the product lifecycle data in a systematic and integrated manner to facilitate the creation and delivery of suitable services during the lifecycle of a product. It also contained an intelligent data unit which maintains the options for recycling and reuse of the products. Kuo [39] proposed a collaborative design platform to support waste electrical and electronic equipment recycling. A collaborative design platform was constructed and collected the needed information using computer-aided design (CAD), enterprise resource planning (ERP), and product lifecycle management (PLM) systems. Rahimifard et al. [40] developed a Computer Aided Recycling Process Planning (CARPP) system. This system determines the bespoke EOL recycling process plan for individual WEEE. The different plans could be stored in an operational database and applied to similar products families. It is able to support designers, manufacturers, and recyclers.

At the transportation stage, a WEEE transportation network was proposed by means of an integrated solution approach [41]. The methodological steps regard the following topics: data collection techniques, vehicle routing methods and heuristic procedures for creating different system scenarios, and simulation modelling for obtaining solutions satisfying technical performance measures. Achillas et al. [42] developed a decision support system for the optimal location of electrical and electronic waste treatment plants. Optimising reverse logistics network was also developed to support policy-making in the case of Electrical and Electronic Equipment [43]. Che [44] proposed an optimisation mechanism to balance and detect supply chain problems considering WEEE/RoHS directives.

Despite the IT technologies supporting sustainable WEEE design, transportation and decision making, there is still a lack of an integrated system that manages the whole WEEE lifecycle and coordinates recovery process collaboratively. Thus in WR2Cloud, the capability offered by the cloud recovery/recycling system is different from conventional solutions. In the cloud, recovery resources and capabilities are packaged as service modules and published, for example quoting-as-a-service and warehouse-as-a-service. In this way, the users do not directly interact with the recovery and recycle activities. Instead they are supported by the everything-as-a-service (XaaS) model that is deployed by the service providers in the cloud.

In this system information management needs support from manufacturers, retailers and end users. A key differentiating factor of this new system from typical production activities is that the WR2Cloud is able to integrate product knowledge and data throughout the lifecycle of products. An integrated and unified data sharing/management mechanism is an important prerequisite for recovery and recycling services. With the help of cloud databases, cloud participants are able to retrieve and stream WEEE data dynamically. The business model for this cloud system is shown in Fig. 13.3. In a conventional WEEE recovery system, the user is the recovery stakeholder who works with the industrial process directly. In WR2Cloud, the users of the system are the consumers of the EEE. Participants that are related to recovery and recycling service act as the service providers in the cloud, e.g. collecting and recycling service providers. The role of the recyclers and remanufacturers changes from end users to providers at the back end, while the consumers interact with the cloud at the frontend. It provides a clear classification and interact mechanism for different stakeholders in the cloud.

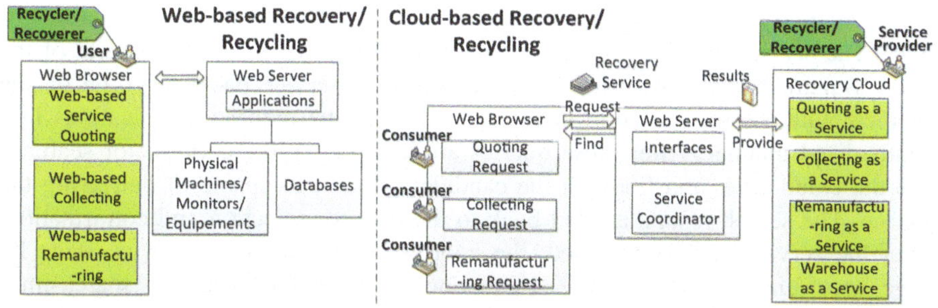


**Fig. 13.3** WR2Cloud business model

From the perspective of a business model, the traditional web-based system supports the customer with the whole infrastructure, which forms a *one-to-one* business model. In the cloud, multiple service objectives are achieved by the whole cloud. The customers may or may not need to know the identities of the providers or their whereabouts or vice versa. It forms a *many-to-many* model. Service requests and results are transferred by the coordinator mechanism between users and cloud. Thus it forms a "request-find-provide" procedure for the recycling/recovery business.

## 13.2.2 WR2Cloud: System Framework

To meet the requirements mentioned above, a three layer system is introduced to support WEEE recovery/recycling activities (see Fig. 13.4). In the WR2Cloud, component recovery and recycling facilities and capabilities are provided as cloud service packages in the cloud layer. The outlines and specifications of these services

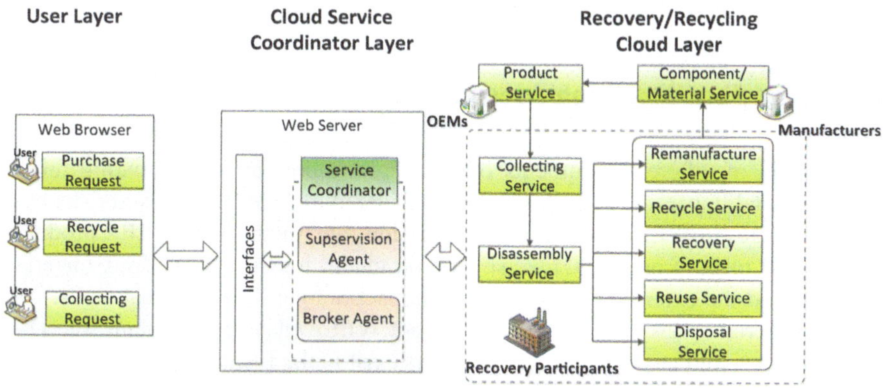


**Fig. 13.4** Three-layer WR2Cloud

are maintained in the cloud database and published at the cloud service coordinator layer. The service coordinator acts as the neutral supervisor or orchestrator of the cloud system. When a user needs one or multiple service package(s), he or she interacts with the service coordinator and provides the request specifications in a standardised manner. After these queries are interpreted, the coordinator searches into the cloud layer and finds all the candidates that are capable of delivering the service. The search results are fed back to the user for selection and confirmation. After the service plan is finalised, the coordinator executes the service stage by stage according to the schedule. In this way, WR2Cloud offers a search-find-provide service loop.

Based on these detailed descriptions, the service coordinator is able to search for appropriate service solutions, to organise optimised service combinations, and to execute service tasks. As a Service-Oriented Architecture (SOA), the WR2Cloud is able to coordinate the input/output flow through service packages and offer them as a concrete virtual service combination. At the user layer, the end-users are able to access the system from their local web browsers over the network. Different interfaces are developed to support their needs in recovery and recycling, e.g. registration, updating, service querying, resource tracking, etc.

As mentioned above, web-based WEEE systems aim to connect multiple modules via the network. It mainly supports the business within one organisation and there is a lack of orchestration mechanism to coordinate the process across different enterprises. In WR2Cloud, WEEE recovery/recycling processes are integrated in terms of cloud manufacturing services. A coordinator mechanism is developed to deliver remanufacturing resources and capabilities as services packages. Users are able to work with the shared service pool regardless the boundaries between different stakeholders. Compared with conventional cloud manufacturing systems, WR2Cloud extends the service scope from typical production processes to WEEE recovery services. Moreover, supporting technologies are developed to assist the WEEE remanufacturing, i.e. standardised data management and product tracking mechanism. These technologies also combat the data exchange and product management difficulties observed in current manufacturing clouds.

## 13.3 Product Tracking Mechanism

Comparison of different treatments of WEEE is important for evaluating related environmental impacts. The factors to be considered include benefits and risks from different stages in WEEE management, emission rate (quantity of WEEE produced or expected to be produced), reverse logistic chain, toxicity (hazardous materials content), and the extent of environmental impact among different treatments.

The materials present in the EEE or future WEEE play an important role [14]. A wide variety and large quantity of products exist in the EEE context (Table 13.1), and their internal composition is also complex. First it is important to understand that the materials content depends on the kind of equipment under consideration.

For example, for large domestic appliances, there is a considerable quantity of metals. However, in the case of small appliances, there are individual components such as cartridges, batteries, cables, printed circuit boards, ferrous and nonferrous fractions thus a wider range of materials is present and typically in smaller quantities. With the help of the cloud-based environment, it is possible to document different types of EEE products and maintain the knowledge database of their compositions.

Environmental impacts of WEEE need to be evaluated before decision making. It has to be admitted that both positive and negative impact factors exist during the recovery and recycling processes. For example, component recovery (also called remarket process) adds new value to the WEEE (especially UEEE), to bring the used products back to working order and this can be considered a positive factor. However, these recovery activities could also bring negative impacts to the environment occasionally, for example due to the chemicals and energy used in the product recovery process. In the recycling (material recovery) cases, when the products are reversed back to their raw materials, the energy and resources used for its original manufacture are lost. Moreover, energy and resource is required to enable this return to raw material thus energy and resource are lost twice over; and even more energy and resource would be needed to turn the raw material into a useable form. Thus it is important to assess the environmental impacts of different strategies of component recovery or material recycling before they are processed [14]. It is well known that inappropriate recycling procedures output high environmental impact, such as the air pollution from burning and dismantling activities, ashes from incineration, fly ashes and bottom ashes with high concentration of dioxins, Pb (industrial soils), PBDEs (urban soils), and also leaching potential [4].

Yet the systematic recycling process offers important environmental benefits in saving natural resources. Another issue to consider is the transportation of WEEE, which leads to high cost and negative environment impacts as well because of the energy and resources consumption during transportation. One solution could be locating processing plants in close proximity, for example recyclers and remanufacturers. Thus products that are not suitable for remanufacturing could be put through recycling (and vice versa) without much resource spent on transportation. Finally, the highest negative environmental impact is the landfill of WEEE without processing, which implies disposal of toxic materials without treatment. Therefore, the cloud-based approach is expected to be able to evaluate and optimise the WEEE recovery and recycling processes at a high level and also deliver sustainable service strategies.

### *13.3.1 WR2Cloud*

In the WR2Cloud, the information management and data sharing is supported by a standardised environment. Data models are developed to describe the important

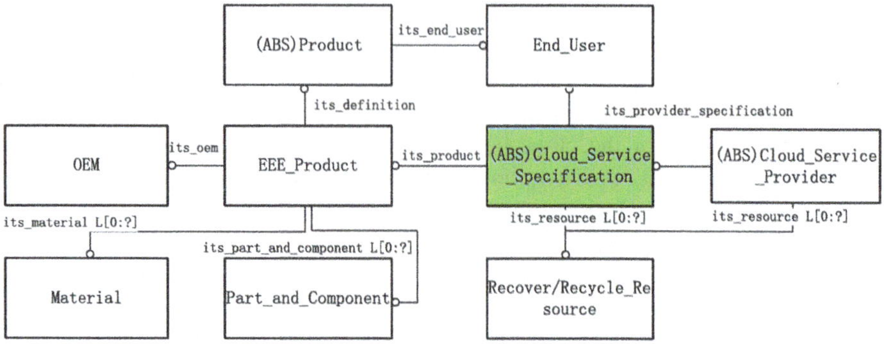


**Fig. 13.5** WEEE data model set

elements throughout the recovery and recycling chain, e.g. end-user, service provider, service model, etc. Uniformed Application Protocol Interfaces (APIs) are utilised to connect these models and provide an interoperable solution for the processes and interactions in the cloud (Fig. 13.5). In this SOA, the data description methodology is also built based on the service-oriented principle. The core of the data model set is the Cloud Service data model that documents what kind of recovery or recycling service is requested and how it is archived. The EEE data model set is established based on current ISO 10303 standards [45, 46] and extensive developments on Cloud Manufacturing [23, 24]. The top entity of cloud recovery and recycling service is defined as "project". For one service case or task, it is maintained as a project that is supported by related recovery and recycling resources and service providers.

For WEEE, its knowledge maintenance starts from EEE product registration at the BOL phase. The Original Equipment Manufacturer is responsible for establishing and maintaining the details and specifications as discussed above. The components/sub-components information is integrated with the product data model. In WR2Cloud, the data is documented at both component and material levels of the product. This information is especially valuable for the recovery and recycling processes afterwards. For instance, special element treatments are organised by the service coordinator based on the hazard element list of the WEEE product model, and the disassembly service can be assigned specifically for the valuable components.

After the EEE reaches the MOL stage, the end-users play an important role in updating the status of EEE or registering WEEE after its service ends. In practice, the WR2Cloud not only takes care of the WEEE process at the EOL phase, it also supports the EEE maintenance throughout the lifecycle of the product. The maintenance can be categorised in three groups, i.e. reactive maintenance, preventive maintenance and predictive maintenance [47].

- Reactive maintenance, or breakdown maintenance, can be described as a fire-fighting approach which allows the equipment/products to work till failure. With the integrated information sharing environment, the users are able to easily report faults or breakdowns via the cloud platform. Customised maintenance solutions can be quickly organised based on the existing product specifications in the cloud database, e.g. warranty status, model, customer location, etc.
- Preventive maintenance is often referred to as use-based maintenance. It is comprised of maintenance activities that are undertaken after a period of time or amount of use. In the cloud-based system, the service provider is also able to interact with customers actively, e.g. reminding them of key component expiration and offering safety check/maintenance based on the usage/duration data in the cloud.
- Predictive maintenance is frequently referred to as condition-based maintenance. For costly or important EEE products, the maintenance can be initialised with additional monitoring/diagnostic data, e.g. noise, temperature, corrosion, and so forth. Predictive maintenance reduces the possibility of the breakdowns on critical devices and parts.

### 13.3.2 'Cloud + QR'-based Tracking Methodology

To further improve the performance and portability of WR2Cloud, the data integration mechanism can be supported by the Quick Response Code (QR Code). QR code is a type of matrix barcode or two-dimensional barcode that is an optically machine-readable label [48]. The labels can be attached to products or even components inside. The capacity of QR code has been improved greatly in recent years. The latest version is able to obtain up to 1852 characters with high error correction. In this chapter, an LCD television is chosen as the WEEE that is owned by an end-user. The basic information of the product is recorded in its QR code tag that is attached on the back of the product (Fig. 13.6). In this case, 764 characters are utilised to document the information of the device, e.g. the product type, model, OEM and most importantly the unique product reference number. The user is able to scan the code via smart phones, tablets or camera devices and then submit queries to the cloud via web browsers or mobile apps. Based on the serial number, the detailed product specifications can be quickly retrieved from the cloud database. The user then links the latest WEEE status to the product profiles and registers it as WEEE. Related recovery and recycling services can be organised according to the information at both material level and component level. Compared with traditional barcode methods, more pollution and recovery data can be stored in the QR code tag, and additionally, for example, specifications compliant with the Restriction of the Hazardous Substances Directive. In this LCD case the pollutants and recyclable parts are recorded in the QR code that is visible to the end user. In this way, the user is able to understand the environmental, and recovery information in detail and request for the preferred services accordingly.

## 13.4 Implementations and Case Studies

To evaluate the methods mentioned above, a cloud-based remanufacturing system is implemented. At the preliminary phase, the virtual environment is built in the cloud environment which contains 32 computing cores and more than 132 GB memories in total. With the help of extendable cloud resources, customers are able to access and maintain the WEEE cloud without installing or configuring any local applications. The working environment of cloud is capable of virtualising multiple operating environments, i.e. Linux, MS Windows, and UNIX family. Thanks to the platform independency of JAVA, the developments can be deployed across different environments to suit the different needs or requirements of the users.

### *13.4.1 Case Study 1: Cloud WEEE Management at Product Level*

In the QR code management module, the remanufacturing stakeholders are able to generate the code tag and attach it on the product (Fig. 13.7). When the product stops functioning, the customer is able to scan the tag and quickly upload it onto the cloud. The basic information of the WEEE stored in the QR tag is directly interpreted by the system, and further details can be tracked based on the unique product ID kept in the tag.

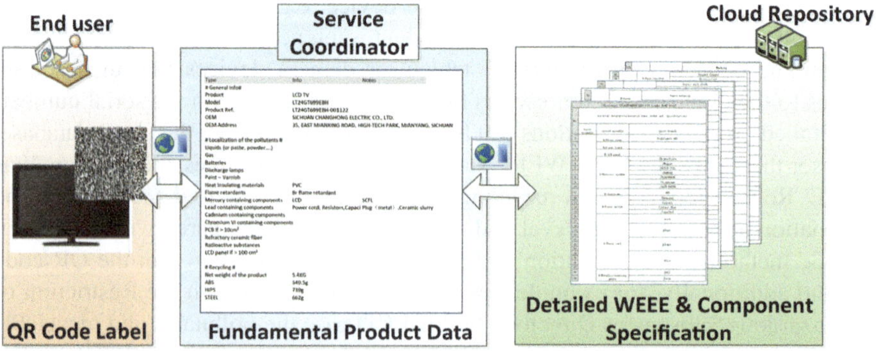


**Fig. 13.6** QR code enabled cloud service

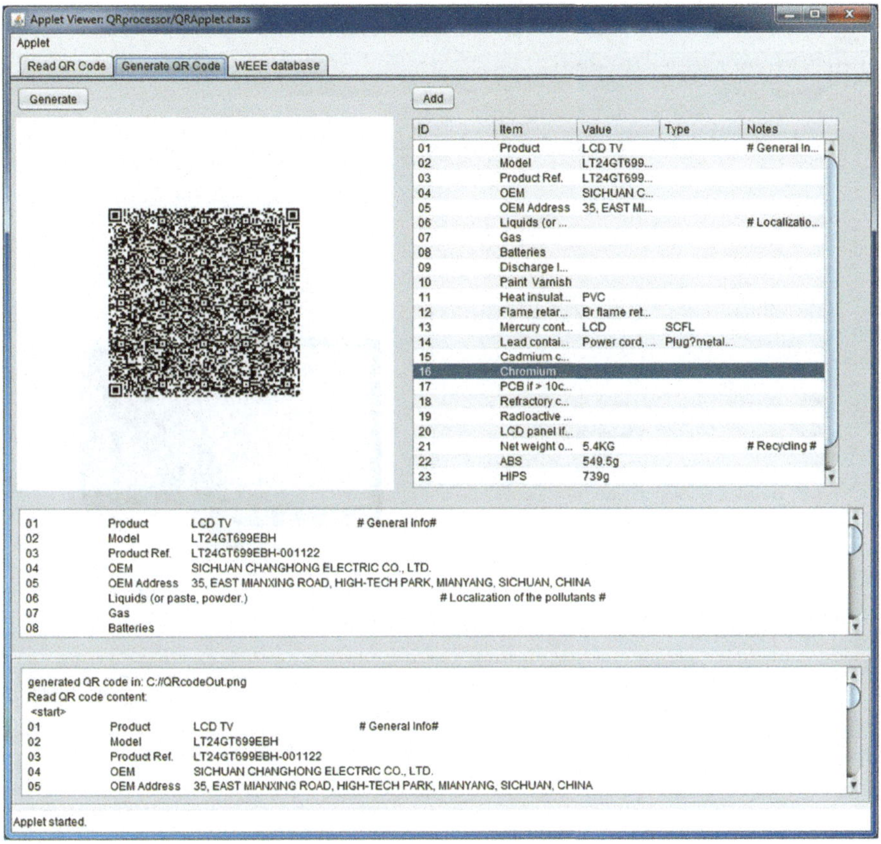


**Fig. 13.7** QR code processors

In the WEEE management module, MySQL databases are also established in the cloud. To better access and maintain the dynamic data of WEEE, a meta-model is adopted in a flat structure. The object-oriented WR2Cloud standard is interpreted into the meta-model. Thus the cloud user is able to understand and maintain the database without the expertise of the standards and schemas. Besides standardised product data, the cloud database also maintains supporting documents, e.g. instructions, designs, disassembly directives, etc.

In the graphical user interface, the user is able to view all the running databases and quickly locate more details of the product specifications based on the unique product ID (see Fig. 13.8). In the study case, the user is able to select from multiple databases, and also maintain the WEEE profile dynamically, including adding, deleting, and updating. In this case, the replacement of the LCD screen is added to the product specifications. When the product needs to be recovered in the future, all the changing and repairing records are integrated and extracted without further efforts.

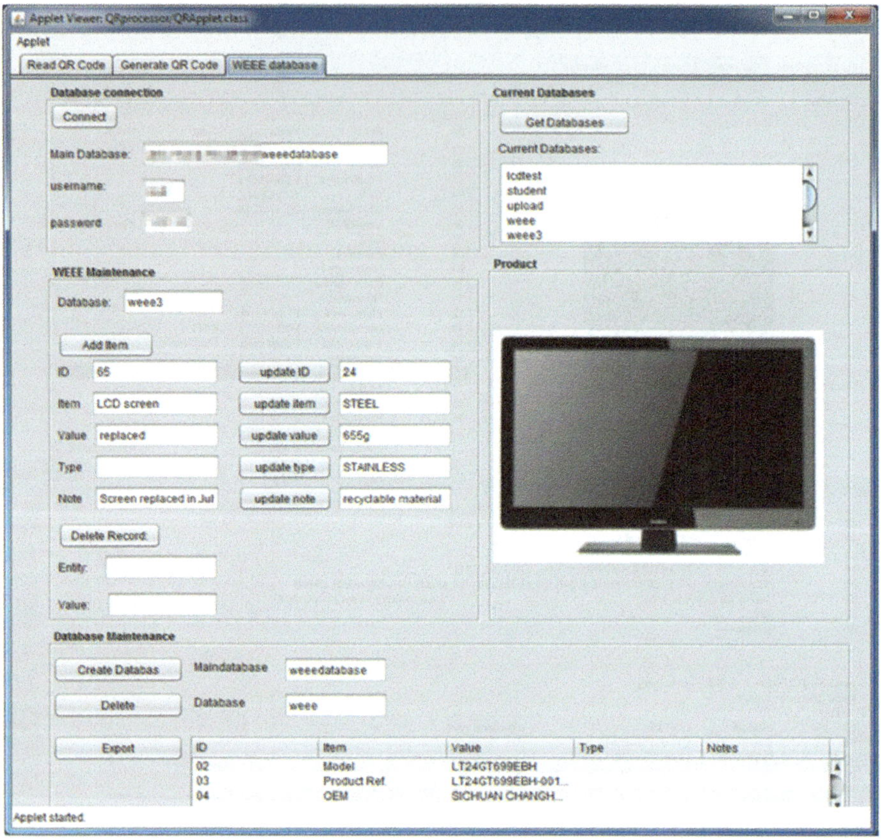


**Fig. 13.8** WEEE management module

### *13.4.2 Case Study 2: Cloud-based WEEE Management at National/International Level*

After the WEEE information is maintained and integrated on the WR2Cloud, this method can be applied at the global level. Take another type of product, Used Lead Acid Batteries (ULAB), as an example: the ULAB case can represent how the information is organised. With the help of WR2Cloud, the decision makers are able to be informed of an item's whereabouts as raw material or product.

In this case study, lead metal and LAB is monitored and managed by the proposed WR2Cloud from a global perspective. The material flow could be explained as follows: currently more than 80% [49] of lead metal productions are used in Lead Acid Batteries (LAB) and their recycling reproduces lead that could be used in future production of LAB, which is called secondary lead production. Another source of lead is from the primary production, which is produced by mining, especially as by-product of Pb-Zn mines. The sum of the primary and secondary production is the total lead metal, which can be utilised by industry for lead acid batteries production. In this way, these two products are interconnected. Furthermore, the secondary production started to act as an important source for lead metal ingot also, due to its high recyclability [50]. Therefore, it is logical to investigate the management of the information about the movements of these two products in order to get the location information of the suppliers and consumers. The major consumers and suppliers in the world are China, USA, UK, Germany, Canada, Japan and India. These countries represent more than 75% of the global lead metal production in 2011. WR2Cloud, in this case, functions as a data and material bank, in order to understand the usage of lead metal. The material flow of lead metal includes production, consumption and export/import, which could be represented by the Physical Trade Balance (PTB). With the help of WR2Cloud, the PTB indicates if a country is a consumer or supplier in principle [51]. Thus, the total consumption of lead in a country is the result from its production plus its PTB. At the international level there are two data sessions maintained in the cloud domain (Fig. 13.9): one corresponds to import/export refined lead as raw material, while the other refers to LAB as a product. The distribution of them during year 2011 is illustrated. In this case WR2Cloud is to monitor and documents these trades.

In conclusion, it can be observed that the biggest producers and consumers of refined lead and LAB are China and USA. However, the quantity that they represent in trade (no more than 10%) is quite small compared with their production, which means that there is a substantial amount of lead located in these two countries. Canada's production of lead is one quarter of that produced in the USA. However, the majority of the production is directly exported to the USA (more than 90%) with no significant material flow to anywhere else.

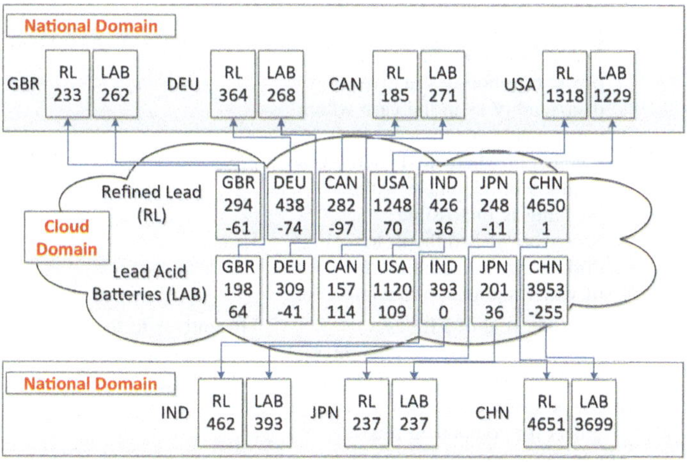


**Fig. 13.9** Lead products cloud

Since all the information from the different countries is properly maintained in WR2Cloud, it is easy to locate the principal markets, producers and consumers. Additionally, it also provides an opportunity to modify the production strategy by considering the financial and environmental impacts. The lifecycle could be completed by using ULAB to produce new LAB that is the currently the principal lead production method. These circumstances would help create regional management that could significantly reduce the environmental impact due to transportation and other informal practices.

## 13.5 Conclusions

During the development of WR2Cloud, interviews were taken with different experts in the fields of component recovery and material recycling within the UK. From these interviews, some barriers for WEEE recovery were identified as follows:

| Barrier | | Solution |
|---|---|---|
| Quantity | The volume of WEEE is huge and the variety is wide | Awareness about WEEE production |
| Recovery processes | – It is difficult to establish universal operations due to the diversity of WEEE in the market<br>– The disassembly process is difficult due to the lack of design considerations for material recovery (recycling) or repairing (product recovery) at EOL. Some joining methods may simplify the manufacturing process and minimise the product size and weight but hinder EOL manipulation by for example preventing non-destructive disassembly as in the case where components are soldered or fused together or where the material is so fragile that dissemble without breaking is impossible<br>– Moreover, it is difficult to differentiate the different materials of which the parts are composed | Incentives to the OEM for design considering component recovery or recycling aspects |
| Lifetime | – The lifecycle of EEE is unpredictable due to different reasons for recovery. Some EEE are destined for recycling and others for component recovery but currently there is no system to | Incentives to the OEM for design considering component recovery or recycling aspects |

(continued)

(continued)

| | | |
|---|---|---|
| | indicate which products or components should be routed towards product recovery<br>– The MOL is often shorter than the design lifetime from OEMs Nondurable materials are widely used in EEE, especially for small electronic devices. These materials may break or sustain other damage during the recovery processes. However, it is difficult to predict in advance the condition of these materials at product EOL and thus set up an appropriate recovery process<br>– The evolution of EEE products is rapid due to the swift changes in market trends. This volatility of EEE products' technology makes it more difficult to find customers for the recovered EEE | |
| Rapid obsolescence | – As the product and process technology for EEE is rapid. Some products are withdrawn because of the arrival of new models/versions as customers may no longer want them because they do not offer the latest functionality. The OEM may prefer to stop their manufacture to make way for the new models in order to compete on a novelty basis with their competitors | Awareness about WEEE production/ recovery |
| Recovery cost | – In some cases, the cost for recovering UEEE is higher than producing or purchasing new alternatives<br>– From the customer's point of view, the price advantage of a recovered product may be little due to labour, resource and facility utilisation required. In other cases, the remanufactured EEE is even more expensive than new ones, despite the environmental benefits<br>– The general preference is directed to new products from the economic point of view<br>– In the case of considering producing competitive UEEE, special facilities are required, which are costly; skilled operators | Incentives to the OEM for design considering component recovery or recycling aspects<br>Incentives for the consumers<br>To consider the EEE, UEEE and WEEE management in a holistic way, as in a hub, where OEM, recovers and recyclers are located closer |

(continued)

(continued)

| | | |
|---|---|---|
| | and technicians are also needed. Currently skilled operators and technicians are not yet available, especially in the component recovery sector<br>– Transportation of WEEE is another gap that is identified. Since WEEE is considered as waste, it contains a large volume of hazardous materials after filtered and centralised. In many countries, specific permission is needed, which increases the cost and effort of transportation | |
| Awareness | – Despite the impacts on environment, it is challenging to convince manufacturers to commercialise WEEE recovering and adopt remanufacturing processes in the current supply chain<br>– The current market still lacks awareness of social and environmental factors<br>– It is also challenging to convince customers to accept recovered EEE, or products containing UEEE components<br>– Since the warranty condition of these products may be changed, it is difficult to persuade consumers to choose recovered EEE or components over premium models. It is necessary to define the warranties respectively, e.g. warranty for the whole product and warranty for the recovered components<br>– Although in some countries the disposal of WEEE is separated from Municipal Solid Waste, the end users still need more assistance for the WEEE recovery, e.g. knowledge of classification and disposal. Especially for recovery services, consumers need to be supported by sufficient information and knowledge regarding the options for component recovery and recycling | Awareness to build customer's purchase motivation as sharing knowledge related to the hazardous materials that could be reduced thanks to the extended lifecycle and the recovery and recycling service |

(continued)

(continued)

| | | |
|---|---|---|
| WEEE chain | – The business model of WEEE recovering needs to be reviewed and improved to encourage more interactions with end users in order to take a bigger market share<br>– It is also difficult to involve end users in the recovery supply chain | Generate a channel where all the stakeholders (manufacturers, collectors, dissemblers, repair shops and remanufacturers) are involved and able to access to the information and find the better path |

In conclusion, the cloud is a promising approach that provides better management and treatment of WEEE. It creates the opportunity of changing waste to valuable UEEE for end-users, with numerous environmental benefits when properly applied. A WEEE recovery/recycling cloud offers an organised platform where not only end-users are able to know how to deal with their used equipment, but also the whole of the supply and resupply chain including manufacturers, recyclers, remanufacturers, repairing shops and collectors are able to interact with the knowledge stored in the cloud.

In this chapter, a cloud-based recovery system is introduced, namely WR2Cloud. The current recovering/recycling resources and capabilities are integrated as cloud services in the service-oriented infrastructure. With the help of the cloud structure, the communication between users and service providers are supported by a distributed, flexible and intelligent network. Data sharing and exchange are further enhanced by the QR code method. It is possible to track and manage the physical flow of WEEEs at both material level and component level. The cloud environment provides a distributed platform to highlight, broadcast and share the advantage of WEEE recovery throughout the EEE supply and resupply chain. With reliable and dynamic information, the effectiveness of WEEE management will be greatly improved. The lead recycling was chosen as a case study to demonstrate how the WR2Cloud could be used in the global management of a substance as raw material and product, especially in order to locate the consumers and producers. Since WR2Cloud can act as the virtual depository of refined lead and LAB due to the fact that they are interrelated for primary and secondary lead production, it would enable the users to decide where would be the most convenient location for the suppliers or consumers. It is expected that, this type of global management could bring environmental and economic benefits.

The interviews undertaken during the development of the WR2Cloud showed barriers related to quantity, recovery processes, and length of lifetime, obsolescence cost, customer awareness and understanding as well as the WEEE supply chain. These issues could be overcome by implementing legislation to ensure better collection channels. Standardised WEEE/UEEE definitions would also assist in creating a filter process before used electrical and electronic products are considered as waste. The commercialisation of WEEE recovery/recycling options could be boosted through awareness, as this would help to gradually increase customers' willingness to purchase recovered WEEE. It also generates a channel in which all participants are involved in the EEE lifecycle. The consumers are able to access the information and find better path for their EEE when it reaches the EOL stage. Hazardous material management is another important consideration. End users should be aware of what is contained in their WEEE and the options of WEEE recovery. The generation of hazardous material can be reduced, thanks to the extended lifecycle and the recovery service from WR2Cloud. Additionally, the recycling of WEEE has an important impact on the environment, because it reduces raw materials utilisation from primary production, which leads to a decrease in the pollution generated from mining and industrial processes.

Future work would include further development of the supervisory mechanism of the recycling and component recovery cloud to integrate and coordinate the services within one package. Quantitative validations can be conducted in particular to compare with conventional approaches in terms of speed, service circle, quality and so forth. Customisation and optimisation solutions can be established with the help of a shared pool of knowledge and resources in the cloud. User-friendly interfaces and mobile phone applications are also needed to evaluate the system and related methods. Mobile apps have particularly become powerful in recent years due to the increase in their mobility and flexibility. It will be helpful for the users to update WEEE data on mobile devices, and to easily link the service case with their location and contact details. As discussed above, the tracking and management of WEEE is difficult due to the lack of data feedback from end users. With the help of the cloud and QR code mechanism, it is possible to establish closed-loop control on the EEE/WEEE flow and achieve a high-level integration of the data and services.

## References


1. US Environmental Protection Agency, *Statistics on the Management of Used and End-of-Life Electronics* (2012). Available http://www.epa.gov/wastes/conserve/materials/ecycling/manage.htm
2. J. Li et al., Regional or global WEEE recycling. Where to go? Waste Manag. **33**, 923–934 (2013)
3. J. Li et al., Heavy metal contamination of surface soil in electronic waste dismantling area: site investigation and source-apportionment analysis. Waste Manag. Res. p. 0734242 X10397580 (2011)
4. A. Sepúlveda et al., A review of the environmental fate and effects of hazardous substances released from electrical and electronic equipments during recycling: Examples from China and India. Environ. Impact Assess. Rev. **30**, 28–41 (2010)
5. X. Yang et al., Pyrolysis and dehalogenation of plastics from waste electrical and electronic equipment (WEEE): a review. Waste Manag. (2012)
6. A. Leung et al., Environmental contamination from electronic waste recycling at Guiyu, southeast China. J. Mater. Cycles Waste Manage. **8**, 21–33 (2006)
7. B.N. Lopez et al., Major pollutants in soils of abandoned agricultural land contaminated by e-waste activities in Hong Kong. Arch. Environ. Contam. Toxicol. **61**, 101–114 (2011)
8. J.-P. Wu et al., Bioaccumulation of polybrominated diphenyl ethers (PBDEs) and polychlorinated biphenyls (PCBs) in wild aquatic species from an electronic waste (e-waste) recycling site in South China. Environ. Int. **34**, 1109–1113 (2008)

9. W. Deng et al., Distribution of PBDEs in air particles from an electronic waste recycling site compared with Guangzhou and Hong Kong, South China. Environ. Int. **33**, 1063–1069 (2007)
10. K. Ni et al., A review of human exposure to polybrominated diphenyl ethers (PBDEs) in China. Int. j hyg. environ. health (2013)
11. G. Zhao et al., Burdens of PBBs, PBDEs, and PCBs in tissues of the cancer patients in the e-waste disassembly sites in Zhejiang, China. Sci. Total Environ. **407**, 4831–4837 (2009)
12. E. Dimitrakakis et al., Small WEEE: determining recyclables and hazardous substances in plastics. J. Hazard. Mater. **161**, 913–919 (2009)
13. F. Ongondo et al., ICT reuse in socio-economic enterprises. Waste Manag **33**, 2600–2606 (2013)
14. V. Goodship, A. Stevels, *Waste Electrical and Electronic Equipment (WEEE) Handbook Cambridge* (Woodhead Publishing, Cambridge, 2012)
15. BS 8887-211. *Design for Manufacture, Assembly, Disassembly and End-of-life processing (MADE)—Specification for Reworking and Remarketing of Computing Hardware* (British Standards Institution, London, 2012)
16. W. Ijomah et al., Development of robust design-for-remanufacturing guidelines to further the aims of sustainable development. Int. J. Prod. Res. **45**, 4513–4536 (2007)
17. M. Sobolewski, Object-oriented service clouds for transdisciplinary computing, in *Cloud Computing and Services Science*, ed. Cham (Springer, Heidelberg New York Dordrecht London 2012), pp. 3–31
18. L. Zhang et al., Cloud manufacturing: a new manufacturing paradigm. Enterp. Inform. Syst. **8**, 167–187 (2014)
19. D. Wu et al., Cloud-based design and manufacturing: a new paradigm in digital manufacturing and design innovation. Comput. Aided Des. **59**, 1–14 (2015)
20. L. Wang, Machine availability monitoring and machining process planning towards cloud manufacturing. CIRP J. Manuf. Sci. Technol. **6**, 263–273 (2013)
21. X.V. Wang, X. Xu, ICMS: a cloud-based manufacturing system, in *Cloud Manufacturing: Distributed Computing Technologies for Global and Sustainable Manufacturing*, ed. by W. Li, J. Mehnen (Springer, London, 2013), pp. 1–22
22. X.V. Wang, L. Wang, From cloud manufacturing to cloud remanufacturing: a cloud-based approach for WEEE recovery. Manuf. Lett. **2014**, 91–95 (2014)
23. L. Wang et al., A cloud-based approach for WEEE remanufacturing. CIRP Ann. Manuf. Technol. **63**, 409–412 (2014)
24. X.V. Wang, X. Xu, An interoperable solution for cloud manufacturing. Robot. Comput. Integr. Manuf. **29**, 232–247 (2013)
25. X.V. Wang, X. Xu, Virtualize manufacturing capabilities in the cloud: requirements, architecture and implementation. Int. J. Manuf. Res. **9**, 348–368 (2014)
26. X. Xu, From cloud computing to cloud manufacturing. Robot. Comput. Integr. Manuf. **28**, 75–86 (2012)
27. L. Wang, Web-based decision making for collaborative manufacturing. Int. J. Comput. Integr. Manuf. **22**, 334–344 (2009)
28. L. Wang, Wise-ShopFloor: an integrated approach for web-based collaborative manufacturing. Syst. Man Cybern. Part C Appl. Rev. IEEE Trans. **38**, 562–573 (2008)
29. Q. Hao et al., Towards a cooperative distributed manufacturing management framework. Comput. Ind. **56**, 71–84 (2005)
30. L. Wang et al., Wise-shopfloor: a web-based and sensor-driven e-shop floor. J. Comput. Inf. Sci. Eng. **4**, 56–60 (2004)
31. L. Wang et al., A Java 3D-enabled cyber workspace. Commun. ACM **45**, 45–49 (2002)
32. H.C. Zhang et al., A web-based system for reverse manufacturing and product environmental impact assessment considering end-of-life dispositions. CIRP Ann. Manuf. Technol. **53**, 5–8 (2004)

33. I. Fidan et al., Intelligent simulation environment for electronics remanufacturing systems, in *Electronics Manufacturing Technology Symposium*, IEEE/CPMT/SEMI 29th International, 2004, 160–164
34. F. Toyasaki et al., The value of information systems for product recovery management. Int. J. Prod. Res. **51**, 1214–1235 (2013)
35. S. Koh et al., Cross-tier ripple and indirect effects of directives WEEE and RoHS on greening a supply chain. Int. J. Prod. Econ. **140**, 305–317 (2012)
36. L. Tompson, S. Chainey, Profiling illegal waste activity: using crime scripts as a data collection and analytical strategy. Eur. J. Crim. Policy Res **17**, 179–201 (2011)
37. A. van Schaik, M.A. Reuter, Dynamic modelling of E-waste recycling system performance based on product design. Miner. Eng. **23**, 192–210 (2010)
38. X. Yang et al., Intelligent products: from lifecycle data acquisition to enabling product-related services. Comput. Ind. **60**, 184–194 (2009)
39. T.C. Kuo, The construction of a collaborative-design platform to support waste electrical and electronic equipment recycling. Robot. Comput. Integr. Manuf. **26**, 100–108 (2010)
40. S. Rahimifard et al., Recycling process planning for the End-of-life management of waste from electrical and electronic equipment. CIRP Ann. Manuf. Technol. **58**, 5–8 (2009)
41. R. Gamberini et al., On the integration of planning and environmental impact assessment for a WEEE transportation network—a case study. Resour. Conserv. Recycl. **54**, 937–951 (2010)
42. C. Achillas et al., Decision support system for the optimal location of electrical and electronic waste treatment plants: a case study in Greece. Waste Manag **30**, 870–879 (2010)
43. C. Achillas et al., Optimising reverse logistics network to support policy-making in the case of Electrical and Electronic Equipment. Waste Manag **30**, 2592–2600 (2010)
44. Z. Che, Using fuzzy analytic hierarchy process and particle swarm optimisation for balanced and defective supply chain problems considering WEEE/RoHS directives. Int. J. Prod. Res. **48**, 3355–3381 (2010)
45. ISO, ISO 10303 -1: *Industrial Automation Systems and Integration—Product Data Representation and Exchange—Part 1: Overview and Fundamental Principles* (ISO, Geneva, 1994)
46. ISO 14649-1. Industrial Automation Systems and Integration—Physical Device Control—Data Model for Computerized Numerical Controllers—Part 1: Overview and Fundamental Principles, 2003
47. L. Swanson, Linking maintenance strategies to performance. Int. J. Prod. Econ. **70**, 237–244 (2001)
48. ISO/IEC 18004: Information Technology—Automatic Identification and Data Capture Techniques—QR Code 2005 bar code Symbology Specification, (2006)
49. CHR, Global end-use breakdown of the lead consumption. Pers. Commun. (2014)
50. International Lead Association ILA, *Lead Recycling, Sustainability in Action*. Lead Action 21 (Oct 2014). Available http://www.ila-lead.org/UserFiles/File/ILA9927%20FS_Recycling_V06.pdf
51. M. Dittrich, S. Bringezu, The physical dimension of international trade: Part 1: direct global flows between 1962 and 2005. Ecol. Econ. **69**, 1838–1847 (2010)

# Chapter 14
# Big Data Analytics for Scheduling and Machining

## 14.1 Introduction

### *14.1.1 Algorithms Used in Big Data Analytics*

Methods employed in big data analytics are basically developed from the traditional approaches for data analytics. The principal methods are summarised as follows.

- Cluster analysis: It captures the natural structure of data. Originated in anthropology in 1932 and introduced to psychology in 1938 [1, 2], it was then used for trait theory classification of personality psychology in 1943 [2].
- Factor analysis: It is a statistical method used to describe variability among observed and correlated variables in terms of a potentially lower number of unobserved variables called factors [3].
- Analysis of correlation and dependence: In statistics, dependence or association is a statistical relationship between two random variables or two sets of data. Correlation is any of a broad class of statistical relationships involving dependence, though it was usually used to refer to the linear relationship between two variables. Similarly, dependent phenomena include the correlation between the parental physical statures and the offspring, and the correlation between the demanded product and price of product.
- Regression analysis: It is a statistical process for estimating the relationships among variables. It includes many techniques for modelling and analysing several variables, when the focus is on the relationship between a dependent variable and one or more independent variables.
- A/B testing: It is a term of a randomised experiment with two variants, *A* and *B*, which are the control and variation in a controlled experiment [4].
- Data mining: It is an interdisciplinary subfield of computer science. The computational process discovers patterns in large datasets involving methods at the intersection of AI, machine learning, statistics, and database systems [5].



L. Wang and X. V. Wang, *Cloud-Based Cyber-Physical Systems in Manufacturing*,
https://doi.org/10.1007/978-3-319-67693-7_14

**Table 14.1** Most popular tools used in big data analytics [6]

| No. | Tool | Alone (%) | User ratio (%) | |
|---|---|---|---|---|
| | | | 2014 | 2013 |
| 1 | RapidMiner | 35.1 | 44.2 | 39.2 |
| 2 | R | 2.1 | 38.5 | 37.4 |
| 3 | Excel | 0.1 | 25.8 | 28.0 |
| 4 | SQL | 0.1 | 25.3 | n/a |
| 5 | Python | 0.9 | 19.5 | 13.3 |
| 6 | Weka | 0.4 | 17.0 | 14.3 |
| 7 | KNIME | 10.6 | 15.0 | 5.9 |
| 8 | Hadoop | 0 | 12.7 | 9.3 |
| 9 | SAS base | 0 | 10.9 | 10.7 |
| 10 | Microsoft SQL server | 0 | 10.5 | 7.0 |

### *14.1.2 Tools Used in Big Data Analytics*

The tools used in big data analytics have been collected by K.Dnuggets™ News, a popular site on business analytics, data mining, big data, and data science etc. [6]. As shown in Table 14.1, 1453 of 3285 voters mentioned that they used RapidMiner, while 35% of RapidMiner users indicated that RapidMiner was the single software that they used.

## 14.2 Background Information

### *14.2.1 Related Works on Scheduling*

The research on scheduling started with static scheduling useful for the development of shop-floor scheduling systems for mass production. A mathematical model was developed by Manne [7]. Laporte et al. [8] analysed two integer linear programming formulas to address job sequencing and tool switching problems, in which branch-and-cut and branch-and-bound algorithms were developed and compared. Asokan et al. [9] used adaptive genetic algorithm and particle swarm optimisation to obtain optimal schedules and storage assignments.

The concept of integration of both process planning and scheduling was developed by Khoshnevis and Chen [10]. In their method, the feasible process for a given feature of a part was not found in the shop, in case of which, the real-time system was then applied to generate process plans and schedules together. Within the context, Chen and Khoshnevis [11] also presented some methods for the integrated system and the performance of the algorithm. Tan and Khoshnevis [12] further extended the approach. Mohapatra et al. [13] proposed an improved controlled elitist non-dominated sorting genetic algorithm to reduce scheduling objectives, for instance makespan, cost, idle time and efficiency for the integration of process planning and scheduling. Freitag and Hildebrandt [14] proposed a simulation-based multi-objective hyper-heuristic to develop optimisation dispatching rules for complex manufacturing systems. Li et al. [15] pointed out that the integration of process planning and scheduling would be developed towards multi-objectives, dynamic and hybrid algorithm application. Rajkumar et al. [16] applied greedy randomised adaptive search procedures algorithm to the integration of process planning with production scheduling, with regard to the process problems having multi-objectives of makespan, maximum workload, total workload, tardiness and total flow time. In addition, scheduling was also carried out for some of special objectives, e.g. sustainable development. Gahm et al. [17] suggested developing energy-efficient scheduling for manufacturing companies.

However, the static scheduling does not have the capacity to handle the situation with growing products of both small batch and wide variety, in particular unexpected faults. Therefore, dynamic scheduling method was developed, where the decisions can be made with rapid response automatically. Real-time scheduling of a manufacturing system involves scheduling and revised scheduling [18]. Vieira et al. [19] proposed a rescheduling method based on a wide variety of experimental and practical approaches. In the method, two common strategies were introduced, dynamic scheduling and predictive–reactive scheduling. Ham et al. [20] proposed a three-stage flexible job-shop scheduling method to deal with unpredictable system disturbances. Iwamura et al. [21] introduced an estimation of future status based real-time scheduling approach for holonic manufacturing systems (HMS). In their method, the future status of an HMS is predicted by applying a neural network model based simulation model. In addition, an agent-based service-oriented architecture was presented for real-time distributed shop-floor scheduling [22]. Semi-Markov decision models were also applied to real-time scheduling by Yih and Thesen [23]. Ant colony optimisation was applied to two dynamic job scheduling by Zhou et al. [24]. Within real-time scheduling, real-time decision making is a key. There are many algorithms. By constructing a decision tree, Metan et al. [25] proposed a new scheduling system for selecting dispatching rules in real-time. The proposed scheduling system was developed by combining the techniques of simulation, data mining, and statistical process control charts. Bayesian algorithm was used to discover priority dispatching rules from large amounts of structured or unstructured data for the single machine scheduling problem [26]. In addition, real-time monitoring methods were also developed. Kohn et al. [27] proposed repair-control of manufacturing systems using real-time RFID information. Through applying RFID on shop floor, the real-time information of objects including operators, machines and materials, can be automatically captured, bound and synchronised with manufacturing orders [28, 29]. According to the literature, dynamism, flexibility and adaptability are the important features in modern scheduling, and a scheduling system should be able to perform task rearrangement in case of unexpected events.

### 14.2.2 Related Works on Machining Optimisation

Nowadays, machine tool selection is basically out of the scope of process planning due to limited machine tools on a shop floor; however, in the cloud manufacturing environment, there are so many machine tool resources that can be selected. For machining condition generation and cutting tool selection, there are two common ways: (1) in most of reported process planning methods, cutting tool is regarded as a standard machining resource, where machine tool and cutting tool optimisation is not considered; and (2) machining conditions are optimised after tool selection [30]. The three decisions on machine selection, cutter selection and machining parameter assignment are made sequentially during process planning, hindering the loss of both machining accuracy and efficiency.

Machining process optimisation started with mathematical model based methods that were popular in the 1990s. Chua et al. [31] proposed a series of mathematical formulations to optimise cutting conditions and to reduce operation time. Yeo [32] developed a multi-pass optimisation methods for a CNC lathe, in which near-optimum solutions were obtained. Akturk and Avci [33] presented a hierarchical method for a CNC machine tool. In their method, the mathematical models were established regarding system characterisation, to minimise the total production cost. Lazoglu and Liang [34] developed a model of the mathematical characterisation of cutter-workpiece interaction to plan machining operations and to optimise cutting parameters.

Experimental methods were applied for some specific aims. Yang et al. [35] applied a Taguchi experiment method to optimise the cutting parameters regarding the groove difference and the average roughness. Chen et al. [36] proposed an experimental plan of a four-factor D-optimal design to obtain the optimal spindle speed, feed rate, cutting depth, and the status of lubrication concerning vibration and surface roughness in precision turning. The cutting tool geometries were optimised in the high feed rate experiment considering surface roughness and cutting force [37]. Zhang et al. [38] proposed an experimental optimisation method in turning of hardened steel.

The machining optimisation problem, regarded as a search problem, was treated based on search algorithms in which tabu search (TS) is a popular one. A constraint-based TS approach was applied in optimisation of the processes of selecting machining resources, determining setup plans and sequencing operations for process planning of a prismatic part [39]. Taiber [40] proposed a set of modified algorithms from the field of combinatorial search problems, gradient projection method named as von Rosen, branch and bound algorithm, and shortest common super sequence algorithm, etc. The method was to assist a human planner fulfil machine tool and cutter selection, determination of setup and process sequence, definition of tool paths and optimisation of cutting parameters. In addition, many other search algorithms were applied, e.g. throughput profit [41], simulated annealing (SA) [42], search heuristics based on SA and TS [43], branch and fathoming algorithms [44], iterative approach [45], evolutionary strategy based optimisation [46], and harmony search [47].

Genetic algorithm (GA), a most basic evolutionary algorithm, was a popular method used for optimisation. Swarm intelligence based optimisation algorithms are also popular in process planning, e.g. particle swarm optimisation (PSO), ant colony optimisation (ACO) and honey bees mating optimisation. Moreover, expert systems were developed to utilise the machining knowledge. In order to minimise the cost of machining, Gupta et al. [48] developed an expert system based model to sequence operations among a set of machines, select cutting tools, and determine process parameters. Data classification methods, e.g. decision trees and artificial neural networks (ANN), were also applied. Sluga et al. [49] developed a decision tree based method to predict tool features, cutting geometry and cutting parameters to improve and automate the tool selection and determination of cutting parameters. Monostori et al. [50] used a general ANN based process to satisfy various requirements.

In general, process planning is treated as an NP-hard problem. Hybrid methods are therefore applied to relax the limitation of one singe algorithm. Li et al. [51] developed a genetic algorithm and simulated annealing approach to optimising process plans for prismatic parts. Wong et al. [52] proposed a fuzzy expert system and GA to sequence machining process. A hybrid GA and intelligent search method was also reported [53] and applied to optimise machine tool, cutting tool and tool access direction for each operation. Moreover, Petrovic et al. [54] utilised PSO algorithm and chaos theory to optimise process plans, in which PSO was used in early stages of the optimisation process by implementing ten different chaotic maps that enlarged the search space and provided diversity.

From the literature, optimisations in machining have been focused on specific cases and processes due to the limitations of traditional physical and experimental based methods in terms of high-dimensional data optimisation and high-accuracy optimisation. Nowadays, hybrid methods and big data analytics popularly used in many other areas show promise of providing high-accuracy solution strategies, in particular, for optimisation problems of high-dimensional data. Big data analytics combining hybrid algorithms for integrated optimisation of cutting tools and machining conditions are explained hereafter.

### *14.2.3 Big Data Analytics Application*

Although limited in achievements, big data analytics shed lights in fault prediction. In big data research, many reported work have focused on applications of big data in production lifecycle and supply chain management [55]. Zhang et al. [56] proposed a big data-based analytics for product lifecycle, supply chain management and maintenance of complex product, where big data analytics and service-driven patterns were used. Zhong et al. [57] reviewed the state of big data technology used in services and manufacturing supply chain management, including six aspects of challenges, opportunities and future perspectives: data collection, data transmission, data storage, processing technologies, big data-enabled decision-making models, and big data interpretation and applications. Babiceanu and Seker [58] reviewed relevant research and indicated that big data analytics will be used for cyber-physical manufacturing systems. Woo et al. [59] developed a big data analytics platform for manufacturing systems, in order to create prediction models specific for the target machine tools. In summary, big data analytics has been used widely and shows the potential of fault prediction in shop-floor scheduling [60].

## 14.3 Big Data Analytics for Shop-Floor Scheduling

### *14.3.1 Big Data Analytics Based Fault Prediction in Scheduling*

The main aim of a manufacturing shop floor is to finish and deliver products timely and in good quality. The main elements on the manufacturing shop floor can be divided into several categories, as shown in Fig. 14.1, according to their functions and properties.

- Physical operation level: human, robots, machine tools, forklift, automatic guided vehicle (AGV), and accessories (fixture and cutting tool)
- Physical monitor and control level: monitoring equipment, control equipment, and computers (connecting with the controller of robots or machine tools)

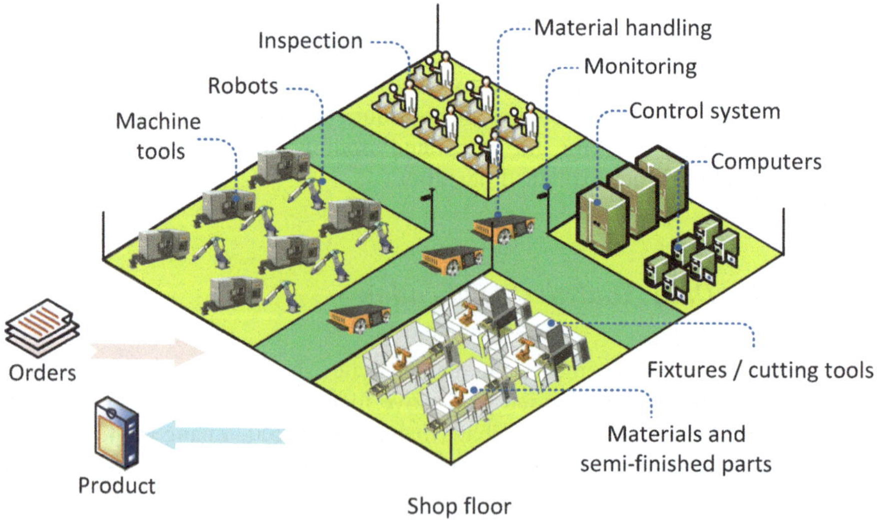


**Fig. 14.1** Main elements on a manufacturing shop floor

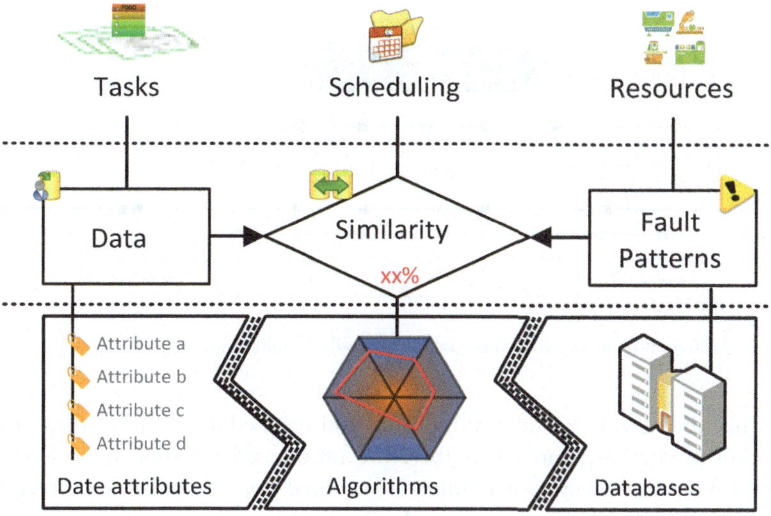


**Fig. 14.2** Concept of fault prediction in scheduling

- Product level: materials (including blanks), semi-finish products, and products
- Cyber level: nets and database

On a real manufacturing shop floor, the main tasks include task scheduling, material handling, machining and inspection. Basically, the tasks of scheduling consider currently available equipment where cost and time are the main concerns. However, the condition prediction of machines is not included in the considerations due to the limitation of traditional techniques.

As shown in Fig. 14.2, the planned tasks are compared with the mined fault patterns from shop-floor databases, as a result of which, the similarity used to make a decision can be obtained. Real-time data are collected as well and compared with the mined fault patterns by combining with the data attributes of the ongoing tasks. Consequently, the similarity as a reference for rescheduling the remaining tasks can be obtained. Within the context, the historical data related to the shop floor are collected in advance of prediction and used in data analysis for fault patterns generation. Tasks are represented by relevant data attributes, and then the similarity or difference are calculated by comparing the mined fault patterns with the planned task. Moreover, the risk probability of the planned task can be obtained, which provides a reference for the final scheduling decision making.

### *14.3.2 System Architecture*

This section addresses two challenges: (1) to avoid any mismatch of machining tasks on machines before scheduling, and (2) to prevent faults from happening

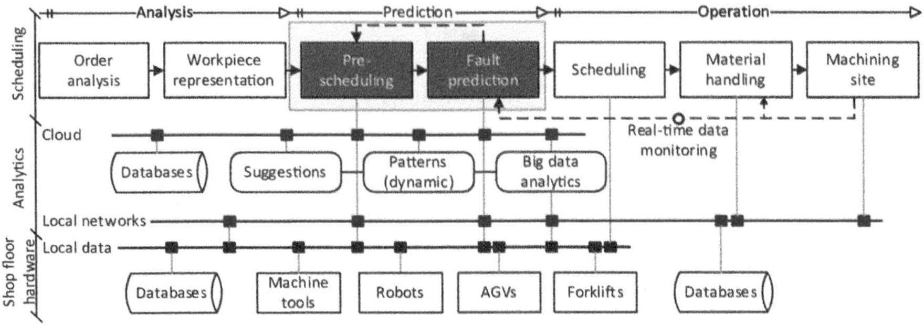


**Fig. 14.3** Architecture of big data analytics in shop-floor scheduling

during machining via real-time monitoring and rescheduling. A big data analytics based fault prediction approach is therefore introduced for shop-floor scheduling. Figure 14.3 depicts the system architecture of three steps in horizontal direction and three levels in vertical direction. The *Analysis* step consists of order analysis and workpiece representation. The *Prediction* step contains pre-scheduling and fault prediction. These two steps are introduced before scheduling in order to predict potential machining faults due to machine conditions and machining status via real-time monitoring. In the last *Operation* step, scheduling, material handling and machining take place. During machining, real-time data are collected and passed to the fault prediction module for fault predictions. Moreover, if there are still unpredicted errors during machining, the error message is sent to the pre-scheduling module, and the unfinished task is re-arranged for defect prevention.

In the analytics level, big data analytics is carried out resulting in the patterns used by fault prediction are discovered. Finally, suggestions if any are given to the shop floor operations. In the shop-floor hardware level, equipment such as machine tools, robots, AGVs and forklifts are connected by a local area network. For the sake of page limit, only fault prediction and big data analytics are further explained.

#### 14.3.2.1 Fault Prediction

Figure 14.4 shows the workflow of fault prediction, including initialisation, pre-scheduling, prediction, termination criteria judgment, and error probability evaluation. Initial parameters are set up during initialisation, e.g. maximum number of calculation (generation) and the probability. Then the task plans are generated (by using the same procedures as in normal scheduling) in the pre-scheduling process, based on which the error probabilities in statistics level are calculated for the planned task in the prediction process. If the error probabilities are greater than the initial one $P_I$, the high-risk tasks will be rescheduled. However, the task plans may still be undecided after the number of generations reaches the initially-defined threshold value $N_I$. In such a case, the values of the initial parameters must be properly adjusted.

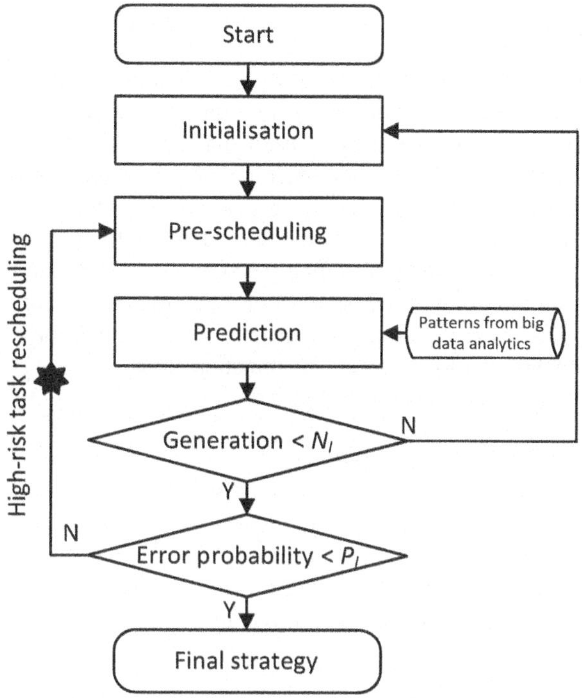


**Fig. 14.4** Workflow of fault prediction

Among the processes, the fault prediction is the key, where potential error patterns are mined by big data analytics based on the databases of the shop floor. Together, they provide a reference for scheduling decision making. The potential patterns from big data analytics are as follows:

- Potential faults of machine and workpiece: it refers to the fault probability of a class of machine tools or a single machine tool before task scheduling, and potential errors during machining regarding real-time data.
- Maintenance state of machine and workpiece: maintenance state of machine tool is related to its usage time, basically. However, if an unsuitable task is scheduled for the machine, the machine should be maintained in advance.
- Machining quality of workpiece and machine tool: machining quality is coupled with the state of the machine tool, e.g. stiffness, repeatability, and stability. Therefore, if the state of the machine tool is not suitable to machine a workpiece, the task should not be assigned to the machine.

#### 14.3.2.2 Data Treatment

Considering the size of data, data should be divided and stored in different databases, i.e. local data, local network data and cloud data, as shown in Fig. 14.5.

Local data are stored in distributed computers, including both historical data and real-time data from monitoring systems, and covering the information of machining features, machine tools, processes, operators, quality measurement and time. Local data analysis considers the patterns of each single machine tool with respect to the special characteristics of each machine tool. The local network data refer to the historical data covering machine tool class, parts, time and technical capability (one kind of evaluation based on process, operator and machining quality in the local data level). Local network data analysis focuses on the patterns of machine tool classes with respect to their performances. The cloud data also refer to the historical data involving time, shop floor level, manufacturing capacity and tasks. Cloud data analysis is to obtain the pattern of a shop floor, which is used to evaluate the manufacturing capacity of the shop floor.

#### 14.3.2.3 Data Attributes

The attributes of big data include the information regarding machine tools, workpiece, machining processes, machining time, machining results and operators (see Fig. 14.5). The details of those information cover all the factors affecting the manufacturing tasks. A sample data sheet is depicted in Table 14.2. The details of the data attributes include:

1. Data attributes of workpiece: workpiece information includes the quantity of parts, part geometries, part materials and machining requirements, etc. Here, the part geometries and quantity can be represented by machining features and their quantity. Part materials refer to the material profiles with respect to the material machinability, e.g. hardness, brittleness, and plasticity, etc., so that every material can be represented by a series of parameters.
2. Data attributes of machining time: machining time refers to the duration of machining period which is related to the usage of a machine tool, and also the machining time, where machining time includes two cases: (1) uncovered shop floor: environmental factors change over time, e.g. the temperature in the

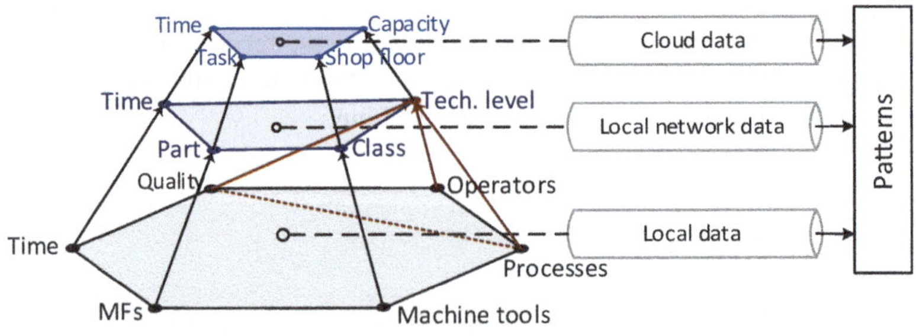


**Fig. 14.5** Data stored and processed at three levels

**Table 14.2** A sample data sheet

| No. | Workpiece | | | | | | | |
|---|---|---|---|---|---|---|---|---|
| | Quantity | MFs | Volume per MF ($cm^3$) | Hardness (HRC) | Yield strength (MPa) | Density ($g/cm^3$) | Roughness (Ra) | Accuracy (mm) |
| 000001 | 20 | Face | 100 | 35 | 690 | 7.9 | 1 | ±0.02 |

| Time | | Machine tool | | | | | |
|---|---|---|---|---|---|---|---|
| When | Duration (min) | No. | Type | Structure | X\Y\Z error (mm) | A\B\C error (mm) | XYZ power (kw) |
| 14:30 | 30 | M15 | 4-axis | XYZBC | 0.0005\0.0008 \0.0007 | 0.0\0.0015 \0.001 | 2.8\3.0\3.5 |

| Machine tool | | | | Machining process | | | | |
|---|---|---|---|---|---|---|---|---|
| Spindle power (kw) | Maintenance date | Fault date | Fault type | Cutting speed (m/min) | Feed (mm/ min) | Cutting depth (mm) | Cutting width (mm) | Fluid type/ pressure |
| 26 | 2016-03-05 | 2016-02-28 | A | 50 | 2000 | 0.2 | 20 | F01/ 30 bar |

| Machining process | | | | | | | | |
|---|---|---|---|---|---|---|---|---|
| Tool type | Tool mate. \coat | Tool radius (mm) | Tool length (mm) | Entrance angle (°) | Rake angle (°) | Flank angle (°) | X\Y\Z cutting force (N) | X\Y\Z vibration ($m/s^2$) |
| IT | K2\ TiCN | 30 | 50 | 60 | 10 | 10 | 400\400 \300 | 40\35\20 |

| Machining result | | | | | Human factor | |
|---|---|---|---|---|---|---|
| Error (mm) | Ra (μm) | White layer (μm) | Dark layer (μm) | Hardened layer (μm) | Operator | Level |
| 0.01 | 0.8 | No | No | No | 0001 | T02 |

morning is lower than in the afternoon; (2) closed shop floor: the controlled environmental factors include temperature, humidity, and dust content, etc.

3. Data attributes of machine tools: machine tool information usually involves the number of machine tools, machine tool types, machine tool structures, power of spindle, linear axis and rotational axis, errors of each axis, and energy consumption of each component.
4. Data attributes of machining processes: machining processes refer to the cutting conditions, cutting tools, cutting accessories and physical data of cutting process. Cutting conditions include cutting parameters, cutting fluid and its pressure. Cutting tools involve tool types, tool materials, tool coats, and tool geometrical parameters. Cutting accessories include fixture types, and their feature parameters. The physical data consist of cutting force, cutting vibration, and cutting temperature, etc.
5. Data attributes of machining results: machining results mean the machining qualities in terms of geometrical errors and surface integrity. The geometrical errors include geometric and dimensional errors against the nominal tolerances. Surface integrity indicates surface roughness, surface morphology and

subsurface layer qualities (e.g. white layer, dark layer, grain deformation layer, and residual stress).

6. Data attributes of human factors: human factors refer to operators, especially the workers who perform the machining tasks manually. The skills of the operators are the core factors, closely related to the machining results.

#### 14.3.2.4 Data Cleansing

High-quality data increase the accuracy of prediction. The data quality depends on a set of quality criteria [61]: (1) validity: data constraint is the key part and there are many types of constraints like data type, range, mandatory, uniqueness, set-membership, foreign-key, regular expression patterns, and cross-field validation; (2) decleansing: error detection and syntactically removing; (3) accuracy: it refers to the degree of conformity of a measured value to a standard or a true value; (4) completeness: it mentions all the required measures that should be known; (5) consistency: the degree to which a set of measures are equivalent across all systems; and (6) uniformity: the degree to which a set of data measures are specified using the same units of measures in all systems.

Data cleansing, also called data cleaning, is the process of correcting or removing inaccurate records from databases. The main processes of data cleansing consist of data auditing, workflow specification, workflow execution, post-processing and controlling [61]. Here, in the data auditing process, the statistical methods are applied to detect anomalies and contradictions in the databases. In order to gain information about the existing anomalies in the data collection, detection and elimination are carried out by a sequence of operations on the data, in workflow specification. After the data cleansing workflow is executed, the results are inspected to verify the correctness during post-processing and controlling. The methods for data cleansing are parsing, data transformation, integrity constraint enforcement, duplicate elimination and statistical methods.

#### 14.3.2.5 Data Integration

Databases are built for each individual machine tools, where the real-time data from the sensors embedded in the machine tools are collected and stored in the databases, together with the historical data. Then the local databases are obtained for each machine class by integrating the machine tool databases. Based on the machine class databases, a shop-floor database can be generated. In this process, data integration is the key operation from databases to data warehouse, and it includes combining data residing in different sources and providing users with a unified view of these data [62]. Data integration takes place with increasing frequency as the volume and the need to share existing data explodes.

In the process of data integration, a specific rule related to the shop floor should be followed, i.e. the unique patterns of individual machine are kept for scheduling of the machine, and the shared patterns of the machine classes are used for high-level scheduling.

#### 14.3.2.6 Process Operators

Among big data analytic methods, data mining is used widely and can fully satisfy the requirements of fault prediction for shop-floor scheduling. Generally, there are two types of approaches used in data mining: classification and clustering. In classification, the task is to assign instances to pre-defined classes, whereas in clustering, the task is to group related data points together without labelling them. Classification is considered an instance of supervised learning, while clustering is considered an instance of unsupervised learning. Therefore, classification is suitable for error prediction on shop floor. There are many algorithms used as classifiers, e.g. decision tree [63], Naïve Bayesian classifier [64], Bayesian network [65], artificial neural network [66], support vector machine (SVM) [67], frequent pattern [68], lazy learner [69], genetic algorithm [70], rough set [71], and fuzzy set [72].

There are other important processes for implementing big data analytics based fault prediction, e.g. accuracy and error measures for classifier and predictor, accuracy evaluation, ensemble methods (for accuracy improvement), and model selection [73]. By following the processes carefully, the accuracy of fault prediction can be guaranteed.

### *14.3.3 A Simplified Case Study*

A simplified proof-of-concept case is illustrated to show the process of the proposed method. This case only focuses on the fault prediction at the beginning of scheduling. The processes of the fault prediction regarding real-time monitoring and maintenance state are similar to the prediction at the beginning of scheduling, at least at the big data analytics level.

On a shop floor, there are four machine tools, M1–M4. Two machining tasks have been arranged to the shop floor (Fig. 14.6). Two machining features (MFs, one for each task) are selected together with material hardness, required tolerance and roughness, and delivery deadline (see Table 14.3).

Due to the fact that meaningful big data are not available to the authors at the time of performing the case study, a hypothetic data sheet is generated randomly (Fig. 14.7a) to test the system. The generated data are imported to RapidMiner where the decision tree operators are applied (Fig. 14.7b). The running process and results are depicted in Fig. 14.7c. According to the results, if Task 1 is arranged to

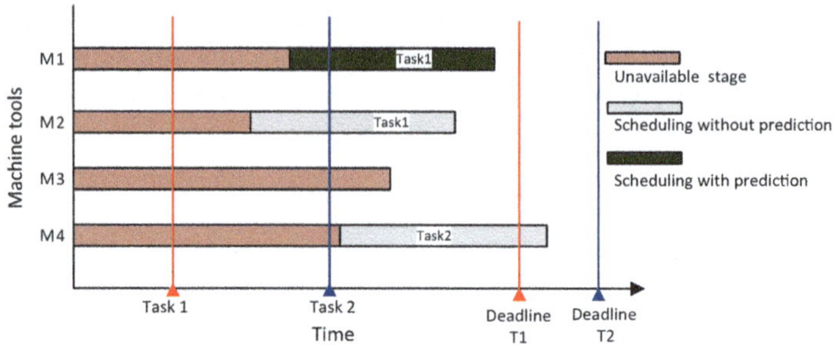


**Fig. 14.6** Gantt chart of two machine tools

**Table 14.3** Details of two machining tasks

| Task | MF | Hardness (HRC) | Tolerance (mm) | Roughness (Ra) | Deadline |
|---|---|---|---|---|---|
| 1 | Side | 31 | 0.005 | 1.2 | T1 |
| 2 | Slot | 45 | 0.05 | 0.8 | T2 |

M2, the error probability would be over 50% (indicated by the red line in Fig. 14.7c), while it is feasible if Task 2 is arranged to M4 (see the blue line in Fig. 14.7c). Therefore, the risk of the current plan is too high, hindering the timely delivery. Machine tool carrying out Task 1 is thus rescheduled to M1 as shown in Fig. 14.6. According to the prediction result, the error probability drops to 0% for the new arrangement (shown in the green line in Fig. 14.7c).

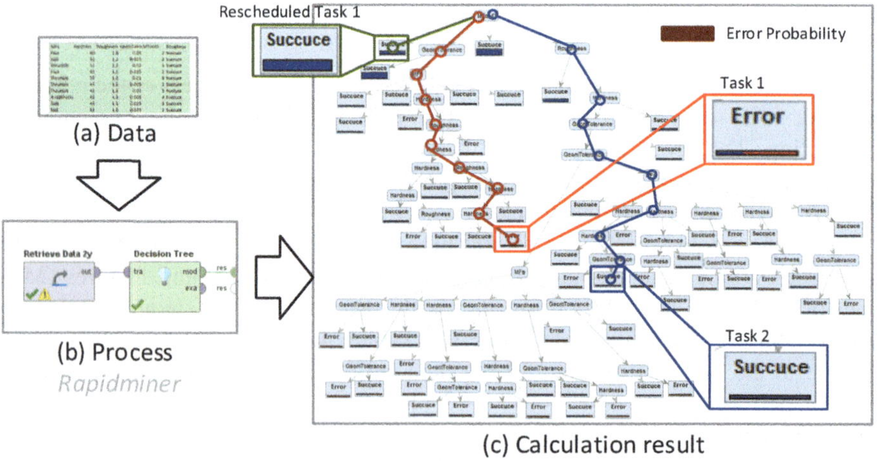


**Fig. 14.7** Result of classification using decision tree

## 14.4 Big Data Analytics Based Optimisation for Machining

### *14.4.1 Analysis of Machining Process*

In manufacturing industry, orders are made to selected manufacturers, produced by machine tools, and then delivered [74], as shown in Fig. 14.8. A manufacturer concerns machining efficiency, cost and quality, whereas in machining, the right decisions of cutting conditions and cutting tools are the major objectives. A cutting tool refers to tool geometries, tool quality, tool material, and the match between the geometry and the material. Machining conditions include cutting parameters, cutting conditions, and tool position. Performance related force, heat and deformation consist of chip control, dynamics, wear, fracture, etc., and they relate the machining conditions to the cutting tool. There are some connections, e.g. geometric constraint, cutting force distribution, and cutting layer, etc. among cutting tool, machining condition, and performance factors, as shown in ❶–❻ of Fig. 14.8, and their details are described as follows:

- Geometric constraint ❶: it refers to the geometric interferences constraining the process of cutting, such as, the description of relationships between tool geometries and chips.
- Cutting force distribution ❷: it is related to cutting vibration and cutting stability, and used to establish the relationship of cutting force with cutting tool, cutting parameters and tool positions.

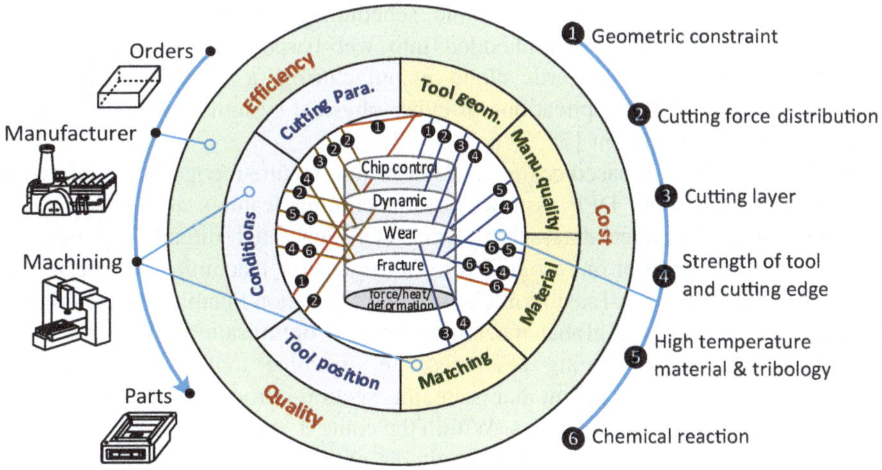


**Fig. 14.8** Relationship between cutting tool and machining condition

- Cutting layer ❸: it concerns material deformation and cutting area analysis within the generations of chip and machined surface, e.g. the relationship between tool geometries, cutting parameters and tool wear.
- Strength of tool and cutting edge ❹: it is to describe the strength of tool and cutting edge in cutting process, which refers to the safety of cutting tool and workpiece. For instance, it can be used to descript the relationship between tool fracture, cutting parameters, tool geometries and tool materials.
- High temperature material and tribology ❺: it is related to material properties and friction with another material in high temperature, and it deals with the interaction between tool wear, tool fracture and cutting conditions.
- Chemical process ❻: it refers to some of chemical reactions which may take place during machining to avoid unsuitable patterns related to tool material, workpiece, cutting fluid, and atmosphere.

## 14.4.2 Enriched Distributed Process Planning (DPP)

### 14.4.2.1 DPP Concept

As described in Chap. 5, the system architecture of DPP consists of supervisory planning, execution control and operation planning, as shown in Fig. 14.9. In this design, the execution control module is placed in-between the supervisory planning and operation planning modules, and looks after jobs dispatching (in the unit of setups) based on up-to-date monitoring data, availability of machines and scheduling decisions [75, 76]. Distribution is a key feature of DPP. Combining with web-based knowledge sharing, dynamic scheduling, real-time monitoring and remote control, DPP can be embedded into web-based environment, which is named Web-DPP [77]. Towards cloud manufacturing, a Cloud-DPP was also developed as one of the applications of cyber-physical systems for more complex manufacturing environment [78, 79].

Within DPP, feature-based design and machining feature recognition are beyond the scope of this book. DPP assumes that machining features are already made available in the product data—they are either generated directly by using a feature-based design tool or recognised by a third-party machining feature recognition solution. However, Execution Control and Operation Planning in the original DPP do not consider the global machining process optimisation due to the complexity of relevant machining resources, i.e. machine tool, cutting tool and machining conditions. Targeting that issue, this section introduces an enriched DPP combining it with big data analytics. Within the context, the suitable machine tool is selected in Execution Control, and the suitable cutting tool and machining conditions will be selected and optimised accordingly.

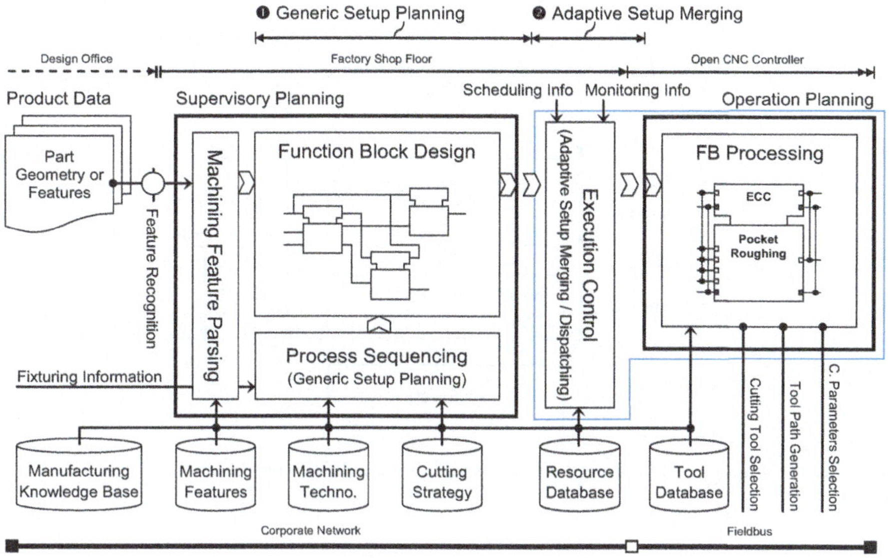


**Fig. 14.9** System architecture of DPP, adapted from Ref. [76]

#### 14.4.2.2 Enriched DPP

Machine tool selection, cutting tool selection and machining condition decision are the major processes after Supervisory Planning in the original DPP (Fig. 14.9). They are treated separately by using the existing methods (experimental- and physical-based), as shown in Fig. 14.10, where their relationships are ignored, resulting in the feasible sets of those elements are bounded. Targeting this issue, the present method is to address the whole process from customer orders to final parts, and to develop a generalised system in which those processes are merged together, as shown in Fig. 14.10. The optimisation can thus be treated as an integrated one towards the globally best solution.

### *14.4.3 Solution Strategy of Enriched DPP*

#### 14.4.3.1 Problem Transformation

The machining process is transformed into a statistic problem in the enriched DPP. The available parameters of machine tool, cutting tool and machining condition form the original solution space $S_o$ with multi-dimensional data, as shown in Fig. 14.11.

The optimisation in the enriched DPP includes initialisation, optimisations ①, ② and ③. Initialisation specifies the workpiece and machining requirements which

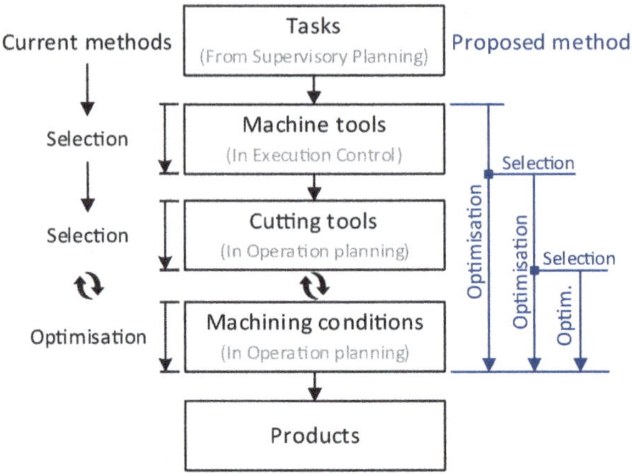


**Fig. 14.10** Differences in cutting tool and machining conditions optimisation

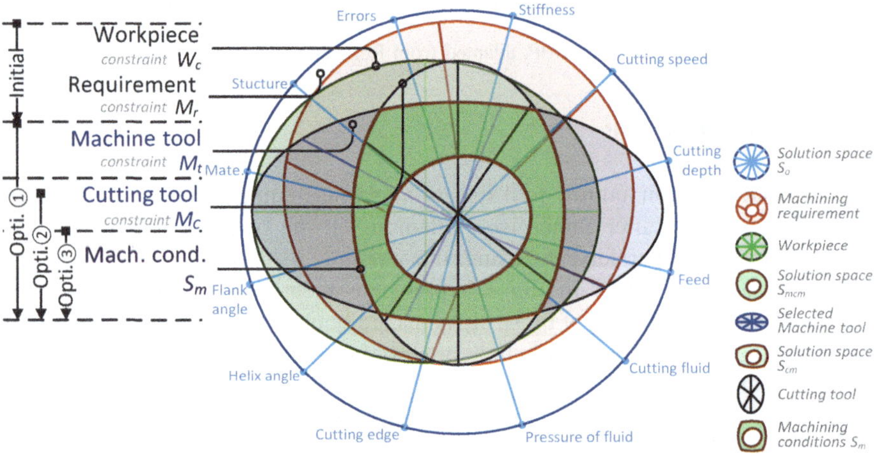


**Fig. 14.11** Optimisations of machining process and their solution spaces

are the two constraints to the original solution space, i.e. workpiece constraint $W_c$ and machining requirement $M_r$. Current solution space $S_{mcm}$ can be calculated by $S_{mcm} = S_o \cap W_c \cap M_r$. Then the global optimisation ① is carried out within space $S_1$, as a result of which machining tools, machining conditions and cutting tools are determined. The results provide a reference for selecting machine tools. Once the optimal machine tool is chosen, the machining conditions and tool are decided simultaneously. On the contrary, a substitute machine tool as a constraint $M_t$ bounds solution space $S_{mcm}$. Then, the current feasible set $S_{cm}$, $S_{cm} = S_{mcm} \cap M_t$, defines the updated solution space for execution of optimisation ② for selecting cutting tool and machining conditions. Similar to the selection process of machine tool, the selected cutting tool, as a constraint $M_c$, bounds the feasible set $S_{cm}$, resulting in the machining condition solution space $S_m$, $S_m = S_{cm} \cap M_c$.

#### 14.4.3.2 A Hybrid Algorithm for Enriched DPP

Hybrid algorithms can enhance optimisation performance by relaxing the limitations of each single algorithm. This section introduces an optimisation method combining three algorithms, as shown in Fig. 14.12, i.e. evolution based algorithms (EBA) or swarm intelligence based algorithms (SIBA), a neural network (NN) based model trained by big data, and analytic hierarchy process (AHP) based weight decision. Within the context, a global optimisation is carried out by EBA or SIBA, and those algorithms refer to several steps, e.g. parameter selection (sample), operation, and criterion. A NN model trained by big data, supervised learning, is employed to obtain individual objective fitness with high accuracy, which plays a key role as oppose to the existing methods. AHP, based on expert questionnaires, is applied to calculate weights for multi-objective optimisation, and its pairwise comparison matrix is established by big data analytics in statistics level, in case that the higher accuracy can be obtained by comparing with the questionnaires.

### *14.4.4 A Simplified Case Study*

A simplified case study is chosen to show the process of the reported method as a proof of concept. A set of hypothetic data generated in-house are applied to train the BPNN model used to calculate the fitness of GA.

#### 14.4.4.1 Data Attributes

The attributes of big data include the information regarding workpiece, machining requirements, machine tools, machining processes, and machining results. Also the

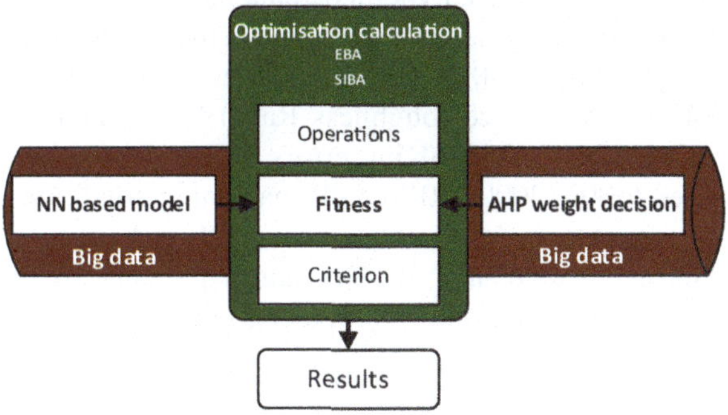


**Fig. 14.12** Optimisation processes

details of such information cover all the factors affecting the part machining, as shown in Table 14.4, including:

- Data attributes of workpiece: workpiece information includes the quantity of parts, part geometry, and part materials, etc. Here, the structural part geometry and quantity are represented by machining features and their quantity, whereas the curved or freeform parts are described by tool paths. Part materials refer to the material profiles with respect to the material machinability, e.g. hardness, brittleness, toughness and plasticity, etc., so that each material can be represented by a series of parameters.
- Data attributes of machining requirements: machining requirements refer to the designed machining quality in terms of tolerances and surface integrity, e.g. dimensional tolerance, geometrical tolerance, roughness, white layer, and dark layer.
- Data attributes of machine tool: machine tool information usually involves the number of machine tool, machine type, machine structure, structure and power of spindle, linear axis and rotational axis, errors of each axis, energy consumption of each component, and other related parameters. In this way, each machine tool can be represented by a set of parameters.
- Data attributes of machining process: machining process refers to machining conditions, cutting tools, accessories and physical data of the cutting process. The machining conditions include cutting parameters, cutting fluid and its parameters. The cutting tool involves tool type, tool material, tool coating, and tool geometrical parameters. The cutting accessories include the fixture types and their parameters. The physical data of the cutting process consist of cutting force, cutting vibration, etc.
- Data attributes of machining results: machining results mean the machining quality in terms of geometrical errors and surface integrity against the machining requirements. Such data are obtained through inspections of machined parts both qualified and unqualified.

#### 14.4.4.2 Data Generation

The attributes of the hypothetic data include: (1) workpiece consisting of machining feature and material information, including Poisson's ratio (0.25, 0.45), Young's modulus (100, 250 GPa), and hardness (20, 70 HRC); (2) machining requirements referring to surface roughness Ra (0.4, 6.3) and white layer depth (0.3, 10 μm); (3) machine tool involving errors of linear axis (1, 10 μm) and maximum spindle speed (3000, 20000 RPM); (4) machining conditions including cutting parameters, i.e. cutting speed (30, 150 m/min) and feed rate (0.05, 0.5 mm/z); (5) cutting tool referring to rake angle (0, 30°) and flank angle (6, 20°); and (6) machining results, covering the machine requirements, including surface roughness and white layer

**Table 14.4** A sample data sheet*

| No. | Workpiece | | | | | | | |
|---|---|---|---|---|---|---|---|---|
| | Geometry | | | Material | | | | |
| | Quantity | MFs | Volume per MF | Poisson's ratio | Young's modulus | Elastic modulus | Shear modulus | Density |
| 00001 | 20 | Side | 100 | 0.3 | 500 | 690 | 100 | 7.9 |

| | | | Machining requirement | | | | | |
|---|---|---|---|---|---|---|---|---|
| | | | Geometry | Surface | | Subsurface | | |
| CTE/10–6/K 20 °C | UTS | Hardness | Accuracy | Roughness | Morphology | White layer | Dark layer | Hardened layer |
| 18 | 300 | 35HRC | ±0.02 | 0.8 | No | 5 | 20 | 15 |

| | Machine tool | | | | | | | |
|---|---|---|---|---|---|---|---|---|
| Grain deformation | No. | Type | Structure | X\Y\Z error | A\B\C error | XYZ power/kw | Spindle power | Maintenance date |
| 10 | M1 | 4– | XYZB | 0.0005\0.0003\0.0003 | ~\0.0015\~ | 2.8\3.0\3.5 | 26 | 2016-03-05 |

| | Machining process | | | | | | | | | |
|---|---|---|---|---|---|---|---|---|---|---|
| | Cutting parameter | | | | Cooling | | Tool material | | Tool basic geometry | |
| Fault | Cutting speed | Feed rate | cutting depth | cutting width | Fuild | Presure | Tool material | Coat | Tool radius | Tool length |
| No | 50 | 0.15 | 0.2 | 15 | 1 | 20 | K2 | TiCN | 30 | 50 |

| Tool basic geometry | | Tool angles | | | Cutting process | | Time |
|---|---|---|---|---|---|---|---|
| Tool radius | Tool length | Entrance angle | Rake angle | Flank angle | X\Y\Z cutting force | X\Y\Z vibration | Duration |
| 30 | 50 | 60 | 10 | 10 | 400\400\300 | 40\35\20 | 20 |

| Machining result | | | | | | |
|---|---|---|---|---|---|---|
| Geometry | Surface | | Subsurface | | | |
| Accuracy | Roughness | Morphology | White layer | Dark layer | Hardened layer | Grain deformation |
| 0.01 | 0.8 | No | 4.8 | 21 | 16 | 9 |

*A part of data attributes

depth, and labelled for training the BPNN model. The selected data are bounded within the limited feasible sets, and they include 10,000 row data generated randomly using Matlab®.

#### 14.4.4.3 A Detailed Algorithm

Figure 14.13 depicts a hybrid algorithm of BPNN and GA, where the BPNN model trained by big data is employed to calculate the fitness, and GA is applied to perform the optimisation computation. The workpiece and machining requirements are set firstly. **S1:** Their parameters combining the target factors are divided into two classifications: (1) proceeded to GA calculation processes, and (2) passed to the trained model. **S2:** The former, as a population, is treated in the GA process, generating a new population. Then the latter filters the data which is used to train the BPNN model. The population generated in S1 is decoded into a set of parameters which are imported to the trained model, obtaining a calculation fitness. **S3:** The fitness is compared with the termination criterion, and once the computation is finished, the results provide a reference for machine tool selection, optimisations of cutting tools and machining conditions.

If the selected machine tool is not the optimal one, the parameters of the machine tool are directed sent back to the parameter classification process, and then the steps (S1), (S2) and (S3) are performed again, finally resulting the optimised results of cutting tools and machining conditions.

#### 14.4.4.4 Optimisation Processes and Results

A batch of parts is planned to be produced by a manufacturer, and its material parameters consist of 0.3 (Poisson's ratio), 200 GPa (Young's modulus), and 40 HRC (hardness). Its machining requirements involve Ra 1.6 (surface roughness) and 7 μm (white layer depth).

The hypothetic data filtered by the machining requirements filter, are applied to train the BPNN model whose parameters consist of 1 hidden layer, 10,000 training cycles, 0.001 error epsilon, and 0.2 learning rate. In this case, Matlab® is used for the computation, as a result of which the mean square error is stabilised around 0.375, as shown in Fig. 14.14, when the number of iteration is reaching approximately $4 \times 10^4$ (due to that these is no pattern in the random data).

Then the GA computation is performed to optimise machine tools, cutting tools and cutting parameters. The parameters of GA consist of (1) generations: 1000; (2) population size: 100; (3) crossover probability: 0.8; (4) mutation probability: 0.01; and (5) computation accuracy: 0.001. The fitness of machining requirements is stable after approximately 650 generations, as shown in Fig. 14.15a, together with the parameters of machine tool, cutting tool and cutting parameters, detailed as

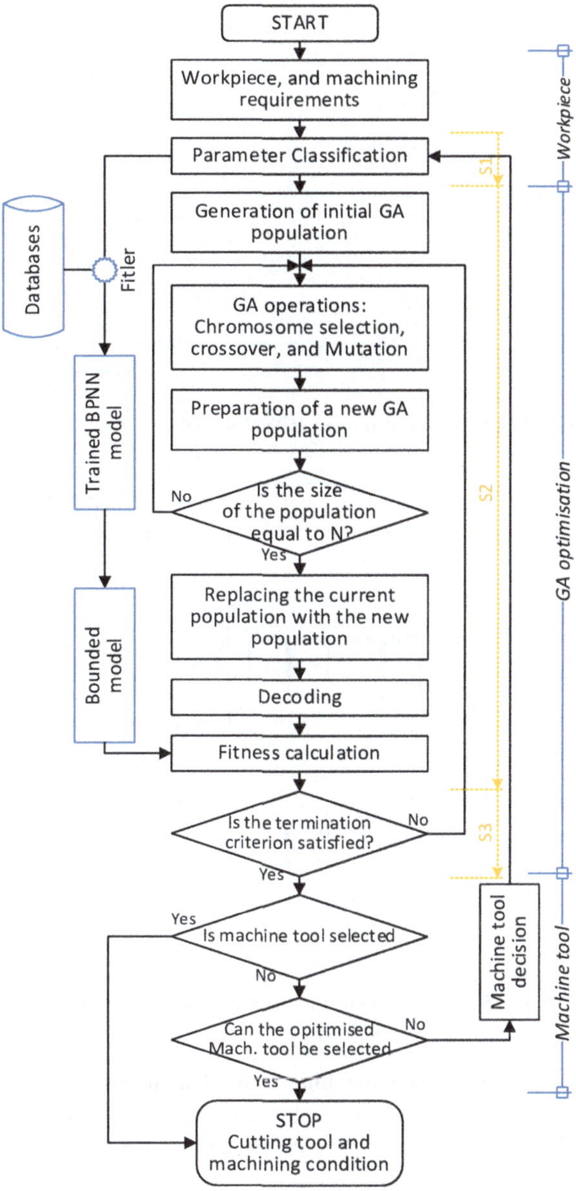


**Fig. 14.13** A hybrid algorithm combining GA and BPNN

follows: (1) machine tool: 10 μm (errors of linear axis) and approximately 4000 RPM (maximum spindle speed) (in Fig. 14.15b); (2) cutting tool: 30° (rake angle) and 6° (flank angle) (in Fig. 14.15c); and (3) cutting parameters: 30 m/min (cutting speed) and 0.05 mm/z (feed rate) (in Fig. 14.15d).

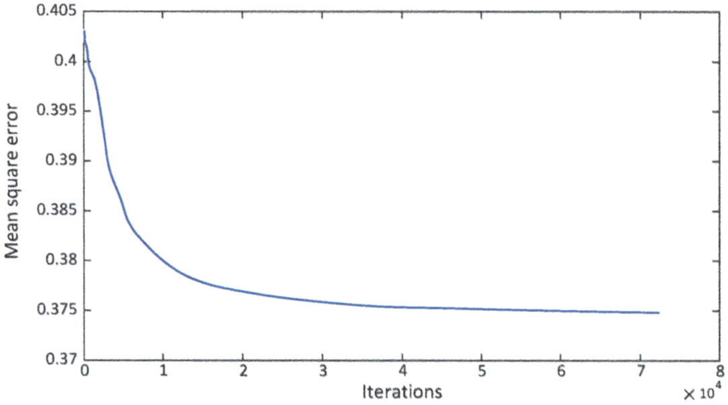


**Fig. 14.14** Errors of trained BPNN model

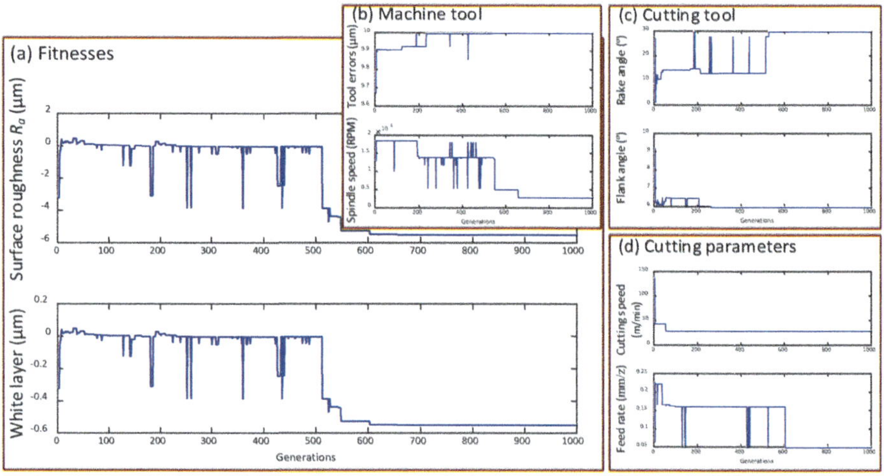


**Fig. 14.15** Calculation results including machine tool, cutting tool and cutting parameters

The optimised machine tool (if available), cutting tool geometries and cutting parameters are determined, simultaneously. In case of no optimal machine tool, an alternative machine is selected, and its parameters are imported to GA. A machine tool, for instance, is equipped with a linear axis of 3 μm in error and a maximum spindle speed of 8000 RPM. Based on the selected machine tool and the initial workpiece parameters, the GA calculation is performed again. Figure 14.16a depicts the fitness of machining requirements after approximately 90 generations. The optimised results of cutting tools and cutting parameters consist of: (1) cutting tool: 28° (rake angle) and 20° (flank angle) (as shown in Fig. 14.16b); and (2) cutting parameters: 30 m/min (cutting speed) and 0.3 mm/z (feed rate) (in

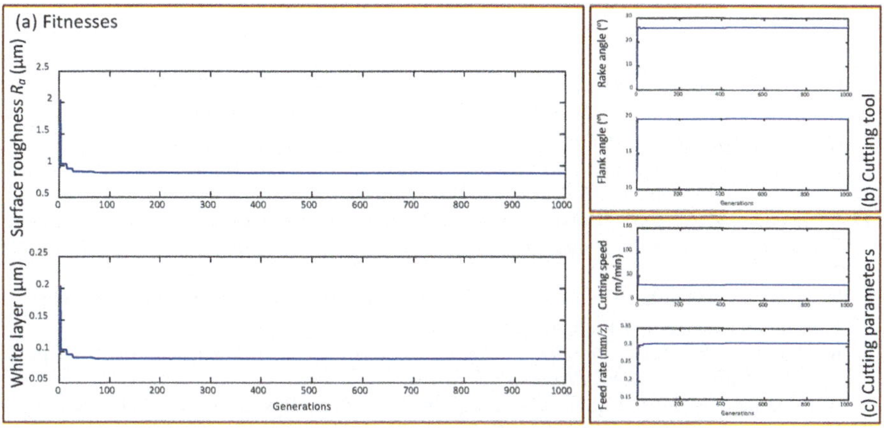


**Fig. 14.16** Calculation results including cutting tool and cutting parameters

Fig. 14.16c). The subsequent processes, i.e. cutting tool selection and machining condition optimisation, are the same in the statistics level, and are omitted in this chapter for the sake of page limit.

## 14.5 Conclusions

Big data analytics can identify hidden patterns from a large amount of historical data to help predicting potential defects of machining. This is useful for scheduling and task assignment to available machines. Based on the past experience and know-how, it helps prevent machining defects and guarantee the product quality.

Different from the conventional optimisation methods in which machine tool, cutting tool and machining condition are decided separately, this chapter also introduces a big data analytics based approach for machining process planning in an enriched DPP, combining the optimisations of the three as an entire package. Within the context, each machining resource is represented by the data attributes and is regarded as a constraint bounding the solution space. Big data analytics then specifies the relationship among those attributes. A hybrid method combining NN and AHP is employed for the optimisations. This approach is validated by a simplified case in which a GA and BPNN hybrid algorithm is implemented based on a set of hypothetic data for proof of concept.

The future work of applying big data analytics in factory shop floors includes: (1) detailed machining resource representation; (2) data collection from real-world machining environment for data filtering and cleansing; and (3) real-time decision making that requires more efficient algorithms for big data analytics. The results of these will bring big data analytics closer to industrial practice.

## References


1. R.C. Tryon, *Cluster Analysis: Correlation Profile and Orthometric (factor) Analysis for the Isolation of Unities in Mind and Personality* (Edwards Brothers, 1939)
2. D.B. Kenneth, *Numerical Taxonomy and Cluster Analysis* (SAGE Publications, 1994)
3. D.J. Bartholomew, F. Steele, J. Galbraith, I. Moustaki, *Analysis of Multivariate Social Science Data* (Taylor & Francis, 2008)
4. J. Manyika, M. Chui, B. Brown, J. Bughin, R. Dobbs, C. Roxburgh, A.H. Byers, *Big Data: The Next Frontier for Innovation, Competition, and Productivity* (McKinsey Global Institute, 2011)
5. D. Hand, H. Mannila, P. Smyth, *Principles of Data Mining* (The MIT Press, Cambridge, Massachusetts, London, England, 2001)
6. K. Dnuggets, *What Analytics, Data Mining, Data Science Software/Tools You Used in the Past 12 Months for a Real Project Poll* (2016)
7. A.S. Manne, On the Job-Shop Scheduling Problem. Oper. Res. **8**(2), 219–223 (1960)
8. G. Laporte, J.J. Salazar-GonzÁLez, F. Semet, Exact algorithms for the job sequencing and tool switching problem. IIE Trans. **36**(1), 37–45 (2004)
9. P. Asokan, J. Jerald, S. Arunachalam, T. Page, Application of adaptive genetic algorithm and particle swarm optimisation in scheduling of jobs and AS/RS in FMS. Int. J. Manuf. Res. **3** (4), 393–405 (2008)
10. B. Khoshnevis, Q.M. Chen, Integration of process planning and scheduling functions. J. Intell. Manuf. **2**(3), 165–175 (1991)
11. Q. Chen, B. Khoshnevis, Scheduling with flexible process plans. Prod. Plan. Control **4**(4), 333–343 (1993)
12. W. Tan, B. Khoshnevis, Integration of process planning and scheduling—A review. J. Intell. Manuf. **11**(1), 51–63 (2000)
13. P. Mohapatra, A. Nayak, S.K. Kumar, M.K. Tiwari, Multi-objective process planning and scheduling using controlled elitist non-dominated sorting genetic algorithm. Int. J. Prod. Res. **53**(6), 1712–1735 (2014)
14. M. Freitag, T. Hildebrandt, Automatic design of scheduling rules for complex manufacturing systems by multi-objective simulation-based optimization. CIRP Ann.—Manuf. Technol. **65** (1), 433–436 (2016)
15. X. Li, L. Gao, C. Zhang, X. Shao, A review on Integrated Process Planning and Scheduling. Int. J. Manuf. Res. **5**(2), 161–180 (2010)
16. M. Rajkumar, P. Asokan, T. Page, S. Arunachalam, A GRASP algorithm for the integration of process planning and scheduling in a flexible job-shop. Int. J. Manuf. Res. **5**(2), 230–251 (2010)
17. C. Gahm, F. Denz, M. Dirr, A. Tuma, Energy-efficient scheduling in manufacturing companies: a review and research framework. Eur. J. Oper. Res. **248**(3), 744–757 (2016)
18. P.M. Khodke, A.S. Bhongade, Real-time scheduling in manufacturing system with machining and assembly operations: a state of art. Int. J. Prod. Res. **51**(16), 4966–4978 (2013)
19. G.E. Vieira, J.W. Herrmann, E. Lin, Rescheduling manufacturing systems: a framework of strategies, policies, and methods. J. Sched. **6**(1), 39–62 (2003)
20. M. Ham, Y.H. Lee, S.H. Kim, Real-time scheduling of multi-stage flexible job shop floor. Int. J. Prod. Res. **49**(12), 3715–3730 (2011)
21. K. Iwamura, N. Okubo, Y. Tanimizu, N. Sugimura, Real-time scheduling for holonic manufacturing systems based on estimation of future status. Int. J. Prod. Res. **44**(18–19), 3657–3675 (2006)
22. C. Wang, H. Ghenniwa, W. Shen, Real time distributed shop floor scheduling using an agent-based service-oriented architecture. Int. J. Prod. Res. **46**(9), 2433–2452 (2008)
23. Y. Yih, A. Thesen, Semi-Markov decision models for real-time scheduling. Int. J. Prod. Res. **29**(11), 2331–2346 (1991)

24. R. Zhou, H.P. Lee, A.Y.C. Nee, Applying Ant Colony Optimisation (ACO) algorithm to dynamic job shop scheduling problems. Int. J. Manuf. Res. **3**(3), 301–320 (2008)
25. G. Metan, I. Sabuncuoglu, H. Pierreval, Real time selection of scheduling rules and knowledge extraction via dynamically controlled data mining. Int. J. Prod. Res. **48**(23), 6909–6938 (2010)
26. V.K. Jain, S. Premalatha, N. Baskar, Implementation of supervised statistical data mining algorithm for single machine scheduling. J. Adv. Manag. Res. **9**(2), 170–177 (2012)
27. W. Kohn, V. Brayman, J. Littleton, Repair-control of enterprise systems using RFID sensory data. IIE Trans. **37**(4), 281–290 (2005)
28. C. Wang, P. Jiang, *Manifold learning based rescheduling decision mechanism for recessive disturbances in RFID-driven job shops* (J. Intell, Manuf, 2016)
29. T. Qu, H.D. Yang, G.Q. Huang, Y.F. Zhang, H. Luo, W. Qin, A case of implementing RFID-based real-time shop-floor material management for household electrical appliance manufacturers. J. Intell. Manuf. **23**(6), 2343–2356 (2012)
30. X. Xu, L. Wang, S.T. Newman, Computer-aided process planning—A critical review of recent developments and future trends. Int. J. Comput. Integr. Manuf. **24**(1), 1–31 (2011)
31. M.S. Chua, H.T. Loh, Y.S. Wong, Optimization of cutting conditions for multi-pass turning operations using sequential quadratic programming. J. Mater. Process. Technol. **28**(1–2), 253–262 (1991)
32. S.H. Yeo, A multipass optimization strategy for CNC lathe operations. Int. J. Prod. Econ. **40** (2–3), 209–218 (1995)
33. M.S. Akturk, S. Avci, An integrated process planning approach for CNC machine tools. Int. J. Adv. Manuf. Technol. **12**(3), 221–229 (1996)
34. I. Lazoglu, S.Y. Liang, Analytical modeling of force system in ball-end milling. Mach. Sci. Technol. **1**(2), 219–234 (1997)
35. Y.K. Yang, M.T. Chuang, S.S. Lin, Optimization of dry machining parameters for high-purity graphite in end milling process via design of experiments methods. J. Mater. Process. Technol. **209**(9), 4395–4400 (2009)
36. C.C. Chen, K.T. Chiang, C.C. Chou, Y.C. Liao, The use of D-optimal design for modeling and analyzing the vibration and surface roughness in the precision turning with a diamond cutting tool. Int. J. Adv. Manuf. Technol. **54**(5–8), 465–478 (2011)
37. W. Ji, X. Liu, L. Wang, S. Sun, Experimental evaluation of polycrystalline diamond (PCD) tool geometries at high feed rate in milling of titanium alloy TC11. Int. J. Adv. Manuf. Technol. **77**(9–12), 1549–1555 (2015)
38. J. Zhang, S.Y. Liang, D. Yen, Optimisation of the finish hard turning process for hardened 52100 steel with PCBN tools. Int. J. Manuf. Res. **2**(4), 428–447 (2007)
39. W. Li, S.K. Ong, A.Y.C. Nee, Optimization of process plans using a constraint-based tabu search approach. Int. J. Prod. Res. **42**(10), 1955–1985 (2004)
40. J.G. Taiber, Optimization of process sequences considering prismatic workpieces. Adv. Eng. Softw. **25**(1), 41–50 (1996)
41. S.Y. Wei, C.C. Lo, C.A. Chang, Using throughput profit for selecting manufacturing process plan. Comput. Ind. Eng. **32**(4), 939–948 (1997)
42. G.H. Ma, Y.F. Zhang, A.Y.C. Nee, A simulated annealing-based optimization algorithm for process planning. Int. J. Prod. Res. **38**(12), 2671–2687 (2000)
43. D.H. Lee, D. Kiritsis, P. Xirouchakis, Search heuristics for operation sequencing in process planning. Int. J. Prod. Res. **39**(16), 3771–3788 (2001)
44. D.H. Lee, D. Kiritsis, P. Xirouchakis, Branch and fathoming algorithms for operation sequencing in process planning. Int. J. Prod. Res. **39**(8), 1649–1669 (2001)
45. D.H. Lee, D. Kiritsis, P. Xirouchakis, Iterative approach to operation selection and sequencing in process planning. Int. J. Prod. Res. **42**(22), 4745–4766 (2004)
46. Y.C. Wang, T. Chen, Modelling and optimization of machining conditions for the multi-pass dry turning process. *Proc. Inst. Mech. Eng. Part B-Journal Eng. Manuf.* **222**(11), 1387–1394 (2008)

47. K. Abhishek, S. Datta, S.S. Mahapatra, Multi-objective optimization in drilling of CFRP (polyester) composites: application of a fuzzy embedded harmony search (HS) algorithm. Measurement **77**, 222–239 (2016)
48. D.P. Gupta, B. Gopalakrishnan, S.A. Chaudhari, S. Jalali, Development of an integrated model for process planning and parameter selection for machining processes. Int. J. Prod. Res. **49**(21), 6301–6319 (2011)
49. A. Sluga, M. Jermol, D. Zupanic, D. Mladenic, Machine learning approach to analysis. Comput. Ind. **37**(3), 185–196 (1998)
50. L. Monostori, Z.J. Viharos, S. Markos, Satisfying various requirements in different levels and stages of machining using one general ANN-based process model. J. Mater. Process. Technol. **107**(1–3), 228–235 (2000)
51. W.D. Li, S.K. Ong, A.Y.C. Nee, Hybrid genetic algorithm and simulated annealing approach for the optimization of process plans for prismatic parts. Int. J. Prod. Res. **40**(8), 1899–1922 (2002)
52. T.N. Wong, L.C.F. Chan, H.C.W. Lau, Machining process sequencing with fuzzy expert system and genetic algorithms. Eng. Comput. **19**(2–3), 191–202 (2003)
53. M. Salehi, A. Bahreininejad, Optimization process planning using hybrid genetic algorithm and intelligent search for job shop machining. J. Intell. Manuf. **22**(4), 643–652 (2011)
54. M. Petrovic, M. Mitic, N. Vukovic, Z. Miljkovic, Chaotic particle swarm optimization algorithm for flexible process planning. Int. J. Adv. Manuf. Technol. **85**(9–12), 2535–2555 (2016)
55. A.K. Choudhary, J.A. Harding, M.K. Tiwari, Data mining in manufacturing: a review based on the kind of knowledge. J. Intell. Manuf. **20**(5), 501–521 (2008)
56. Y. Zhang, S. Ren, Y. Liu, S. Si, A big data analytics architecture for cleaner manufacturing and maintenance processes of complex products. J. Clean. Prod (2016)
57. R.Y. Zhong, G.Q. Huang, S. Lan, Q.Y. Dai, X. Chen, T. Zhang, A big data approach for logistics trajectory discovery from RFID-enabled production data. Int. J. Prod. Econ. **165**, 260–272 (2015)
58. R.F. Babiceanu, R. Seker, Big data and virtualization for manufacturing cyber-physical systems: a survey of the current status and future outlook. Comput. Ind. **81**, 128–137 (2016)
59. J. Woo, S.J. Shin, W. Seo, *Developing a big data analytics platform for increasing sustainability performance in machining operations* (Flexible Automation and Intelligent Manufacturing, Seoul, Korea, 2016), pp. 1–8
60. W. Ji, L. Wang, Big data analytics based fault prediction for shop floor scheduling. J. Manuf. Syst. **43**, 187–194 (2017)
61. H. Müller, *Problems, methods, and challenges in comprehensive data cleansing* (Humboldt-Universität zu Berlin zu Berlin, Berlin, Germany, 2005)
62. M. Lenzerini, Data integration: a theoretical perspective, in *Proceedings of the Twenty-first ACM SIGACT-SIGMOD-SIGART Symposium* (2002), pp. 233–246
63. J.R. Quinlan, Simplifying decision trees. Int. J. Man Mach. Stud. **27**(3), 221–234 (1987)
64. S. Russell, P. Norvig, *Artificial Intelligence: A Modern Approach* (Prentice Hall, Egnlewood Cliffs, New Jersey, 1995)
65. T.D. Nielsen, F.V. Jensen, *Bayesian Networks and Decision Graphs* (Springer Science & Business Media, 2009)
66. M.T. Hagan, H.B. Demuth, M.H. Beale, O. De Jesús, *Neural Network Design* (PWS Publishing Company Boston, Boston, MA, US, 1996)
67. C. Cortes, V. Vapnik, Support vector networks. Mach. Learn. **20**(3), 273–297 (1995)
68. R. Agrawal, R. Srikant, Fast algorithms for mining association rules. Proc. 20th Int. Conf. Very Large Data Bases **1215**, 487–499 (1994)
69. D.W. Aha, Lazy learning, in *Lazy Learning*, ed. by W.A. David (Kluwer Academic Publishers, 1997), pp. 7–10
70. M. Melanie, *An Introduction to Genetic Algorithms* (MIT Press, Cambridge, Massachusetts, London, England, 1999)

71. Z. Pawlak, *Rough Sets: Theoretical Aspects of Reasoning about Data* (Springer, Netherlands, 1991)
72. L.A. Zadeh, Fuzzy sets. Inf. Control **8**(3), 338–353 (1965)
73. J. Han, J. Pei, M. Kamber, *Data Mining: Concepts and Techniques* (Elsevier, 2011)
74. W. Ji, L. Wang, Big data analytics based optimisation for enriched process planning: a methodology. Procedia CIRP **63**, 161–166 (2017)
75. L. Wang, W. Shen, DPP: An agent-based approach for distributed process planning. J. Intell. Manuf. **14**(5), 429–439 (2003)
76. L. Wang, H.Y. Feng, N. Cai, Architecture design for distributed process planning. J. Manuf. Syst. **22**(2), 99–115 (2003)
77. L. Wang, Web-based decision making for collaborative manufacturing. Int. J. Comput. Integr. Manuf. **22**(4), 334–344 (2009)
78. L. Wang, Machine availability monitoring and machining process planning towards Cloud manufacturing. CIRP J. Manuf. Sci. Technol. **6**(4), 263–273 (2013)
79. L. Wang, Cyber manufacturing: research and applications, in *Proceedings of the TMCE*, Budapest (2014), pp. 39–49

# Chapter 15
# Outlook of Cloud, CPS and IoT in Manufacturing

## 15.1 Introduction

In the past decades, advancements in Web-/Internet-based systems and applications have opened up the possibility for industries to utilise the cyber workspace to conduct efficient and effective daily collaborations from anywhere in distributed manufacturing environments [1]. For example, remote robot control becomes relevant not only in rescue operations but also in cyber and/or cloud manufacturing environments where distant operations can be done quickly and economically. Among many recently emerging technologies, CPS is treated as the main thread and summarised below together with Cloud and IoT concepts.

The term *Cyber-Physical Systems* (CPS), was coined in the US in 2006 [2], with the realisation of the increasing importance of the interactions between interconnected computing systems and the physical world. CPS can be characterised as a thematic subject as opposed to a disciplinary topic. Multidisciplinary areas such as mechatronics, robotics and CPS typically start as themes, and then eventually evolve into disciplinary areas [3]. It is interesting to note that mechatronics was adopted and promoted from electrical or mechanical engineering disciplines whereas CPS has initially been driven from computer science and electrical engineering directions. It is currently not fully clear to whether CPS will evolve into a discipline in itself.

Correspondingly, there are multiple definitions of CPS, for example the early one from 2008 [4]: CPS are integrations of computation and physical processes. Embedded computers and networks monitor and control the physical processes, usually with feedback loops where physical processes affect computations and vice versa. In other words, CPS use computations and communication deeply embedded in and interacting with physical processes so as to add new capabilities to physical



L. Wang and X. V. Wang, *Cloud-Based Cyber-Physical Systems in Manufacturing*,
https://doi.org/10.1007/978-3-319-67693-7_15

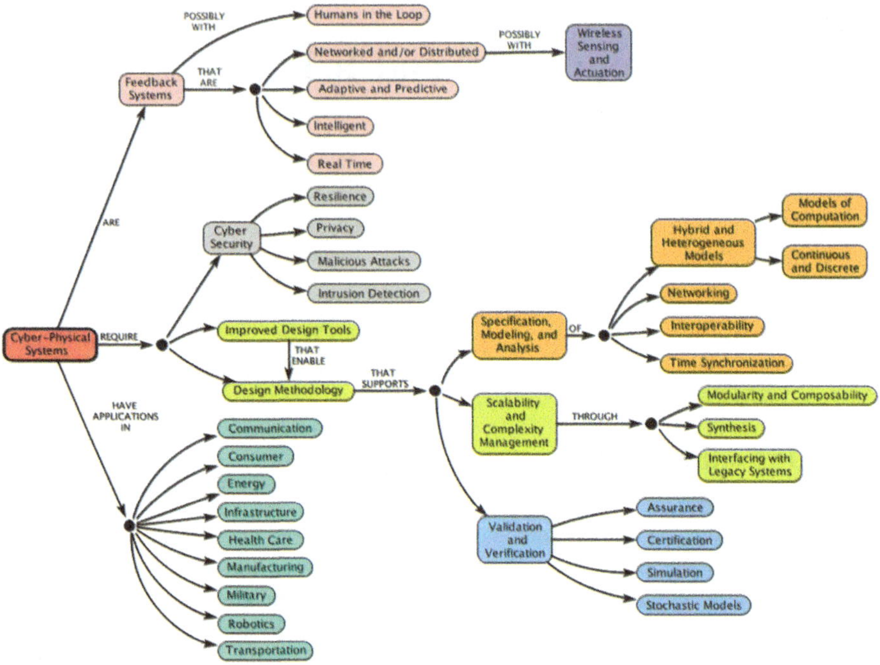


**Fig. 15.1** A concept map of CPS [5]

systems. A CPS may range from minuscule (a pace maker) to large scale (a national power grid). Figure 15.1 illustrates schematically a concept map of CPS.

In recent years, research and applications of CPS have been active in such areas like transportation, smart home, robotic surgery, aviation, critical infrastructure, defence, etc. CPS also positively affected manufacturing in form of cyber-physical production systems (CPPS) in process automation and control [6]. Due to the huge application potentials of CPS in manufacturing and yet the lack of common understanding of CPS in manufacturing sector, there is a need to systematically document the current status and the latest advancement of CPS with future trends clearly identified. The remainder of this chapter is therefore organised as follows. The drivers, barriers and initiatives of CPS are presented in Sect. 15.2, followed by their characteristics and requirements in Sect. 15.3. Section 15.4 provides examples of CPS in cloud manufacturing in connection with IoT, which highlight key characteristics of such systems. Section 15.5 identifies the future trends and R&D directions of Cloud, CPS and IoT, before concluding the key findings of this chapter in Sect. 15.6.

## 15.2 Drivers, Barriers and Initiatives

Depending on the formalities of the definition, CPS may be viewed as having been present in the manufacturing sector for quite some time. For example, embedded controllers, sensor systems, collaborating robots and others may represent the early technologies that contributed to the development of CPS. The advent of advanced communication technologies has now brought new possibilities, as will be shown in the next sections.

The needs of industry have been clear and strong for most stakeholders. The Time-to-Volume and Time-to-Market aspects of most producing companies demand for very rapid product introductions to markets and, if successful, a very quick production increase. As products become more complex, with greater levels of miniaturisation and with embedded electronics, the production needs to adapt just as quickly as possible. If one adds that most manufacturing companies would like to offer personalised products (and services), then the immense variety of products sets obvious challenges.

Recently, there have been new drivers surfacing, mostly related to the need of society to attain sustainability: reuse of end-of-life equipment, reuse of materials, energy efficiency of production systems, self-organising and self-maintenance, as well as online customers support. The main barrier to a full exploitation of CPS technologies, however, remains a rather conservative industry which operates under incredibly tight margins, thus not allowing for major uncertainties at strategic level. The CPS technology must, therefore, find transitional technologies through which the truly novel ideas may be gradually introduced at the shop-floor level, without incurring major investments. The CODESYS solutions by the German software company 3S (Smart Software Solutions) [7] may be viewed as such a transitional step. Secondly, the use of architectural approaches, which is fundamental in industrial CPS applications, needs to be addressed as a global issue instead of a local one. As reported in [8], most manufacturing engineers are mechanical experts, the notion of abstract architectural work does find resistance, and hence the need to bridge the gap between the disciplines, as proven in several recent European projects (IDEAS, EUPASS, GRACE).

Specific to manufacturing, integration is the key that can be facilitated by CPS. Manufacturing industry involves multi-sector activities with a quite broad range of stakeholders. Typical industrial processes include the fabrication of parts, assembly, packaging, transport, quality control, and many more. These activities are operated by producers, system integrators, sub-suppliers, logistics/supply chain experts, and many others. At the heart of these activities is the production of the product itself. Manufacturing has been a rather conservative industry as the costs are very high and have to be minutely monitored to ensure final product quality. This results in production systems that consist of an enormous variety of equipment, ranging from vision systems and sensors to robots and conveyors, including metrology equipment, different controllers, different levels of users, and so forth. Such variety is also affected by the multitude of stakeholders (in sectors of machinery, control systems, robots, etc.). The challenge in manufacturing is the integration of the equipment such that all levels of production may communicate, and CPS shows the promise of potential applications in manufacturing.

Current aspects such as the absence of tailored software approaches, under-performing controllers and limited protocols are addressed by current initiatives. Initially, the non-deterministic nature of multi-agent control was often cited as being the major drawback, but new paradigms that limit this to almost deterministic levels have reverted this barrier (also see Holonic [9], Changeable [10] and Evolvable Systems [11] for more details).

The initial thrust towards a new way of controlling manufacturing and assembly systems was given in [12], in which the idea of industrial agents was proven to hold. This led to Service-Oriented Architectures (SOA), where Rockwell Automation [13] and Schneider Electric [14] developed interesting solutions. In later years a great deal of work has been carried out by academia [15, 16], which has consolidated this approach. In 2011, the first self-organising assembly system was demonstrated at FESTO in Germany, this approach being based on Agent Oriented Architectures (AoA). The advent of AoA architectures meant that the swarm approaches of computer science had now been taken up at industrial level [17] (see Fig. 15.2). An ambitious effort of the swarm approaches is the UC Berkeley led TerraSwarm project addressing design methodology, techniques and

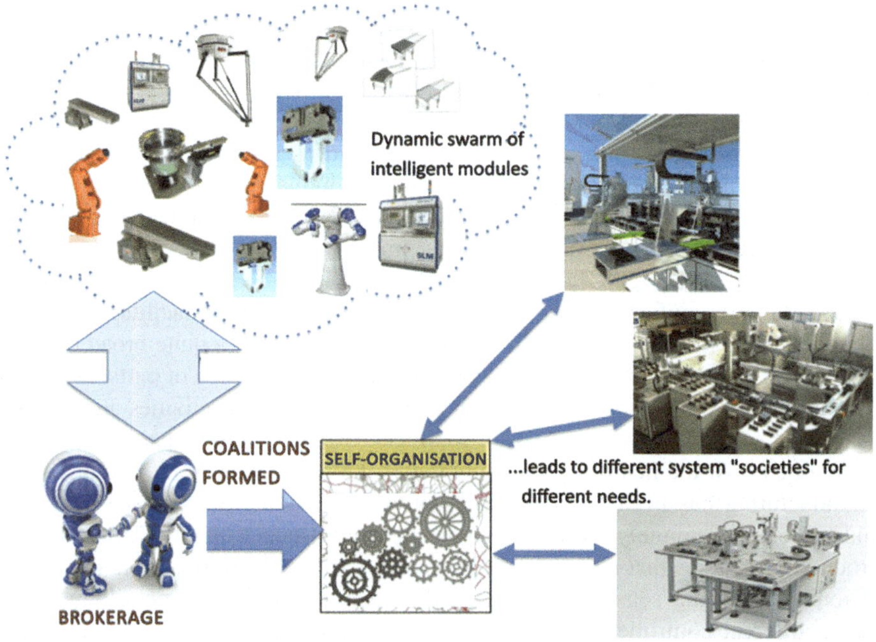


Fig. 15.2 Swarms of modules creating "system societies" [17]

platforms for pervasive integration of smart, networked sensors and actuators into a connected world [18].

The above examples have recently been expended to include ARM processors that enable devices to communicate with one another via embedded communication systems (e.g. BLE, Wi-Fi, etc.). It is therefore safe to say that CPS, even in its more stringent traditional definition, is now part of many manufacturing systems. To date there are several initiatives that cater for the CPS development. These include Advanced Manufacturing Partnership 2.0 [19], Industrial Internet [20] and CPS [21] in the USA, Industry 4.0 [22] in Germany, ECSEL (EU, with ARTEMIS), Factories of the Future [23] in EU, and even the less-known Japanese "Monozukuri" which stands for Coopetition. Other initiatives on this front include Wise-ShopFloor for web-based sensor-driven e-shop floor [24, 25] and the ongoing Cyber-Physical European Roadmap and Strategy [26].

With the above in mind, we will now address a number of pertinent questions that emerge when discussing any cyber-physical systems: (1) What is new with CPS and what characterises does CPS have? (2) How does the term CPS relate to other concepts such as IoT, big data and systems of systems? (3) How does CPS relate to manufacturing? and (4) What are the challenges in dealing with CPS? These questions are answered by analysing the characteristics and requirements of CPS in the next section.

## 15.3 Characteristics and Requirements

Industrial representatives rightfully pointed out that CPS, according to the previous definition, are indeed not new but already existing and manifested by for example existing industrial distributed control systems. Indeed, the case can be made that it is becoming easier to identify non CPS, due to the increasing digitalisation and penetration of embedded systems [27].

The increasing connectivity and capabilities of computational systems, largely driven by the consumer market, however drive the creation of entirely new systems, characterised by:

- Deployment of CPS in mass-products for use in all kinds of applications in society, for example exemplified by smartphone enabled services;
- Opportunities for and introduction of new cross-domain applications, exemplified by intelligent transportation systems that integrate among other communications, mobility, entertainment and safety-related services;
- New cross-domain collaboration typically calling for new business models and interoperability standards; and
- Increasing openness, adaptability and autonomy.

These trends and traits of CPS led the CyPhERS project to carry out a characterisation of CPS, attempting to capture the evolving scope of CPS, from traditionally closed systems, with single jurisdiction, limited adaptability and autonomy. Defining such characteristics would be helpful beyond definitions, because definitions of CPS tend to be very general; instead the characterisation helps to identify various types of CPS. The following aspects of CPS have been identified (elaborated from CyPhERS D2.2 [26]):

1. Deeply embedded versus IT dominated. Traditional embedded systems are represented by resource limited and dedicated computer systems, tightly integrated with the physical processes that they are interacting with (compare for example with an automotive engine controller, directly mounted on the combustion engine). With the increasing connectivity and capabilities of computing systems, there is now rather a grey zone in what encompasses "embedded" versus IT systems. The two types of systems are increasingly becoming connected and a new design choice is where to place functionality (online, embedded) or as part of a remote system, possibly with real-time capabilities. As an example, consider a modern engine controller (e.g. aircraft or automotive application) where the engine controller will be connected to an IT system for remote diagnosis and maintenance. A particular CPS can nevertheless be characterised w.r.t. its "embedded" versus "IT" dominance.
2. Single-domain versus cross-domain. A traditional embedded system often represents a single domain application, compare for example with a refrigerator and temperature control in a building. New cost-efficient communication leads to opportunities for new services that cut across existing domains, or potentially for creating new CPS domains. The smart home and its connection to the electrical grid represent an example of this trend.
3. Open versus closed. A traditional embedded system represents a system which is not connected to other computing systems. The difficulties of diagnosing, maintaining and upgrading widely deployed embedded systems provide strong driver towards more open systems. Another driver is provided by the ability to provide new collaborative services.
4. Automation levels and types. Autonomy can be considered as the ability to operate without constant human supervision/intervention. Automation has traditionally been introduced to relieve humans of dirty, dull, and dangerous operations [28]. Driven by environmental, resources efficiency and safety considerations, autonomy is now moving to all kinds of domains and applications (compare for example with integrated transportation systems to increase transport efficiency).
5. Governance, referring to the entities responsible for dependable and efficient system operation. The division of responsibility will correlate with the system of system nature; for example, a car OEM will be primarily responsible for the functioning of a car, but responsibilities also lie on the driver and road operators. In an intelligent transportation system, responsibilities will be divided even further.
6. Distributed versus centralised control. The increasing connectivity implies that most CPS already constitute distributed computer systems (or are likely to

become so), implying that control will be more or less decentralised. Control in this context refers to the decision making within the distributed system. A CPS will thus as a whole, or in its various parts, be characterised by the degree to which control is centralised/decentralised. Note that "controllers" may well include both humans as well as computerised control.

7. Single jurisdiction versus cross-jurisdiction. This aspect refers to applicable standards and legislation. Generally, the more open and cross-domain a CPS becomes, the more complicated the jurisdiction is. It can be noted that many existing CPS already face this challenge, for example a truck, where "body builders" will add features such as cranes and pumps to a truck platform, implying that a number of standards and laws are applicable. The aspect has a number of implications referring to responsibilities, liability and business models.
8. Adaptability under uncertain conditions. A typical CPS will face varying contexts, in terms of for example environmental conditions, system load and failures. Making a CPS adaptable implies that it has ability to cope with such varying contexts within given bounds, potentially providing benefits in terms of reduced maintenance costs and increased availability. Enhanced adaptability will on the other hand increase the system complexity, providing interesting and important design trade-offs.
9. Human in/outside the loop. Traditional CPS come in two types; those that are more or less fully autonomous (i.e. act independently of humans, but may be triggered by human inputs; for example, a stability controller in a car), and those with a much closer interaction with humans, including shared control. An example of the latter includes gear control in a car where the driver in an automated gearing system can choose to relay on the computer control or override it. In shared control, it becomes crucial to clarify who is in control at any point in time and making sure that unintended control does not take place.
10. Degree of integration. Connectivity paves the way for various types of integration. A CPS, in a certain context and application domain, will have a certain degree of horizontal and vertical integration. Horizontal integration refers to integrating services and functions of similar type (at the same level of abstraction), for example referring to integration of factory floor sensors and actuators. Initial integrations of this type were made in the 1970s in the manufacturing domain. Vertical integration refers to integration across system hierarchies, considering for example smart buildings by integrating energy meters and heating/ventilation devices with building control, up to entire buildings, and further towards local energy distribution and power systems of cities. Extended levels of integration are likely to cut across domains and jurisdictions, thus involving several non-technical challenges.

Several of the proposed aspects refer to a scale or degree, e.g. of openness, shared control, and automation levels. For some of these scales, existing reference models could be used to quantify the aspect, such as proposed in [29]. The aspects may not be fully orthogonal, for example automation is related to adaptability, and the number of domains involved may relate to single-/cross-jurisdiction. Mobility relates to many aspects when comes to manufacturing applications [30]. Moreover, CPS are generally characterised as software intensive systems, in which software provides a major part of the investment and value as part of an integrated system. The connectivity provides opportunities to extend traditional products with additional services.

The increasing connectivity and related opportunities have given rise to multiple terms that provide different perspectives to the enabling technology and the connected society. In the following we briefly contrast CPS with Systems of Systems (SoS), Internet of Things (IoT), Cloud, Big Data, and manufacturing related terms such as Industry 4.0 initiatives.

### *15.3.1 Systems of Systems (SoS)*

The SoS term has a background in the defence domain, but now is increasingly used in and across domains e.g. automotive, rail, aerospace, maritime and logistics [31]. Coordination and collaboration are the keywords for SoS. SoS have the following characteristics [32]:

- Operational and Managerial Independence of Elements—corresponding to different jurisdiction and autonomy in the CPS characterisation.
- Evolutionary Development—systems that are "never" finished, for example a transportation system, also implying legacy.
- Emergent Behaviour—where uncertainties in behaviours and interactions make it impossible to fully anticipate all SoS behaviours.
- Geographical Distribution of Elements.

Most (if not all) SoS will indeed be systems of CPS! We can therefore conclude that most SoS constitute a special class of CPS.

### *15.3.2 Internet of Things (IoT)*

The term IoT was coined in 1999 by Kevin Ashton [33], referring to wireless communication abilities integrated with sensors and computing, thus enabling uniquely identifiable *things* to provide data over the Internet with limited or no human interaction. IoT can be seen as a bottom-up vision, an enabling technology, which can be used to create a special class of CPS, i.e. systems including the Internet. A CPS does not necessarily include the Internet. Some visions of the IoT go beyond basic communication, and consider the ability to link "cloud" representations of the real things with additional information such as location, status, and business related data.

Given the above discussions, it follows that IoT systems will be CPS systems. A given CPS does not however necessarily need to involve the Internet. CPS thus constitute a larger class of systems.

### *15.3.3 Cloud Manufacturing (CM)*

Recently, cloud computing has changed the way of thinking of both IT service providers and their customers. It offers business and application models that deliver infrastructure, platform, software and applications in forms of services [34], which provide different levels of services of cloud applications compared against standalone ones. Inspired by the success of cloud computing, the *cloud* technology has been extended to the manufacturing contexts, leading to the innovation of various cloud manufacturing systems. Cloud manufacturing implies an integrated cyber-physical system that can provide on-demand manufacturing services, digitally and physically, at the best utilisation of manufacturing resources [35]. It aims at offering a shared pool of resources such as manufacturing software, manufacturing facilities, and manufacturing capabilities. However, cloud manufacturing is more than simply deploying manufacturing software applications in the cyber cloud. Besides data storage and virtual machines, the physical resources integrated in the manufacturing cloud are able to offer adaptive, secure, scalable and on-demand manufacturing services over the Internet of Things, including work-cells, machine tools, robots, etc.

Figure 15.3 illustrates the idea of cloud manufacturing with cloud computing as its core. The additional services of cloud manufacturing around cloud computing connect with physical machines, robots and even factories in the real world.

### *15.3.4 Big Data*

Big data, one of the hottest buzzwords of the era, refers to analytics based on large data collections. Advancements in computing and memory performance, together with networking (not the least social networks) have made it possible to gather unprecedented amounts of data. CPS and IoT enable further enormous amounts of data related to physical systems to be made available for analysis, and thus pave the way for new applications of big data in the future. Big data is relevant to non-technical systems and IT systems, but becomes even more interesting when applied in the context of CPS due to the implications of physicality in terms of capabilities, technical risks and costs. Nevertheless, in order not to lose the focus of this chapter we keep the big data aspect short.

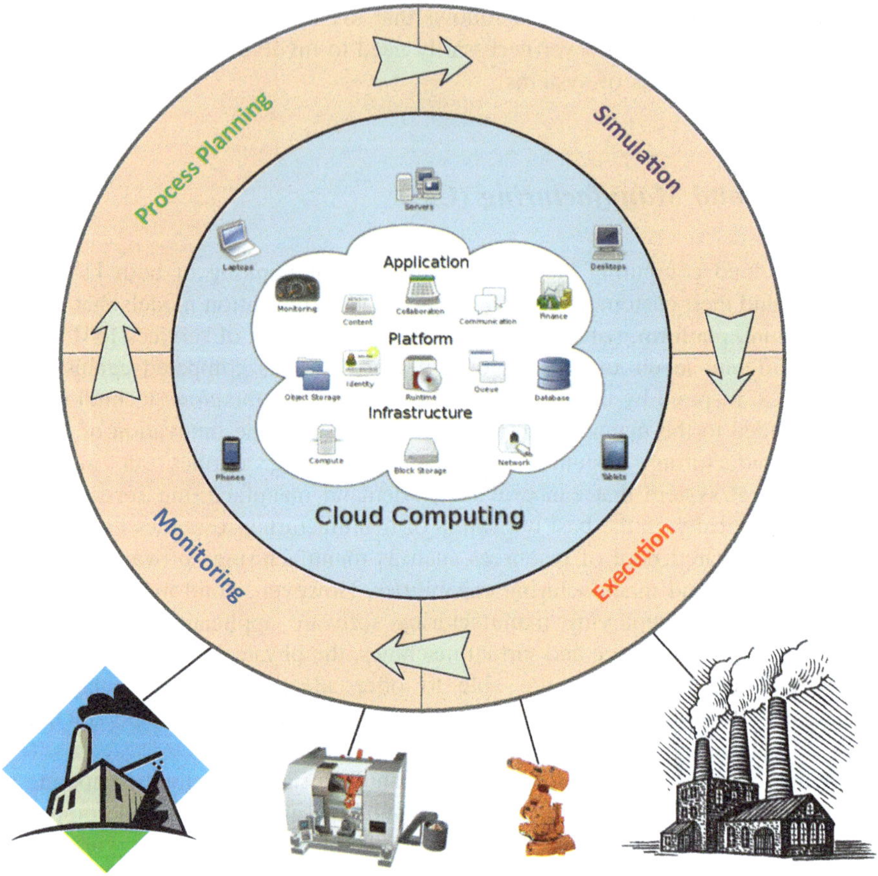


**Fig. 15.3** Cloud manufacturing services connected with the physical world

## *15.3.5 Industry 4.0*

Industry 4.0 is a large German initiative [22] that emphasises the extension of traditional manufacturing systems to full integration of physical, embedded and IT systems including the Internet. It highlights three features for implementation: (1) horizontal integration through value networks, (2) end-to-end digital integration of engineering across the entire value chain, and (3) vertical integration and networked manufacturing systems. The implementation recommendations call for actions in eight key areas including standardisation and reference architecture; managing complex systems; safety and security; work organisation; professional training; regulations, and resource efficiency.

### *15.3.6 Challenges*

While CPS provides a huge amount of opportunities, it also brings and emphasises several challenges (see [36]). A key challenge refers to competence provision and being able to bridge the gaps between previously more or less isolated domains (such as embedded systems and IT systems). Bridging these gaps will require an emphasis on technical as well as non-technical aspects. Among non-technical aspects we would like to emphasise education, life-long learning and work organisation. Among technical aspects we would like to highlight complexity management, interoperability (including technical standards), and the development of adequate technical platforms.

In connecting the physical, embedded and IT systems, a particular (technical as well as organisational) challenge becomes that of security.

The issue of security is most probably a sub-domain that could make or break future advances of CPS technology in industry. The addition of cloud computing only enhances the importance of this aspect, and will have to bring about a harmonisation of security control levels as well as regulations.

Unknown to many industrial engineers, an entire industrial PLC network can be easily accessed by a single search engine (such as SHODAN). This has prompted the US Department of Homeland Security to issue a clear warning that potential hackers had an almost free entry to almost any industrial site. This was further demonstrated in an article posted in Ars Tecnica, Dan Goodin [37] in which the popular CODESYS platform was shown to be extremely vulnerable.

Typical examples include:

- Higgins [38] reported on backdoor exploits that targeted Siemens PLCs allowing the capture of passwords and ability to manipulate the PLC code.
- In 2012, the Saudi Aramco oil and natural gas company had 30,000 computers on their corporate network infected and damaged by a piece of malware called Shamoon [39].
- StuxNet [40] was used to selectively target Iranian nuclear facilities.

The major issue at stake is that security cannot be an add-on. It has to be well developed and integrated within the reference architectures and systems from the very start of the design process. In the meantime there will be only ad hoc solutions, some of which quite capable at limiting damages: see Dunlap [41] and Schuett [42]. Security is moreover closely related to safety; both these system level properties have to be considered in conjunction (where security essentially protects the systems from humans (as attackers), and where safety protects humans from the systems). Cost-efficient safety will not be feasible without considering new security threats arising due to increasing openness and levels of integration.

If CPS and/or cloud computing are to be truly successful and exploitable by industry, the security equation thus needs to be seriously addressed. This is a task for both engineers and legislators, as pointed out by Williams [43].

### 15.3.7 Discussions

The authors believe that there has been some confusion between the terms discussed in this section. Characterisation is the best constructive remedy that we can provide to counter this confusion. As seen from the above descriptions, all the terms rely strongly on networked software-intensive systems, while emphasising different aspects of the corresponding systems. IoT mainly refers to technology and information (from bottom up) while big data emphasises data analysis (regardless of sources; connecting and collecting information from the physical world will definitely create much more data!). CPS instead emphasises interactions between physical and cyber parts, including humans, whereas SoS emphasis interactions within large-scale evolutionary systems. The terms provide different perspectives and from the previous discussion it is clear that there is no overall encompassing term today. CPS however covers a larger scope compared to IoT, embedded systems and mechatronics, and will become increasingly important in the context of SoS and big data (see CyPhERS deliverable D2.1 [44]).

## 15.4 Representative Examples of CPS in Manufacturing

As described previously, CPS applications in manufacturing are not new although the term of CPS was not used explicitly in early applications. In the following we provide four examples of CPS in manufacturing. The examples reflect many of the CPS characteristics as described in Sect. 15.3, in particular increasing openness, autonomy, distributed control, adaptability and degree of integration (refer to the characteristics numbered 3, 4, 6, 8 and 10 in Sect. 15.3).

### 15.4.1 Example 1: Service-Oriented Architecture

One of the first service-oriented architectures which was effectively deployed in industry and is still active at present is the Ford Motor Company application to the Valencia assembly plant (Ford Transit models). Developed by a system integrator (IntRoSys SA), this approach left all current PLCs as they were. Specific agent technology was embedded as "wrappers" to the PLCs according to the architecture given in Fig. 15.4.

As shown in Fig. 15.4, the IMASA approach is based on three different agents: (1) a Product Agent (PA) that formulates and dispatches the workflow, (2) a Knowledge Manager Agent (KMA) that performs a check of the physical feasibility of the proposed workflows, and (3) a Machine Agent (MA) that translates the workflows into specific machine instructions.

Fig. 15.4 IntRoSys multi-agent system architecture (IMASA) [17]

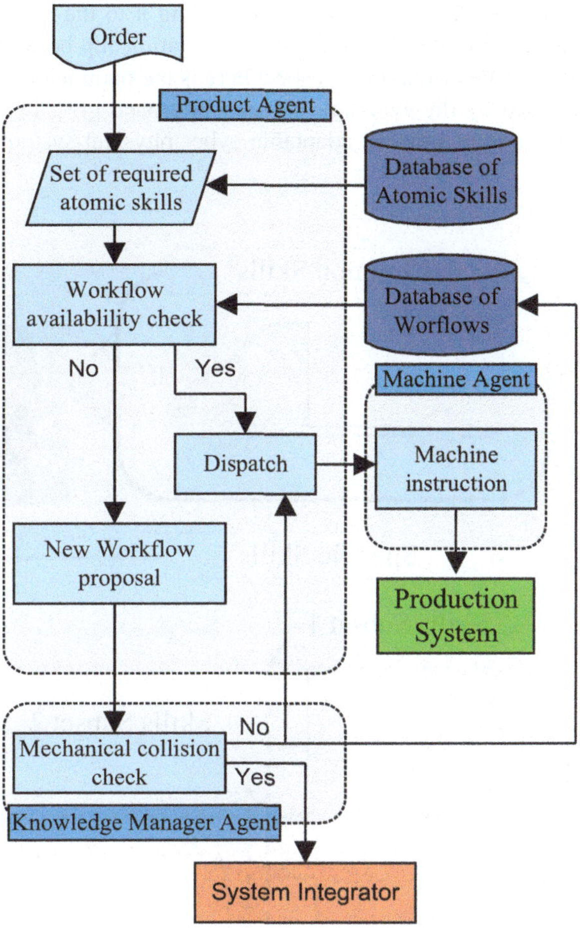


The PA receives the order and identifies the Atomic Skills required. The following step is a match of such required Atomic Skills with a database of workflows already executed in the past. If the order can be executed with an existing workflow, then such a workflow is dispatched to the MA, else the PA elaborates a new workflow that is sent for a feasibility check to the KMA. If the KMA does not detect any problem, the examined workflow is sent back to the PA that dispatches it to the MA. The newly found feasible workflow is also included in the database of existing workflows. Vice versa, in case of problems with the proposed workflow, the MA warns the System Integrator for the necessity of a human intervention to sort out the related order. Finally, all the dispatched workflows are processed by the MA that sends the necessary machine instructions to the production system.

This IMASA agent architecture is open in the sense that if a new atomic skill (for example a new process for a new variant, or a safety routine) is required, it can be integrated in the system without modifying the existing code but simply by coding it independently and eventually adding it to the related database (see Atomic Skill earlier). Figure 15.5 illustrates the relationship between common skills and specific skills. The fundamental aspect here is the parameterisation of common processes, as defined by the evolvable systems approach [45]. In this context, agents and PLCs function as a highly adaptable cyber-physical system.

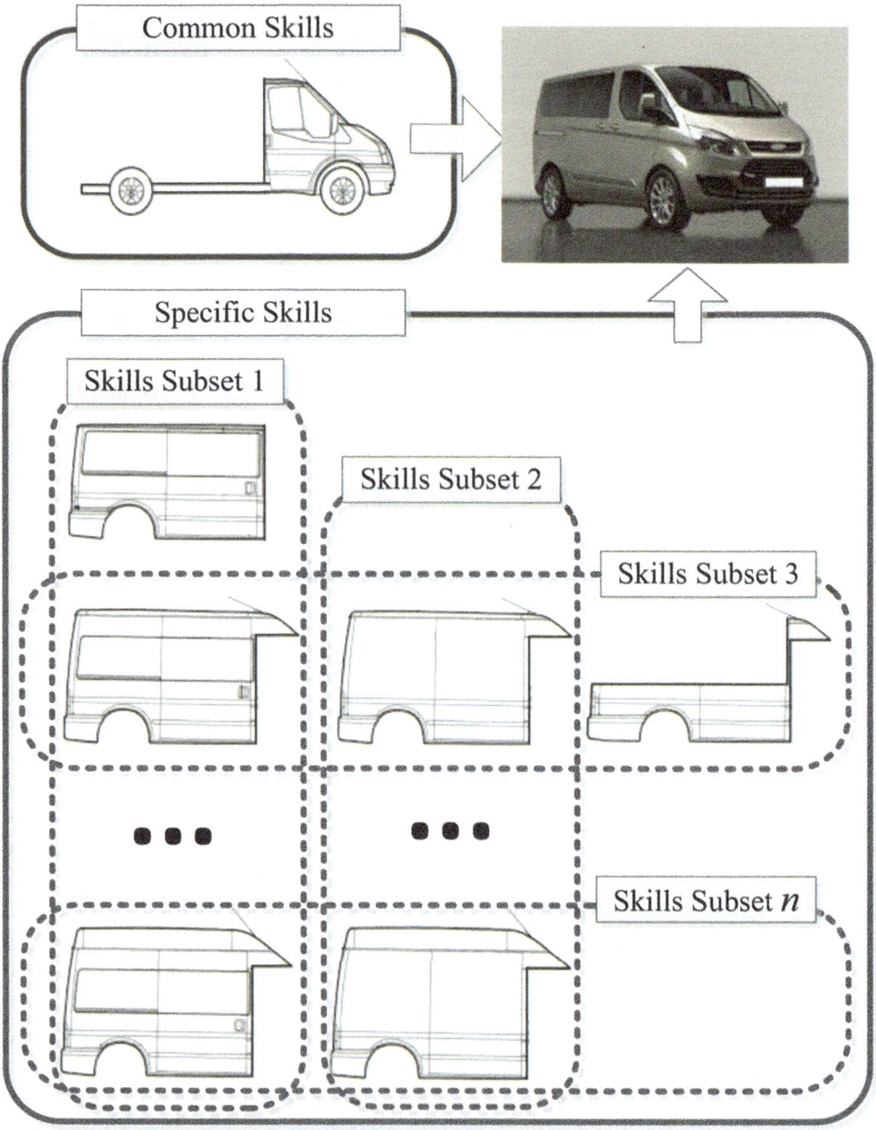


**Fig. 15.5** Common skills and specific skills [17]

### 15.4.2 Example 2: Cloud Manufacturing

Cloud-DPP (Cloud-based Distributed Process Planning) is an EU-funded project and particularly a joint research effort between KTH and Sandvik, Sweden, aiming for cloud-based distributed and adaptive process planning in a shared cyber workspace [46]. As depicted in Fig. 15.6, the four system modules close the loop of information flow. Based on real-time status/information of machines as well as their availability and capability, it is possible for the Cloud-DPP to generate machining process plans adaptively to changes through well-informed decision making [47]. This is accomplished by linking sensors embedded/attached to each machine to a manufacturing cloud in the cyber workspace, and delivering process plans in form of function blocks [48] to the machine controller on the physical shop floor for execution. By properly dividing process planning tasks and assigning them to the cloud and embedded in function blocks, adaptive process planning and machining become possible.

### 15.4.3 Example 3: Adaptive Manufacturing Systems

Another example of CPS in manufacturing is the FESTO pre-industrial system, MiniProd, which was demonstrated in January 2011. It ran with a multi-agent control setup (Agent-oriented Architecture), could be reconfigured on-the-fly, and consisted of modules self-configured thanks to their embedded controllers. The

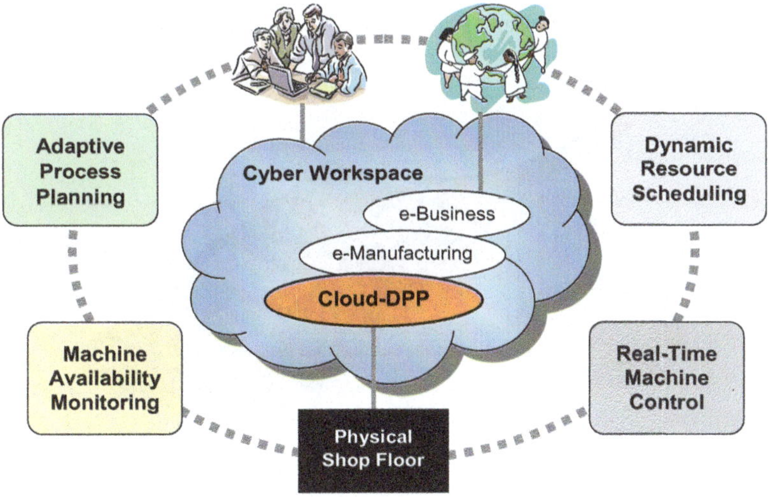


**Fig. 15.6** Cloud-DPP in a shared cyber workspace

MiniProd system relied on many years research [16, 49] and the following developments:

- A simple and effective mechatronic architecture
- Control boards developed for multi-agent applications
- An elaborated and well-structured methodology
- Industrial commitment.

The mechatronic architecture is, first of all, an architecture that considers the control demands from an embedded-system point of view. That is, each assembly system module is an entity with its own controls, hence mechatronic. The difficulty was in creating an architecture out of which an effective control structure could be instantiated for any assembly system layout.

The final mechatronic architecture is based on four basic agents:

- Machine Resource Agent
- Coalition Leader Agent
- Transportation System Agent
- Human Machine Interface Agent.

The second main development has been the one of commercial control boards capable of running the multi-agent setup; these are as follows:

- run on WinCe6
- implemented CrEme™, a Java Virtual Machine (NSI.com)
- implementation of 24V I/Os, Ethernet, CAN and RS232/RS485 connections
- runs CoDeSys V3
- implementation of different drivers (CAN, Ethernet, RS232/RS485)
- implementation of I/Os, Stepper/Frequency-count in FPGA and SW.

ELREST provided the project with several alternatives, out of which Combo211 shown in Fig. 15.7 was chosen. This required some developments:

**Fig. 15.7** Combo211 control board [17]

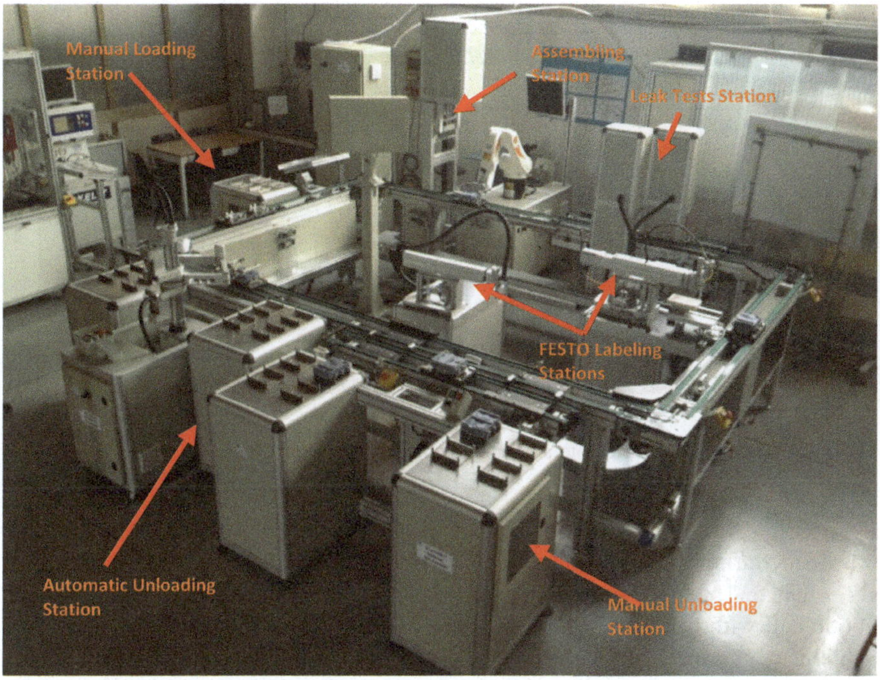


**Fig. 15.8** Electronic control unit assembly system [17]

- Combo200 series runs on WinCe6
- Implemented CrEme™, a Java Virtual Machine (NSI.com)
- Fits to the above needs of the four agents and supports JADE
- Implementation of 24 V I/Os, Ethernet, CAN and RS232/RS485 connections.

The project currently intends to develop three variants of these control boards, depending on the required granularity and number of agents/module. The assembled products are an ECU (electronic control unit) for a commercial vehicle. Figure 15.8 illustrates the layout of the consolidated assembly system.

### *15.4.4 Example 4: Model-Driven Manufacturing Systems*

A 3D model-driven robot-in-the-loop approach is presented in [50, 51] for remote assembly in a cloud environment, where an off-site operator can manipulate a physical robot instantly via virtual robot control in cyber-workspace. Instead of video image streaming, 3D models are used to guide the operator during remote assembly to meet real-time constraint over the Internet. The 3D models of the parts to be assembled are generated based on a sequence of snapshots of the parts

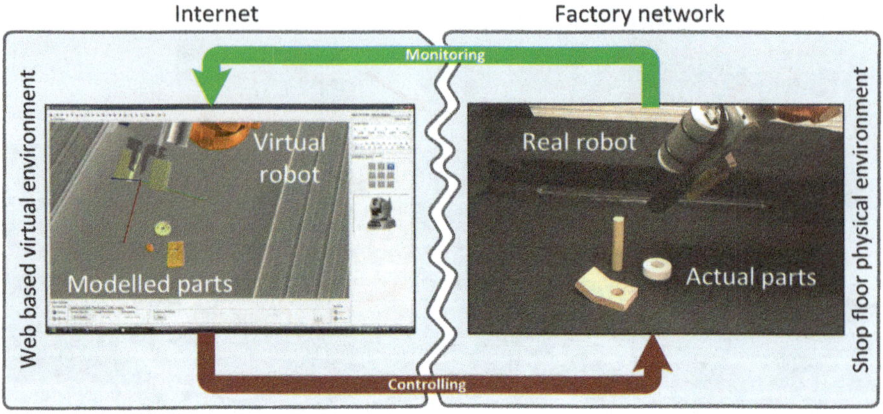


**Fig. 15.9** 3D model-driven remote assembly as a CPS

captured by a robot-mounted camera at the initiation phase. The camera is then turned off during robotic assembly to save network bandwidth for better overall performance. The generated 3D models are integrated with the 3D model of a real robotic cell with built-in kinematic models.

In this context, the robot is treated as a manipulator which mimics the operator's assembly operations from distance. As shown in Fig. 15.9, three simple parts are chosen for a proof-of-concept case study to validate the functionality of the 3D model-driven remote assembly system. The remote operator assembles the 'parts' (3D models) using the 3D robot model in the cyber world, whereas the real robot mimics the virtual robot and assembles the actual parts simultaneously in the physical world. During remote assembly, only the robot control commands are transmitted from the virtual robot to the real one instantly and automatically, without extra robot programming at runtime. Facilitated by the cyber-physical system, it thus enables virtual-to-real remote component assembly in real-time and paves the way towards factories of the future.

## 15.5 Future Research Directions

Whereas the above sections may help predict foreseeable future research directions for CPS in manufacturing, it is never an easy task to do so in any capacity. Technologies in the domain of IoT, CPS and SoS have made a sizeable impact on their applications in the last decade in smart grid and transportation. It is though not quite the case in manufacturing, with an exception of research lab based approaches that have been reported in this chapter and in the literature. Nevertheless, one can safely expect CPS research to continue in the direction of integrating IoT, cloud technology and SoS in an Internet-based environment, particularly in manufacturing settings. Such a trend echoes the regime in which current manufacturing firms function. This regime is featured as collaborative and distributive. Hence, there is a need for the ability to share manufacturing data and information between different stakeholders at different locations, seamlessly and collaboratively. CPS will continue to play a critical role in serving the need to share the data and information between cyber and physical worlds.

It is clear that there is no "silver bullet" to address the aforementioned need. The following research directions are listed based on the authors' literature analysis.

- *Self-organising manufacturing* rests on smart sensor networks and adaptive event-driven control. It also relies on the machine level ability of communication and cognition among constituent manufacturing equipment (or device controllers). Multi-agent systems for cognition and negotiation combined with CPS for communication and execution are important elements towards self-organising manufacturing. Legacy manufacturing process and performance can be integrated and maintained in a cloud-based knowledge repository. Combined with intelligent controllers, the future manufacturing processes can be improved continuously. Research in this area deals with what information is transferred, how it is used and how uncertainty is dealt with. The impact of every manufacturing process needs to be clarified in order to realise fully self-organising manufacturing. Event-driven control mechanisms at low-level controllers linking to the cyber workspace at high-level provides a holistic approach for self-learning and self-organisation. Thus, a cloud-enabled CPS approach will be the research focus.
- *Context- and situation-aware control* based on multi-dimensional data communications with low-level sensors/actuators and high-level planning systems can be facilitated by the CPS approach. Tedious and error-prone native programming of machines and robots by operators today will be replaced by smart decision algorithms tomorrow, runnable in machine controllers for robust and adaptive control. The research in this area includes closed-loop data analysis, sensor fusion and smart algorithms development. The research focus is on the device level with new interfaces with legacy control devices and new design of next-generation intelligent controllers with networking capability, capable of running algorithms than rigid codes.
- *Symbiotic human-robot collaboration* in a fenceless environment will improve productivity and resource effectiveness by combining the flexibility of humans and the accuracy of machines. CPS enables such human-robot collaboration in areas of dynamic task planning, active collision avoidance, and adaptive robot control. Humans can instruct robots by speech, signs, gestures and their combinations during collaborative assembly. On the other hand, in situ human assistance by, for example, 3D goggles will be possible and feasible. Standardisation is one more effort required to turn human-robot collaboration into a reality in the factories of the future.
- *CPS methodology* providing supporting methods and tools with which we will be able to cost-efficiently develop, operate and maintain SoS with the desired

capabilities and quality attributes. The increasing complexity of heterogeneous manufacturing systems is manifested in many ways, including as nonlinear hybrid systems with behaviours which are hard to predict and verify, and in terms of multiple and variable structures, composed of many interacting parts and properties. Research challenges include (1) development of techniques for efficiently integrating and/or relating multiple models, viewpoints and data sets, (2) CPS design methodology for trustworthy end-to-end services including adaptive/autonomous systems, and (3) platforms for safe and secure CPS design that underpin design methodology, facilitating integration and establishing desired system level properties.

## 15.6 Conclusions

This chapter presents the current status and advancement of cyber-physical systems and their future directions when applied to manufacturing. The characteristics of CPS are outlined together with those of SoS, IoT, Big Data and Cloud technology. Relevant initiatives, e.g. Industry 4.0, AMP 2.0 and Industrial Internet, are also briefly mentioned. In the authors' humble opinion, CPS research and applications will continue in the years to come, not only for the unsolved issues but also for the complex and intriguing nature of the problems that never failed to fascinate and challenge researchers and engineers. This is especially true when CPS are applied in manufacturing sector in the future, where self-organising manufacturing, context-/situation-aware control and symbiotic human-robot collaboration can play an important role in turning today's manufacturing shop floors into factories of the future with enhanced safety and security. The unique features of CPS in networking, communication and integrated device control attribute to the smartness and intelligence of manufacturing in the horizon. When combined with Cloud, IoT and Big Data, CPS will become feasible and practical in smart factories soon.

## References

1. A. Patnaik, S. Biswas, S.S. Mahapatra, An evolutionary approach to parameter optimisation of submerged arc welding in the hardfacing process. Int. J. Manuf. Res. **2**, 462–483 (2007)
2. E.A. Lee, Cyber-physical systems-are computing foundations adequate. Position paper for NSF workshop on cyber-physical systems: research motivation, techniques and roadmap (2006)
3. M. Grimheden, M. Hanson, What is Mechatronics? Proposing a Didactical Approach to Mechatronics. 1st Balt. Sea Workshop on Education in Mechatronics (2001)
4. Lee EA, Cyber physical systems: design challenges, in *11th IEEE international symposium on Object oriented real-time distributed computing (isorc)*, pp. 363–369
5. UC Regents, Cyber-physical systems (2009). http://cyberphysicalsystems.org/

6. L. Monostori, Cyber-physical production systems: roots, expectations and R&D challenges. Procedia CIRP. **17**, 9–13 (2014)
7. CODESYS, Industrial IEC 61131-3 PLC Programming (2017). Available from: https://www.codesys.com/
8. M. Onori, Oliveira J. Barata, Outlook report on the future of European assembly automation. Assem. Autom. **30**, 7–31 (2010)
9. H. Van Brussel, J. Wyns, P. Valckenaers, L. Bongaerts, P. Peeters, Reference architecture for holonic manufacturing systems: PROSA. Comput. Ind. **37**, 255–274 (1998)
10. H.A. ElMaraghy, Changing and evolving products and systems–models and enablers. Chang. Reconfigurable Manuf. Syst. 25–45 (2009)
11. M. Onori, J. Barata, Evolvable production systems: mechatronic production equipment with process-based distributed control. IFAC Proc. Vol. **42**, 80–85 (2009)
12. A.A. Rizzi, J. Gowdy, R.L. Hollis, Distributed coordination in modular precision assembly systems. Int. J. Rob. Res. **20**, 819–838 (2001)
13. F.P. Maturana, D.H. Norrie, Multi-agent mediator architecture for distributed manufacturing. J. Intell. Manuf. **7**, 257–270 (1996)
14. A.W. Colombo, R. Schoop, R. Neubert, An agent-based intelligent control platform for industrial holonic manufacturing systems. IEEE Trans. Ind. Electron. IEEE **53**, 322–337 (2006)
15. L. Monostori, J. Váncza, S.R.T. Kumara, Agent-based systems for manufacturing. Ann. CIRP. **55**, 697–720 (2006)
16. L. Ribeiro, J. Barata, J. Ferreira, *An agent-based interaction-oriented shop floor to support emergent diagnosis*, in 8th IEEE International Conference IEEE Industrial Informatics (INDIN) (2010), pp. 189–194
17. L. Wang, M. Törngren, M. Onori, Current status and advancement of cyber-physical systems in manufacturing. J. Manuf. Syst. **37**(2), 517–527 (2015)
18. TerraSwarm, The TerraSwarm Research Center [Internet] (2017). Available from: https://www.terraswarm.org/index.html
19. The Whitehouse, President Obama Launches Advanced Manufacturing Partnership Steering Committee "2.0" [Internet]. Off. Press Secr. 2013. Available from: https://obamawhitehouse.archives.gov/the-press-office/2013/09/26/president-obama-launches-advanced-manufacturing-partnership-steering-com
20. IIC, Industrial Internet Consortium (2010). http://www.iiconsortium.org/
21. NSF, Cyber-Physical Systems [Internet] (2013). Available from: https://www.nsf.gov/funding/pgm_summ.jsp?pims_id=503286
22. H. Kagermann, J. Helbig, A. Hellinger, W. Wahlster, Recommendations for implementing the strategic initiative INDUSTRIE 4.0: securing the future of German manufacturing industry; final report of the Industrie 4.0 Working Group. Forschungsunion (2013)
23. EU, HORIZON 2010—The EU Framework Programme for Research and Innovation [Internet] (2014). Available from: http://ec.europa.eu/programmes/horizon2020/
24. L. Wang, B. Wong, W. Shen, S. Lang, A Java 3D-enabled cyber workspace. Commun. ACM **45**, 45–49 (2002)
25. L. Wang, Wise-ShopFloor: an integrated approach for web-based collaborative manufacturing. Syst. Man Cybern. Part C Appl. Rev. IEEE Trans. **38,** 562–573 (2008)
26. CyPhERS, Cyber-Physical European Roadmap and Strategy (2013). http://www.cyphers.eu/
27. A.S. Vincentelli, CPS week keynote. Commun. Dur. keynote present by Alberto Sangiovanni Vincentelli Int. Conf. CPS CPS Week 2014
28. L. Takayama, W. Ju, C. Nass, Beyond dirty, dangerous and dull: what everyday people think robots should do, in *Proceedings of 3rd ACM/IEEE International Conferences of Humun Robot Interaction ACM* (2008), pp. 25–32
29. R. Parasuraman, T.B. Sheridan, C.D. Wickens, A model for types and levels of human interaction with automation. IEEE Trans. Syst. Man. Cybern. A Syst. Humans. IEEE **30**, 286–297 (2000)

30. D. Mourtzis, M. Doukas, C. Vandera, Mobile apps for product customisation and design of manufacturing networks. Manuf. Lett. **2**, 30–34 (2014)
31. H.A. Thompson, Input paper analysis of the state-of-the-art and future challenges in the application domain related to WG1. *Deliverable of EU CPSoS Project* (2014). http://www.cpsos.eu/wp-content/uploads/2014/10/D2.1b-Input-Paper.pdf
32. K. Nagorny, A.W. Colombo, U. Schmidtmann, A service- and multi-agent-oriented manufacturing automation architecture: an IEC 62264 level 2 compliant implementation. Comput. Ind. **63**, 813–823 (2012)
33. K. Ashton, That "internet of things" thing. RFiD J. **22** (2011)
34. P. Mell, T. Grance, The NIST definition of cloud computing. NIST Spec. Publ. **800**, 7 (2011)
35. L. Wang, A. Mohammed, M. Onori, Remote robotic assembly guided by 3D models linking to a real robot. CIRP Ann. Technol. **63**, 1–4 (2014)
36. M. Törngren, S. Bensalem, V. Cengarle, J. McDermid, R. Passerone, A. Sangiovanni Vincentelli, CPS: Significance, Challenges and Opportunities. (CyPhERS-Cyber-Physical European Roadmap and Strategy, co-funded by the European Commission under the Seventh Framework Programme, 2014)
37. D. Goodin, Backdoor in computer controls opens critical infrastructure to hackers. Ars Tecnica. http://arstechnica.com/security/2012/10/backdoor-in-computer-controls-opens-critical-infrastructure-to-hackers/
38. K.J. Higgins, Utilities facing brute-force attack threat [Internet]. Available from: https://www.darkreading.com/attacks-and-breaches/utilities-facing-brute-force-attack-threat/d/d-id/1102697
39. Symantec, The Shamoon Attacks [Internet] (2012). Available from: http://www.symantec.com/connect/blogs/shamoon-attacks
40. StuxNet, All about Stuxnet [Internet]. Available from: http://www.stuxnet.net/
41. S.J. Dunlap, Timing-based side channel analysis for anomaly detection in the industrial control system environment (Air Force Institute of Technology, 2013)
42. Schuett CD, Programmable logic controller modification attacks for use in detection analysis (Air Force Institute of Technology, Wright-Patterson Afb Oh Graduate School of Engineering and Management, 2014)
43. Williams PM, Distinguishing internet-facing ICS devices using PLC programming information (Air Force Institute of Technology, Wright-Patterson Afb Oh Graduate School of Engineering and Management, 2014)
44. V. Cengarle, S. Bensalem, J. McDermid, R. Passerone, A. Sangiovanni-Vincentelli, M. Törngren, Characteristics, capabilities, potential applications of cyber-physical systems: a preliminary analysis. D2.1 CyPhERS FP7 Proj. (2013)
45. A. Maffei, M. Onori, Evolvable production systems: environment for new business models. Key Eng. Mater. Trans. Tech. Publ. 1592–1597 (2011)
46. L. Wang, An overview of function block enabled adaptive process planning for machining. J. Manuf. Syst. **35**, 10–25 (2015)
47. L. Wang, Machine availability monitoring and machining process planning towards cloud manufacturing. CIRP J. Manuf. Sci. Technol. **6**, 263–273 (2013)
48. IEC, IEC 61499. *Function Blocks for Industrial-process Measurement and Control Systems-Part 1: Architecture* (IEC, Geneva 2005)
49. P. Ferreira, N. Lohse, S. Ratchev, Multi-agent architecture for reconfiguration of precision modular assembly systems. Precis. Assem. Technol. Syst. 247–254 (2010)
50. L. Wang, R. Sams, M. Verner, F. Xi, Integrating Java 3D model and sensor data for remote monitoring and control. Robot. Comput. Integr. Manuf. **19**, 13–19 (2003)
51. L. Wang, A. Mohammed, M. Onori, Remote robotic assembly guided by 3d models linking to a real robot. CIRP Ann. Manuf. Technol. **63**(1), 1–4 (2014)

# Index





L. Wang and X. V. Wang, *Cloud-Based Cyber-Physical Systems in Manufacturing*,
https://doi.org/10.1007/978-3-319-67693-7

The manufacturer's authorised representative in the EU is Springer Nature Customer Service Centre GmbH, Europaplatz 3, 69115 Heidelberg, Germany. If you have any concerns regarding our products, please contact ProductSafety@springernature.com

Printed and bound by CPI Group (UK) Ltd, Croydon, CR0 4YY
30/09/2026
02225229-0014